Cyber Warfare and the Laws of War

The information revolution has transformed both modern societies and the way in which they conduct warfare. *Cyber Warfare and the Laws of War* analyses the status of computer network attacks in international law and examines their treatment under the laws of armed conflict. The first part of the book deals with the resort to force by states and discusses the threshold issues of force and armed attack by examining the permitted responses against such attacks. The second part offers a comprehensive analysis of the applicability of international humanitarian law to computer network attacks. By examining the legal framework regulating these attacks, Heather Harrison Dinniss addresses the issues associated with this method of attack in terms of the current law and explores the underlying debates which are shaping the modern laws applicable in armed conflict.

HEATHER HARRISON DINNISS is a postdoctoral research fellow at the International Law Centre of the Swedish National Defence College. She has previously taught at the London School of Economics and Political Science and at the School of Oriental and African Studies, University of London. Her research focuses on the impact of modern warfare on international humanitarian law, in particular, on advanced and autonomous weapons systems and the status and use of computer network attacks in the law of armed conflict.

CAMBRIDGE STUDIES IN INTERNATIONAL AND COMPARATIVE LAW

Established in 1946, this series produces high quality scholarship in the fields of public and private international law and comparative law. Although these are distinct legal sub-disciplines, developments since 1946 confirm their interrelations.

Comparative law is increasingly used as a tool in the making of law at national, regional and international levels. Private international law is now often affected by international conventions, and the issues faced by classical conflicts rules are frequently dealt with by substantive harmonisation of law under international auspices. Mixed international arbitrations, especially those involving state economic activity, raise mixed questions of public and private international law, while in many fields (such as the protection of human rights and democratic standards, investment guarantees and international criminal law) international and national systems interact. National constitutional arrangements relating to 'foreign affairs', and to the implementation of international norms, are a focus of attention.

The Series welcomes works of a theoretical or interdisciplinary character, and those focusing on the new approaches to international or comparative law or conflicts of law. Studies of particular institutions or problems are equally welcome, as are translations of the best work published in other languages.

A list of books in the series can be found at the end of this volume.

Cyber Warfare and the Laws of War

Heather Harrison Dinniss

CAMBRIDGE
UNIVERSITY PRESS

CAMBRIDGE UNIVERSITY PRESS
Cambridge, New York, Melbourne, Madrid, Cape Town,
Singapore, São Paulo, Delhi, Mexico City

Cambridge University Press
The Edinburgh Building, Cambridge CB2 8RU, UK

Published in the United States of America by Cambridge University Press, New York

www.cambridge.org
Information on this title: www.cambridge.org/9781107011083

First published 2012

Printed in the United Kingdom at the University Press, Cambridge

A catalogue record for this publication is available from the British Library

Library of Congress Cataloguing in Publication data
Dinniss, Heather Harrison.
Cyber warfare and the laws of war / Heather Harrison
Dinniss.
p. cm. – (Cambridge studies in international and comparative law)
Includes bibliographical references and index.
ISBN 978-1-107-01108-3 (hardback)
1. Information warfare (International law) 2. War (International law)
3. Cyberterrorism. I. Title.
KZ6718.D56 2012
341.6′3–dc23 2012013369

ISBN 978-1-107-01108-3 Hardback

For Marc

Thank you … for everything

Contents

Acknowledgements

The writing of a Ph.D. thesis and the task of turning it into a book necessarily requires assistance from many individuals and there are a number of people who have helped me both in the formation of my ideas and in the writing of this book. First and most importantly, my thanks go to my Ph.D. supervisor, Christopher Greenwood, for his insightful comments and suggestions, and for keeping me focused on the danger of being too technical at the expense of the law. Special thanks are also due to Christine Chinkin, my supervisor during Chris's sabbatical, who also commented on a draft of the manuscript. Her perceptive insights form the basis of my methodology section, a contribution for which I am profoundly grateful. I am indebted to Michael Schmitt and Chaloka Beyani for examining the original thesis and particularly to Mike, who has offered numerous comments and suggestions for this subsequent published edition. He has been a supportive and encouraging mentor, always willing to discuss ideas and offer advice.

Although the media seems full of new reports of cyber warfare (some worthy of the title, many not), at the time that the original thesis was written there were very few examples in the public domain of computer network attacks. My thanks are also due to those specialists in information technology and network design who patiently reviewed my hypothetical examples and advised on the feasibility of particular computer network attack scenarios, in particular Andy Batchelor, Karon James and Michelle Ellis. I am grateful for their assistance, and the fault for any errors or inaccuracies remains entirely mine. Special mention must also go to Andrew Ladley of Victoria University of Wellington, for his encouragement to continue postgraduate study and for the stories of his adventures that proved so inspirational. I would also like to thank all of my colleagues in the law department of the London School of Economics and Political Science; it is often remarked that the drafting of a Ph.D. thesis can be a lonely experience and I am grateful to all my friends for making sure that this was not the case, and for making my time at the LSE so rewarding both personally and academically.

The thesis on which this book is based was funded in part by the generous Freyberg Scholarship provided by the New Zealand Defence Force and the Research Studentships of the London School of Economics and Political Science.

Finally, and most importantly, I would like to thank my family. My parents for their support and encouragement throughout this process; my mother, Ann Dinniss, who painstakingly proofread the original manuscript, and my father, David Dinniss, who sadly did not live to see it completed. Above all, my thanks go to my wonderful husband Marc; without his support, patience, sense of humour, and steadfast belief, the thesis and this subsequent book would not have happened. This book is dedicated to him.

Table of cases

List of abbreviations

ADIL	Annual Digest and Reports of Public International Law Cases
Additional Protocol I *also* API	Protocol Additional to the Geneva Conventions of 12 August 1949, and relating to the Protection of Victims of International Armed Conflicts (Protocol I), 8 June 1977
Additional Protocol II *also* APII	Protocol Additional to the Geneva Conventions of 12 August 1949, and relating to the Protection of Victims of Non-International Armed Conflicts (Protocol II), 8 June 1977
AFJ Log.	*Air Force Journal of Logistics*
AFL Rev.	*Air Force Law Review*
AJIL	*American Journal of International Law*
Am. U. L. Rev.	*American University Law Review*
ASIL Proc.	*American Society of International Law Proceedings*
BFSP	British and Foreign State Papers.
BMJ	*British Medical Journal*
BU Int'l L. J.	*Boston University International Law Journal*
BU J. Sci. & Tech. L.	*Boston University Journal of Science and Technology*
BYBIL	*British Yearbook of International Law*
Case W. Res. J. Int'l L.	*Case Western Reserve Journal of International Law*
Chi. J. Int'l L.	*Chinese Journal of International Law*
CNA	Computer Network Attack
Col. J. Trans. L.	*Columbia Journal of Transnational Law*
Conventional Weapons Treaty	Convention on Prohibitions or Restrictions on the Use of Certain Conventional

	Weapons Which May Be Deemed to Be Excessively Injurious or to Have Indiscriminate Effects, 10 October 1980
CPU	central processing unit
DCS	distributed control system
DDoS	distributed denial of service
EJIL	*European Journal of International Law*
EMP	electromagnetic pulse
ENMOD	Convention on the Prohibition of Military or Any Other Hostile Use of Environmental Modification Techniques, 10 December 1976
FRY	Federal Republic of Yugoslavia
Geneva Convention I	Geneva Convention for the Amelioration of the Condition of the Sick and Wounded in Armed Forces in the Field of August 12, 1949
Geneva Convention II	Geneva Convention for the Amelioration of the Condition of Wounded, Sick and Shipwrecked Members of Armed Forces at Sea of August 12, 1949
Geneva Convention III	Geneva Convention relative to the Treatment of Prisoners of War of August 12, 1949
Geneva Convention IV	Geneva Convention relative to the Protection of Civilian Persons in Time of War of August 12, 1949
GJIA	*Georgetown Journal of International Affairs*
GovExec	Government Executive
Hague Convention 1899	Hague Convention Concerning the Laws and Customs of War on Land
Hague Regulations	Regulations Annexed to the 1907 Hague Convention IV Respecting the Laws and Customs of War on Land
Harv. Int'l L.J.	*Harvard International Law Journal*
Hous. J. Int'l L.	*Houston Journal of International Law*
HPCR	Harvard Program on Humanitarian Policy and Conflict Research
IAEA	International Atomic Energy Agency
ICBL	International Campaign to Ban Landmines
ICJ	International Court of Justice

ICRC	International Committee of the Red Cross
ICTR	International Criminal Tribunal for Rwanda
ICTY	International Criminal Tribunal for the Former Yugoslavia
IFF	Identification Friend or Foe
IMF	International Monetary Fund
IO	Information Operations
IP Address	Internet Protocol Address
IRRC	*International Review of the Red Cross*
Israel Y.B. Hum. Rts	*Israeli Yearbook of Human Rights*
IW	Information Warfare
J. Legal Stud.	*Journal of Legal Studies*
JC&SL	*Journal of Conflict and Security Law*
JSTARS	Joint Surveillance Target Attack Radar System
Keesings	*Keesings Record of World Events*
LOAC	Laws of Armed Conflict
Mil. L. Rev.	*Military Law Review*
NIL Rev.	*Netherlands International Law Review*
NYU J. Int'l L. & Pol.	*New York University Journal of International Law and Politics*
NZLR	*New Zealand Law Reports*
OAS	Organization of American States
OAU	Organisation of African Unity
Ottawa Convention	Convention on the Prohibition of the Use, Stockpiling, Production and Transfer of Anti-Personnel Mines and on Their Destruction.
PMC	Private Military Company
RIAA	Reports of International Arbitral Awards
SCADA	Supervisory Control and Data Acquisition
SCSL	Special Court for Sierra Leone
Stan. J. Int'l L.	*Stanford Journal of International Law*
T. Jefferson L. Rev.	*Thomas Jefferson Law Review*
Tex. L. Rev.	*Texas Law Review*
Tulsa J. Comp. & Int'l L.	*Tulsa Journal of Comparative & International Law*
UAV	Unmanned Aerial Vehicle
UN	United Nations
UNCIO	Documents of the United Nations Conference on International Organisation

UNTS	United Nations Treaty Series
UN Y. B. Int'l L. Comm'n	*United Nations Yearbook of the International Law Commission*
URL	Universal Resource Locator (a web address)
US	United States (of America)
USAF Acad. J. Legal Stud.	*US Air Force Academy Journal of Legal Studies*
Vand. J Transnat'l L.	*Vanderbilt Journal of Transnational Law*
Yale Hum. Rts & Dev. L.J.	*Yale Human Rights and Development Law Journal*
Yale J. Int'l L.	Yale Journal of International Law
ZaöRV	*Zeitschrift für ausländisches öffentliches Recht und Völkerrecht*

1 The world in which we live and fight

On 16 May 1943 one of the most famous missions of World War II, the 'Dambusters' raid, took place. Nineteen Lancaster bombers, modified to carry weapons at the cutting edge of technology, flew over most of southern Germany to attack three hydroelectric dams supplying electricity to German industrial installations in the Ruhr valley. Two of the three targeted dams were breached, causing significant damage; however it was at the cost of eight bomber crews lost during the mission.[1] Fifty-five years later, according to some reports, a twelve-year-old boy hacked into the control system of Arizona's Roosevelt Dam, gaining control of its massive floodgates and the 489 billion gallons of water which it contains.[2] Although the boy was unaware of the fact, federal authorities stated that he could have released the 489 billion gallons of water contained by the dam downstream, causing massive amounts of damage. Such an incident demonstrates the power and possibility of computer network attacks if utilised in an armed conflict; it also illustrates the vulnerability of states that are dependent on critical infrastructures that are not adequately protected against this new method of attack.

[1] In the Möhne and Ruhr valleys 11 factories were totally destroyed, 114 seriously damaged, 25 road and rail bridges were destroyed and throughout the region power, water and gas supplies were seriously disrupted. Communications by road and canal were severely disrupted and for the remainder of the war the Germans had to divert an additional 10,000 troops to guard the dams. National Archives, *Dambusters: The Legacy*, www.nationalarchives.gov.uk/dambusters/legacy.htm (last accessed 15 July 2011).

[2] Barton Gellman, 'Cyber-Attacks by Al Qaeda Feared', *Washington Post* (Washington DC), 27 June 2002, A01. Although there is debate over the veracity of some of the facts of this case, including the year and severity of the attack and the age of the

The most recent example is the Stuxnet worm which is widely believed to have caused significant damage to approximately 1,000 centrifuges at Iran's Natanz nuclear enrichment facility in June 2010.[3] For commentators who had questioned the veracity or significance of previous incidents circulating in the public domain, Stuxnet served as a declaration that cyber warfare had finally come of age. Although widely touted as the first cyber war in the media, the denial of service attacks, which took place against Estonia in 2007 and Georgia in 2008, did not cause physical damage and later came to be seen as criminal attacks. However, the incidents offered concrete examples to analysts and commentators working in the field and focused the minds of national policy makers. In recent years states and international organisations have become increasingly aware of the threats and challenges presented by cyber security and the urgency with which these issues need to be addressed. The national security strategies of states such as the United States and the United Kingdom now reflect the central role of cyber security in their planning;[4] likewise, each of these countries has recently created cyber commands within either the armed forces or intelligence agencies to address the threat.

This book examines the laws governing the resort to force and the conduct of hostilities as they relate to one of the newest forms of warfare, computer network attacks. Computer network attacks are 'actions taken through the use of computer networks to disrupt, deny, degrade, or destroy information resident in computers and computer networks, or the computers and networks themselves'.[5] The book is divided into two parts. Part I addresses the *jus ad bellum*; it examines computer network attacks as prohibited acts and the permitted responses to such acts under international law. Chapter 2 first looks at the qualification

hacker (detailed in Appendix 1), the example serves to illustrate the change in the method of warfare, and the comparative ease of achieving the same effect.

[3] See, for example, Kim Zetter, 'Surveillance Footage and Code Clues Indicate Stuxnet Hit Iran', *Wired* 16 February 2011, Threat Level, www.wired.com/threatlevel/2011/02/isis-report-stuxnet/# (last accessed 10 April 2011); David Albright, Paul Brannan and Christina Walrond, *Did Stuxnet Take out 1,000 Centrifuges at the Natanz Enrichment Plant?* Institute for Science and International Security (2010) (hereafter ISIS Stuxnet Report).

[4] UK Cabinet Office, *A Strong Britian in an Age of Uncertainty: The National Security Strategy* (Cmd 7935 ed., The Stationery Office, London, 2010); White House, *National Security Strategy* (2010), www.whitehouse.gov/sites/default/files/rss_viewer/national_security_strategy.pdf (last accessed 15 July 2011).

[5] US Department of Defense, *Dictionary of Military and Associated Terms, 8 November 2010 (as amended through 31 January 2011)* (Washington DC, 2010).

of computer network attacks as a use of force contrary to Article 2(4) of the UN Charter and examines the theoretical underpinnings of the prohibition against force in international law in order to address some of the specific characteristics of computer network attacks. Chapter 3 then considers when an attack will rise to the level of an armed attack, thus triggering the right of self-defence. The chapter also examines the complex issue of the attribution of attacks, a particular problem for a method of warfare that generally relies on anonymity, and the right of states to act in self-defence against non-state actors. The chapter also addresses other possible responses to computer network attacks that do not allow for self-defence, namely countermeasures, and considers the role of computer network attacks both as threats to the peace and possible responses under the collective security regime of the United Nations. Part II of the book examines the *jus in bello* and works systematically through those areas of the law of armed conflict for which computer network attacks raise issues. Chapter 4 begins by examining the concept of armed conflict and discusses under what circumstances the law of armed conflict will apply to computer network attacks. The remaining chapters examine the themes of participants in conflict, targeting and legitimate military objectives, precautions in attack and defence, measures of special protection, and means and methods of warfare (including the law of weaponry).

The book is not, however, limited to a point-by-point analysis of the current laws of armed conflict. The rise of computer network attacks as both a means and method of warfare is born out of, and in turn has influenced, many different societal and military trends. Therefore any attempt to analyse how the laws of armed conflict should affect this form of warfare must take these trends into account or risk becoming outdated as soon as it is completed. Indeed, with much of the current capacity for computer network attacks remaining classified and the exponential growth of computing and transmission power, any attempt to limit such work to present capacity and ignore trends would be foolhardy at best.[6] The book also takes account of the ongoing debates between experts taking place in relation to the laws applicable in conventional armed conflicts. These debates, such as the current discourse on direct participation in hostilities, the use of civilian contractors, the

[6] Moore's Law states that computing power will double approximately every two years; Nielson's law states that bandwidth for high-end users will double in the same period.

applicability of the laws of armed conflict to counter-terrorist operations, and targeting of dual-use facilities, to name just a few, all form the background to the discussion of the law as it applies to computer network attacks.

1. Computer network attacks

Computer network attacks are actions taken through the use of computer networks to disrupt, deny, degrade, or destroy information resident in computers and computer networks, or the computers and networks themselves.[7] The defining feature of this form of attack is the fact that both the weapon and the target of the attack is the network itself and the information contained on such networks. This feature distinguishes computer network attacks from forms of electronic warfare, which may also seek to disrupt or destroy a network, but instead use electromagnetic energy, usually in hardwired weapons such as electromagnetic pulse (EMP) generators or jammers to achieve their aims.[8] A computer network attack uses computer code to effect its damage and is capable of causing a myriad of effects depending on the target system's function. Although some authors have taken issue with the definition in the past,[9] on the whole it appears that these concerns stem from a narrow interpretation of the concept of 'information' in the context of the definition.[10] Information in terms of computing is any data that reduces uncertainty in the state of a system; it includes rather more than the traditional definition of facts and knowledge required by human beings to change or form an

[7] US Department of Defense, *Dictionary of Military and Associated Terms, 8 November 2010 (as Amended through 31 January 2011)* (Washington DC, 2010).

[8] Other forms include other uses of the electromagnetic spectrum such as radar, radio, optics (laser and infrared devices), high-powered microwaves, as well as warning and counteraction systems. Techniques include signal interception, passive listening, electronic surveillance, radar and radio traffic deception, as well as jamming and electronic interference. Roland Heickerö, 'Electronic Warriors Use Mail Order Equipment', *Framsyn Magazine* April 2005, www.foi.se/FOI/templates/Page____4554.aspx# (last accessed 12 April 2011).

[9] See, for example, Yoram Dinstein, 'Computer Network Attacks and Self-Defense' in M. N. Schmitt and B. T. O'Donnell (eds.), *Computer Network Attack and International Law* (Naval War College, Newport, RI, 1999), 99–119, 102.

[10] Multiple conceptions of the term 'information' appear in the literature surrounding the information revolution, see generally: John Arquilla and David Ronfeldt, 'Information, Power and Grand Strategy: In Athena's Camp – Section 1' in J. Arquilla and D. Ronfeldt (eds.), *In Athena's Camp: Preparing for Conflict in the Information Age* (RAND, Santa Monica, 1997), 141–71, 144.

opinion.[11] Indeed the US military definition of information is 'facts, data or instructions in any medium or form'.[12] Thus the operating code of a computer, its automated processes and applications, as well as the files and data it contains are all information. Once one grasps this extended definition, the range of possible effects of a computer network attack become greatly expanded.

The term computer network attack thus covers a broad range of hostile techniques involving computer code. Such malicious software (malware) can cause extensive disruption, as in the case of the denial of service attacks which hit Estonia, or physical destruction, as with the Stuxnet worm in Iran; both incidents are discussed in detail in Chapter 2. Distributed denial of service (DDoS) attacks, although disruptive, are a fairly unsophisticated method of attack and are usually carried out through the use of a botnet through which a single controller can harness the power of many computers. The attack floods a specific target with requests for service, so that either the target shuts down in the wake of its inability to cope with the incoming messages, or the target is effectively blocked to legitimate requests as the attack exhausts the resources available to the target to handle legitimate requests.

These types of attacks are capable of shutting down websites, servers and backbone nodes; generating massive email and spamming campaigns; and disseminating viruses.[13]

Computer network attacks which cause physical effects, such as the Stuxnet worm, often target the control systems which regulate the most critical infrastructure systems of technologically advanced societies; these systems control power plants, water systems, dams, gas pipelines, chemical plants and reactors, to name a few. These control systems regulate most of the critical infrastructure and have proven particularly vulnerable to attack.[14] Prior to the Stuxnet worm's discovery, there were several incidents to which commentators could point to illustrate the physical threat from computer network attacks, but all required audiences to extrapolate and connect the dots.

[11] For a full definition see 'Information', *A Dictionary of Computing* (Oxford University Press, 2004).

[12] US Department of Defense, *Dictionary of Military and Associated Terms*.

[13] Eneken Tikk, Kadri Kaska and Liis Vihul, *International Cyber Incidents: Legal Considerations*, Cooperative Cyber Defence Centre of Excellence (2010), 112.

[14] Supervisory control and data acquisition (or SCADA) systems and distributed control systems (DCS) are two examples of control systems that are often mentioned in the literature.

In March 2007, researchers from the Idaho National Laboratory launched an experimental cyber attack, hacking into a replica of a power plant's control system and changing the operating cycle of a generator.[15] The attack sent the generator out of control and ultimately caused it to self-destruct, alarming the federal government and electrical industry about what might happen if such an attack were carried out on a larger scale.[16] One of the earliest reported incidents of this kind of computer attack, the so-called 'Farewell Dossier' incident, took place in 1982 during the Cold War; however, it only came to light once the incident was declassified in 1996. Following the theft of technology from Western powers by the Soviet KGB, the CIA of the United States and a Canadian software supplier planted malicious code in the software for a gas pipeline control system which a KGB operative had been sent to steal:[17]

> [T]he pipeline software that was to run the pumps, turbines and valves was programmed to go haywire, after a decent interval, to reset pump speeds and valve settings to produce pressures far beyond those acceptable to the pipeline joints and welds. The result was the most monumental non-nuclear explosion and fire ever seen from space.

Control systems were again compromised in the 1998 Arizona Roosevelt Dam example cited previously. In another domestic example, in 2000 a disgruntled former employee, Vitek Boden, hacked the control system of the water and sewerage treatment plant in Queensland, Australia. Over a two-month period Boden accessed the system forty-six times, gaining complete control of the sewerage and drinking water systems for the region and dumping putrid sludge into the area's rivers and parks.[18] Incidents such as these have made states increasingly aware of the amount of critical infrastructure that is controlled by computers and their vulnerability to computer network attacks. The result

[15] Jeanne Meserve, 'Staged Cyber Attack Reveals Vulnerability in Power Grid', *CNN.com* 26 September 2007, http://edition.cnn.com/2007/US/09/26/power.at.risk/index.html (last accessed 15 July 2011).

[16] *Ibid.* Footage of the generator is available at www.youtube.com/watch?v=fJyWngDco3g (last accessed 23 June 2010).

[17] Thomas C. Reed, *At the Abyss: An Insider's History of the Cold War* (Presidio, New York, 2004), 269; for a first-hand account see Gus W. Weiss, 'The Farewell Dossier: Duping the Soviets' (1996) 35(5) *Studies in Intelligence* 121, https://www.cia.gov/library/center-for-the-study-of-intelligence/csi-publications/csi-studies/studies/96unclass/farewell.htm (last accessed 12 April 2011).

[18] *R* v. *Boden* (2002) QCA 164, Court of Appeal of the Supreme Court of Queensland (Australia); Gellman, 'Cyber-Attacks by Al Qaeda Feared'.

has been the increasing emphasis on cyber attacks as one of the major threats to both the US and UK critical infrastructure in recent reports.[19]

Computer network attacks may come in isolation, but will more probably be used in conjunction with a conventional attack, either to ease the way for the conventional attack or to amplify its effects. In the battlespace they may be used to disable the advance warning systems of an air defence network, allowing an attacker's air force to advance unseen into enemy territory. This happened during Israel's penetration of Syrian air defences on 6 September 2007 in order to bomb a suspected nuclear site at Dayr az-Zawr, without being engaged or even detected.[20] That attack combined electronic attack techniques in the form of brute-force jamming, precision missiles to eliminate the facility itself and, most interestingly, computer network attack techniques. The ability of non-stealthy Israeli aircraft to penetrate Syrian airspace without interference rests in part on technology, carried on board modified aircraft, which allowed specialists to hack into Syria's networked air defence system.[21] As one commentator noted, 'network raiders can conduct their invasion from an aircraft into a network and then jump from network to network until they are into the target's communications loop'.[22] Israel is not the only state to have developed this technology. The US has developed 'Suter' network-invasion capability which uses the EC-130 electronic attack aircraft to shoot data streams, laced with sophisticated algorithms, into enemy antennas.[23] The US version of the system has at the very least been tested operationally in Iraq and Afghanistan, most likely against insurgent communication networks.[24]

[19] US Department of Homeland Security, *National Infrastructure Protection Plan*, US Department of Homeland Security (2009), www.dhs.gov/xlibrary/assets/NIPP_Plan.pdf (last accessed 4 July 2011); UK Cabinet Office, *The National Security Strategy of the United Kingdom: Security in an Interdependent World*, UK Cabinet Office, Cm 7291 (2008).

[20] David A. Fulghum, Robert Wall and Amy Butler, 'Cyber-Combat's First Shot: Attack on Syria Shows Israel Is Master of the High-Tech Battle' (2007) 167(21) *Aviation Week & Space Technology* 28.

[21] David A. Fulghum, Robert Wall and Amy Butler, 'Israel Shows Electronic Prowess', *Aviation Week & Space Technology* 25 November 2007, www.aviationweek.com/aw/generic/story_channel.jsp?channel=defense&id=news/aw112607p2.xml&headline=Israel%20Shows%20Electronic%20Prowess (last accessed 12 April 2011).

[22] *Ibid.* [23] *Ibid.*

[24] David A. Fulghum and Douglas Barrie, 'Israel Used Electronic Attack in Air Strike against Syrian Mystery Target', *Aviation Week & Space Technology*, 8 October 2007,

The computer network attacks launched against Georgia in 2008, although not attributed to the Russian authorities, also demonstrate the utility of attacks launched in support of conventional strikes. In those attacks a high degree of coordination was observed between computer network attacks against targets in specific locations and their conventional bombardment. This approach prevented the Georgian authorities from keeping information flowing during a critical period both to the international media and to its own residents.[25] Other hypothetical examples include the use of computer network attacks to switch off or re-divert calls to an emergency response number after a conventional attack, causing further damage and destruction as emergency responders are grounded. Alternatively, an attack against a satellite control centre or other mission-critical facilities could severely affect a state's war effort, as could intrusion into a system which sends supplies to the front line. These examples are a few of the more commonly cited; many more are possible.

2. Law and war in the internet age

Raymond Ku has noted that with each controversy involving the internet, the law is forced to confront cyberspace on two levels.[26] The first is a consideration of what real space rules and legal regimes should apply to cyberspace. At this level we are asked to translate where possible our existing values and legal principles into values and legal principles applicable to cyberspace.[27] On a second level, providing new laws for cyberspace forces us to examine our pre-cyberworld rules as well as our commitment to the values that form the foundation for those laws.[28] Ku's dual analysis can be applied to the interpretation and promulgation of laws to govern armed conflict using computer network attacks. First, it is necessary to examine the current legal regulation of armed conflict and consider how it can be applied to computer network attacks. However, in order to do that effectively, it is necessary to return to the underlying principles for those laws and determine whether the values they seek to protect are the same for the

www.aviationweek.com/aw/generic/story_channel.jsp?channel=defense&id=news/aw100807p2.xml (last accessed 12 April 2011).

[25] See generally Tikk, Kaska and Vihul, *Cyber Incidents*, 66–89.

[26] Raymond Ku, 'Foreword: A Brave New Cyberworld' (2000) 22 *T. Jefferson L. Rev.* 125, 128.

[27] *Ibid.* [28] *Ibid.*, 129.

societies dependent on information technology who are the victims of such attacks. For example, the laws of armed conflict offer protection to civilian property as a consequence of the principle of distinction. Therefore it is necessary to revisit the reasons *why* we protect civilian property, to determine whether those principles should still apply with respect to digital property, in light of societies' changing conceptions of property as a whole and the importance of digital property to the functioning of information societies.

This need to re-address principles comes at a time when the law of armed conflict, even as it relates to conventional armed conflict, is under greater scrutiny than it ever has been in the past. Increased media attention and the proliferation of non-governmental actors involved in conflict, both as participants and observers, has resulted in the inherent tensions and ambiguities in the laws of armed conflict being forced into stark relief. Ku argues that before we can consistently apply existing law to the challenges posed by cyberspace, we must resolve conflicting values and clarify the latent ambiguities that justify existing legal rules.[29] However, while that may be an ideal solution for application to domestic law issues, such an argument cannot work at the international level. The laws relating to the use of force and the conduct of armed conflict owe their existence to a state of perpetual tension between conflicting values; in respect of the use of force it is the balancing of rights and obligations of states, and, most obviously in the case of the laws of armed conflict, the balance between humanitarian principles and military necessity.[30] Further, it is the very ambiguities that Ku is determined to resolve that allow public international law to function – in some cases consensus may only be reached by allowing for differing interpretations. Simply put, the application of the law to cyberspace – in this case computer network attack technologies – cannot be dependent on the resolution of those conflicts and ambiguities that form an integral part of the functioning of the international system. Some of the tensions that are now becoming apparent are the result of the changing character of warfare, the context in which it is waged, and the societies in which it is conducted. This book

[29] *Ibid.*, 127, citing Lawrence Lessig, *Code and Other Laws of Cyberspace* (Basic Books, New York, 1999), 119.

[30] The classic statement of this latter balance is found in the Preamble of the St Petersberg Declaration, *Declaration Renouncing the Use, in Time of War, of Explosive Projectiles under 400 Grammes Weight (St Petersburg Declaration)*, 29 November/11 December 1868.

sets out the competing approaches and examines their validity for the application of the law to computer network attack where these areas of disagreement occur.

The trends affecting modern armed conflict are happening at a societal level as well as at a military and strategic level; thus an understanding of these developments is required in order to understand the legal complexities arising from this new type of warfare. In fact, Alvin and Heidi Toffler point out: 'What is known as the [revolution in military affairs] therefore, is extremely important, but it is, nevertheless, just one facet of the larger civilisational shift, and it needs to be understood in that context.'[31] This view is shared by British military historian Jeremy Black:[32]

> ... the material culture of war, which tends to be the focus of attention, is less important than its social, cultural and political contexts and enablers. These contexts explain the purposes of military action, the nature of the relationship between the military and the rest of society, and the internal structures and ethos of the military.

That is to say, the context of warfare defines it more than the military technology it utilises.

The same context will be reflected in the laws that govern warfare through the application of the general principles which underpin it. In particular, the laws of armed conflict represent the point of balance or compromise between two dynamic forces: the requirements of humanity on the one hand, and military necessity on the other. It is the dialectical relation between these two forces, in the light of historical experience, which determines the contents, contours and characteristics of the law of armed conflict at any moment in time.[33] As Dinstein points out, these humanitarian concerns are shaped by the global *zeitgeist*, and affect the law though influencing the practice of states and the drafters of treaties.[34] The information revolution, and

[31] Alvin Toffler and Heidi Toffler, 'Foreword: The New Intangibles' in J. Arquilla, *et al.* (eds.), *In Athena's Camp: Preparing for Conflict in the Information Age* (RAND, Santa Monica, 1997), xiii–xxiv, xiv.

[32] Jeremy Black, *War in the New Century* (Continuum, London, 2001), 114, cited in Colin S. Gray, *Another Bloody Century: Future Warfare* (Weidenfeld & Nicolson, London, 2005), 84.

[33] Georges Abi-Saab, 'The Specificities of Humanitarian Law' in C. Swinarski (ed.), *Studies and Essays on International Humanitarian Law and Red Cross Principles in Honour of Jean Pictet* (Martinus Nijhoff, Geneva, The Hague, 1984), 265–80, 265.

[34] Yoram Dinstein, *The Conduct of Hostilities under the Law of International Armed Conflict* (2nd edn, Cambridge University Press, 2010), 4.

societies' resultant dependence on computers, directly affects both what will be militarily necessary in order to achieve the objective of an armed conflict and the values of the society that act as the counterbalance in applying the laws of armed conflict. It does this first by causing us to re-address the possible consequences of an attack, both in terms of what damage to civilian property might mean and in respect of assessment of knock-on effects. Second, the information revolution changes the underlying value structure of society, what is meaningful to us and worthy of protection. For example, in a society that considers internet access to be a fundamental human right, the proportionality analysis under the *jus in bello* may look very different.[35]

The following sections outline and examine some of the trends that are influencing both society and the military, and hence the legal context in which future armed conflicts will take place. This chapter places the emergence of computer network attacks as a means and method of warfare in its broader context, both in terms of the revolution in military affairs and its wider societal context, in order to understand the drivers of modern armed conflict and the values which the laws of armed conflict seek to protect.

3. Societal trends generally

3.1. The information revolution and digitisation

The information revolution is one of the defining characteristics of the current age. Advances in information technology are affecting almost every segment of business, society and government in many, if not all, regions of the world.[36] The entire structure of society, governmental leadership, national identity, production values, organisational structures and even domestic attributes such as family and religion have all been affected with a speed and global impact on a scale never seen

[35] At the time of writing, five countries had declared internet access to be a fundamental human right (Costa Rica, Estonia, Finland, France and Greece). A survey conducted across twenty-six countries by GlobeScan for the BBC World Service in March 2010 also found 79 per cent of people surveyed regard internet access as their fundamental right, regardless of their actual access. 'Internet Access is "a Fundamental Right"', *BBC News*, 8 March 2010, http://news.bbc.co.uk/1/hi/technology/8548190.stm (last accessed 12 April 2011).

[36] Richard O. Hundley, *et al.*, *The Global Course of the Information Revolution: Recurring Themes and Regional Variations* (National Defense Research Institute, RAND, Santa Monica, 2003).

before.[37] In examining the effects of this phenomenon on conflict, four main factors must be taken into account: the ubiquity of information technology, the increasing amount and decreasing cost of information, societies' changing value systems based on information, and finally the effects of increased information on organisational structures within both domestic and international society.

3.1.1. Ubiquity

As more and more information becomes digitised and bandwidth expands, societies have become increasingly reliant on networked and electronic information.[38] Information technology is being integrated into everything from appliances and vehicles to business processes and control systems.[39] More importantly, computer systems regulate air traffic control and other transportation networks, oil and gas pipelines, electricity generating systems and networks, sewerage and water treatment facilities, emergency response services, hospital systems and many other systems considered part of the critical infrastructure of modern states. Dorothy Denning has noted that this trend to ubiquitous computing affects information security in two ways: first there are more targets and more attackers; and second, attacks can have real world consequences.[40] The first point is fairly self-explanatory, the more systems that are networked and run by computers, the more open they become; thus increasing the number of possible targets and the number of attackers who have access to try and crack the system. Denning's second point regarding the real world consequences of such actions is a basic but important one. When computer network attacks were first raised as a possible risk, many were sceptical of the seriousness of the threat, seeing them as little more than 'nuisance' attacks of no real consequence for everyday life.[41] Despite continued over-excitement by the

[37] See generally, Alvin Toffler and Heidi Toffler, *War and Anti-War: Survival at the Dawn of the 21st Century* (Warner Books, London, 1994).

[38] Digitisation refers to the encoding, transformation, and transmission of all information – whether audio, video, graphics or text – into a series of binary numbers i.e. 1s and 0s. See Stephen Saxby, *The Age of Information: The Past Development and Future Significance of Computing and Communications* (Macmillan, London, 1990), 3.

[39] Dorothy E. Denning, 'Cyber-Security as an Emergent Infrastructure' in R. Latham (ed.), *Bombs and Bandwidth: The Emerging Relationship between Information Technology and Security* (Manas Publications, New Delhi, 2004), 25, 33.

[40] *Ibid.*

[41] See, for example, PBS Frontline, *Interview with James Lewis for Frontline: Cyber War! (Interview Conducted on 18 February 2003)*, www.pbs.org/wgbh/pages/frontline/shows/cyberwar/interviews/lewis.html (last accessed 15 July 2011).

media over small-scale cyber incidents, that attitude has largely been revised in the face of increased domestic computer network attacks in the form of cyber crime, the publicised results of research in the area, and the reporting in the public domain of incidents such as the Israeli attack on the suspected Syrian nuclear facility and the Stuxnet worm.

Although they are the most obvious examples, control systems are not the only link between computers and the physical world that may be affected by computer network attacks. For example, civilian vehicles, shipping and air traffic controls are increasingly equipped with navigation systems relying on GPS satellites. The same satellites which control US military positions and precision guided missiles are also relied on by cellular and data networks, financial systems, agriculture, railways and the emergency services.[42] This trend will only continue as the global information environment continues to advance at pace with the development of next generation internet providing broadband, always-on connection from multiple devices in every aspect of personal, business and public life.

3.1.2. Amount

As information technology pervades more of daily life, the sheer amount of information available is increasing at a phenomenal rate. In 1993, there were about fifty websites in the world; by the end of the decade that figure had surpassed five million.[43] One study has estimated that the total amount of new information produced in 2002 was approximately five exabytes, in print, film magnetic and optical storage media; 92 per cent of which was stored on magnetic media, mostly in hard disks, and only 0.01 per cent was stored on paper.[44] The amount of new information produced has more than doubled from the

[42] The Royal Academy of Engineers has warned that the UK has become dangerously over-reliant on GPS signals: Royal Academy of Engineering, *Global Navigation Space Systems: Reliance and Vulnerabilities*, Royal Academy of Engineering (2011), www.raeng.org.uk/gnss (last accessed 8 March 2011).

[43] Douglas McGray, 'The Silicon Archipelago' (1999) Spring *Daedalus* 147–76 cited in Joseph S. Nye, Jr, *The Paradox of American Power: Why the World's Only Superpower Can't Go It Alone* (Oxford University Press, 2002), 42.

[44] Five exabytes (5×10^{18} bytes) is equivalent to half a million new libraries the size of the US Library of Congress print collections; see Peter Lyman and Hal R. Varian, *How Much Information?*, University of California at Berkley (2003), www.sims.berkeley.edu/research/projects/how-much-info-2003/ (last accessed 5 July 2011). The same study estimates that the amount of information available on the surface of the World Wide Web (i.e. fixed web pages) is 170 terabytes, and in depth (i.e. including database-driven websites that create web pages on demand) is 91,850 terabytes. Note that this

estimated two exabytes produced in 1999. 2002 is also the first year that worldwide digital storage capacity overtook total analogue capacity; it is worth bearing in mind that as of 2007 almost 94 per cent of humankind's memory is in digital form.[45] Further, the relative cost of transmitting information has dramatically decreased, removing barriers to entry and allowing almost anyone to add information or utilise available information systems.

3.1.3. Societal change

As Marshall McLuhan once remarked: 'we become what we behold. We shape our tools and thereafter our tools shape us.'[46] It is axiomatic to say that the information revolution is fundamentally changing societies. The dramatic growth in computing and communications technologies, sometimes called the third industrial revolution, is changing the nature of governments and sovereignty, increasing the role of non-state actors, and enhancing the importance of soft power in foreign policy.[47] Indeed it can be argued that the change in the mode of communication has had a substantial effect on the distribution of power within society, on social evolution as a whole and on the values and beliefs of that society.[48] As noted above, all of these societal changes will affect the application and interpretation of the laws governing both the resort to force and armed conflict in varying degrees; two in particular deserve further examination here.[49]

First, the increased interconnection that the internet provides facilitates the spread of power to non-state actors, and allows for the strengthening of groups which share like-minded interests but not geographical boundaries. While this increased interconnection can have the effect of strengthening the state system by harnessing the power of national diasporas, it may also result in individuals turning from

figure does not include email (440,606 terabytes) or instant message (274 terabytes) information sources.

[45] University of Southern California, 'How Much Information Is There in the World?', *ScienceDaily* 11 February 2011.

[46] Marshall McLuhan, *Understanding Media: The Extensions of Man* (1st edn, McGraw-Hill, New York, 1964).

[47] Nye, *Paradox*, 43.

[48] For an exposition of 'medium' theory and its effects on the information revolution see Ronald J. Deibert, *Parchment, Printing, and Hypermedia: Communication in World Order Transformation* (Columbia University Press, New York, 1995); see generally McLuhan, *Understanding Media*.

[49] The increasing importance of intangible property will be examined in the section on the knowledge economy (section 3.3 below).

traditional sources of national identity to competing identifications based on ethnic, religious, or other ties which are not dependent on geographical location. Although such changes in self-identification can have a positive effect, for example by providing support for minorities and lessening social isolation of specific groups, they may also have a fragmentary effect by amplifying the existing divisions in a society to the point of armed conflict. For example, it is well known that Al Qaeda actively recruits using the internet and facilitates transnational identification with, and access to, jihadist ideology online.[50] The denial of service attacks which took place against Estonia in 2007 also illustrate the powerful combination of low access costs, a divisive issue and a nationalistic diaspora. The attacks, a response to the Estonian government's relocation of a Soviet-era statue, were mainly, although not exclusively, conducted from outside Estonia and were carried out by so-called 'patriotic hackers' from 178 different countries.[51]

The second societal change of interest is the effect that the information revolution has on citizens' relationships with the government and the resulting constraint on its power. For example, increased access to information allows people to see events happening around the world. This may lead to an increased humanitarian consciousness regarding human rights abuses and an increased awareness of actions carried out in the public's name. The result is the so-called 'CNN effect', the influence of media footage on foreign policy. With the advent of live satellite feeds, and now social media and citizen journalism in the global communications market, the harsh realities of conflict are beamed into the homes of citizens who see the violence committed in their name. This effect has provided both a powerful tool for governments and a constraint on their behaviour.[52] The events surrounding the shooting

[50] Although this book does not deal with hactivism, it is also worth noting the effect of groups such as the hacker collective Anonymous who have made headlines for organising online protests including computer network attacks around concepts such as freedom of information.

[51] Although some of the overseas IP addresses involved may indicate computers recruited to botnets, the botnets were only used in the second phase of the attacks. Reports indicate that code, target lists and instructions were freely available on Russian language hacker sites. See Tikk, Kaska and Vihul, *Cyber Incidents*, 23.

[52] See generally Steven Livingston, *Clarifying the CNN Effect: An Examination of Media Effects According to Type of Military Intervention*, The Joan Shorenstein Center on the Press, Politics and Public Policy, John F. Kennedy School of Government, Harvard University, R-18 (1997), www.hks.harvard.edu/presspol/publications/papers/research_papers/r18_livingston.pdf (last accessed 15 July 2011); Margaret H. Belknap, 'The CNN Effect: Strategic Enabler or Operational Risk?' (2002) Fall *Parameters* 100.

down of the US Black Hawk helicopter crew and subsequent television footage of the treatment of the body of one of the crew members in Mogadishu, Somalia in 1993, illustrated the powerful effects of the new communications environment.[53] The US government announced a withdrawal from Somalia shortly after the events of October 1993, in large part due to the overwhelming public pressure following the airing of this distressing footage.[54]

The effects of technology and in particular digital media have been widely publicised in the Arab Spring revolutions currently taking place in the Middle East and North Africa. While it is too early to reflect on the full impact or importance of this technology on the social uprisings, it cannot be doubted that it has been instrumental in the extraordinary events unfolding.[55]

3.1.4. Organisational change

One aspect has become increasingly clear; the information revolution is favouring and strengthening networked forms of organisation, often giving them an advantage over hierarchical forms.[56] For example, for businesses engaged in information-intensive enterprises, organisations that are structured as networks are proving more competitive than traditional hierarchies.[57] Secondly, the increase of information available to individuals permits them to bypass hierarchies that have – deliberately or inadvertently – controlled or limited information.[58] This is another factor enabling power to migrate to non-state actors, as they are more easily able to organise themselves into sprawling networks

[53] In October 1993, US Delta Force and Army Rangers launched a mission against Somali warlord General Aideed. Two US Black Hawk helicopters were shot down and the resulting fire-fight and mob action left several US servicemen dead. Following these events, CNN (and other media outlets) aired footage of the body of one of the servicemen being dragged through the streets of Mogadishu to the cheers of the gathered crowd.

[54] The decision to place US troops in Somalia in the first place was also seen by many to be in reaction to footage of starving refuges shown in the media.

[55] For an initial report on how activists are using social media see, for example, Peter Beaumont, 'The Truth About Twitter, Facebook and the Uprisings in the Arab World', *Guardian* (London), 25 February 2011, www.guardian.co.uk/world/2011/feb/25/twitter-facebook-uprisings-arab-libya (last accessed 12 April 2011).

[56] John Arquilla and David Ronfeldt, 'The Advent of Netwar (Revisited)' in J. Arquilla and D. Ronfeldt (eds.), *Networks and Netwars* (RAND, Santa Monica, 2001), 1–24, 1.

[57] Tracy Kidder, *The Soul of a New Machine* (Little, Brown, Boston, 1981), cited in Brian Nichiporuk and Carl H. Builder, 'Societal Implications' in J. Arquilla and D. Ronfeldt (eds.), *In Athena's Camp* (RAND, Santa Monica, 1997) 295, 297–8.

[58] Nichiporuk and Builder, 'Societal Implications', 297.

where every node can communicate with every other node, than can traditional hierarchical state actors.[59] This flexibility and effectiveness has not gone unnoticed by those fighting insurgencies against non-state actors such as the Taliban and Al Qaeda. As retired US General Stanley McCrystal has commented, 'just like their allies in al Qaeda, this new Taliban is more network than army, more a community of interest than a corporate structure … it became clear to me and to many others that to defeat a networked enemy we had to become a network ourselves'.[60] This realisation and subsequent reshaping of the organisational model in response, has a clear impact on the laws of armed conflict, particularly for issues like the knowledge requirement of command responsibility, or the requirement that a combatant must be under a responsible command.[61]

3.2. *Globalisation, interdependency and the changing world order*

Globalisation and the information revolution enjoy a symbiotic relationship, each enhancing the other's effects and capabilities. The advances in technology allow cross-border integration of communications, taxes, movement of money, goods, services and people by reducing or removing the regulatory barriers. Conversely, globalisation is shaping the world in which in the information revolution is playing out.[62] These cultural and societal effects are taking place, for good and ill, across an increasingly networked and interdependent world. Thus the information revolution has spawned the most complex and rapid period of interconnectedness and interdependence in history.[63] But if the information revolution is one of the defining characteristics of the modern age, the current era of globalisation must surely be a second. Globalisation takes many forms and affects the economic, political and societal structures in which we live. Much has been written about its effects on the causes of conflict; from a resurgence of nationalist or tribal groupings, rebellion against perceived cultural imperialism, inadequate living and working conditions created by the 'race-to-the-bottom' in the global labour market, to food riots, globalisation has been cited as a causative factor in the resulting unrest. Although

[59] Arquilla and Ronfeldt, 'The Advent of Netwar (Revisited)', 1.

[60] Stanley A. McChrystal, 'It Takes a Network' (2011) 185 *Foreign Policy* 66–70.

[61] See Chapter 5 below. [62] Hundley, *et al.*, *Global Course*, 4.

[63] As Colin S. Gray points out, this is not the first era of globalisation in history; the Huns, Alexander the Great and the empires of the Romans and the Byzantines were highly interconnected. See Gray, *Another Bloody Century*, 78–9.

far from a fixed definition, globalisation is generally understood as referring to the expansion of networks of interdependence spanning national boundaries that follows the increasingly rapid movement of ideas, money, goods, services and people across these borders.[64]

In the economic sector, globalisation has meant increased transnational production of goods,[65] decreased state control over such bastions of sovereignty as national currency,[66] and the rise of an economy based on knowledge and other intangible assets. Commodities and product markets have also gone global and are no longer heavily subject to the policies of national governments or even cartels of national governments.[67] For example, in 2001 multi-national corporations accounted for 25 per cent of world production, and sales equated to almost half of the world's GDP.[68] Although economic globalisation can be seen as the principal driving force behind contemporary globalisation, it is its effects on the political landscape, namely the form and context of state power, which are of interest in this instance.

Globalisation has empowered new actors and placed some traditional forms of governance beyond the reach of national governments.[69] In a world linked by almost instantaneous communication without regard to national borders, political associations have taken on new allegiances, authority and forms. As Held and McGrew point out, 'the intimate connection between "physical setting", "social situation" and politics, which distinguished political associations from

[64] Hundley, *et al.*, *Global Course*, 49.

[65] For an example of transnational, cross-border production of goods, see Thomas Friedman, 'Global Is Good', *Guardian* (London), 21 April 2005, G2.

[66] Daily turnover on the foreign exchange market now exceeds US$3.2 trillion, leaving state control of currency negligible, and in some cases forcing national governments to adjust their financial and monetary policies to prevent currency devaluation. Bank for International Settlements, *Trienniel Central Bank Survey: Foreign Exchange and Derivitaves Market Activity in 2007*, Bank for International Settlements (2007), www.bis.org/publ/rpfxf07t.pdf (last accessed 12 April 2011); Nichiporuk and Builder, 'Societal Implications', 302.

[67] Nichiporuk and Builder, 'Societal Implications', 304. Although as Nichiporuk and Builder point out, this is not the case with extremely rare resources such as diamonds.

[68] UNCTAD, 2001 cited in David Held and Anthony McGrew, 'Introduction' in D. Held and A. McGrew (eds.), *Governing Globalisation: Power, Authority and Global Governance* (Polity Press, Cambridge, 2002), 1–21.

[69] In addition to the financial and economic governance outlined above, states have also lost their power as the major arbiter of information in society and, in many cases, much of their control over the movement of goods and people across borders; the EU is an example.

pre-modern to modern times, has been ruptured; the new communication systems create new experiences, new modes of understanding and new frames of political reference independently of direct contact with particular peoples, issues or events.'[70] Thus, disparate groups of individuals or small collectives are now capable of exercising political power across the globe by exploiting the communications networking power of the information revolution. A prime example of the new power of these networks is the recognition of the International Campaign to Ban Landmines (ICBL) in bringing about the implementation of the 1997 Ottawa Convention.[71] The empowerment of these new actors is also happening at multiple levels. Intergovernmental organisations such as the World Trade Organisation, International Monetary Fund (IMF) and World Bank wield considerable power through structural adjustment programmes; the European Union and other regional alliances now shape policies for their members; international non-governmental groups such as the ICBL, as well as criminal and terrorist organisations such as Al Qaeda, increasingly affect the global agenda. Correspondingly, the number of these organisations has increased dramatically. At the beginning of the twentieth century there were just 37 intergovernmental organisations and 176 international non-governmental organisations (NGOs);[72] by 2006 that number had grown to 970 intergovernmental bodies and 11,859 non-governmental bodies.[73]

Both globalisation and the growing interdependence of states have also contributed to matters which had traditionally remained in the purview of the state becoming widely accepted as part of the international community's concern (most notably human rights abuses). This shift has made the maintenance of closed societies almost impossible. The interdependence between states is also highlighted by environmental globalisation. The actions of many states feed into the effects of phenomena such as global warming, the effects of which will decrease arable land and fresh accessible water which is essential to the survival

[70] Held and McGrew, 'Introduction', 6.

[71] ICBL was the joint Nobel Peace Prize winner in 1997; another NGO, Médecins Sans Frontières, won the Nobel Peace Prize in 1999.

[72] UIA, *Yearbook of International Organisations* (Brussels: Union of International Associations, 1997) cited in Held and McGrew, 'Introduction', 7.

[73] UIA, *Yearbook of International Organisations* 2003, Appendix 3, Table 1, Available at http://www.uia.org/statistics/organizations/types-oldstyle_2003.pdf (last accessed 5 July 2011).

of all states, not just those with agrarian economies. Increasingly, the availability of water will become a significant cause of conflict.[74]

3.3. *The rise of the knowledge economy*

The third societal trend of note is the rise of knowledge-based economies and the resultant change in attitudes towards intangible property. Alvin and Heidi Toffler argue that at the heart of the information revolution lies a shift in the relationship between tangible and intangible methods of production, and, as a corollary, methods of destruction.[75] They note that although knowledge, in its broadest sense, has always been a factor in the economy, in recent decades it has moved from the periphery to a central position.[76] So much so, that the World Bank has stated:

> For countries in the vanguard of the world economy, the balance between knowledge and resources has shifted so far towards the former that knowledge has become perhaps the most important factor determining the standard of living – more than land, than tools, than labour.[77]

Indeed, the dilemma of measuring and quantifying the knowledge assets of a nation, and hence its capacity for socio-economic growth, is something that academics, economists and accountants have struggled with, because it is treated as a 'residual', something that does not fit the category of tangibles, either industrial or agricultural.[78] However this residual category accounts for more than 70 per cent of most developed nations' economies.[79] This resultant change in the status of knowledge or information assets has been troubling for the legal community as well. Intellectual property rules, the most likely body

[74] Egypt has already threatened military force on a number of occasions when its privileged position on the Nile River has been threatened by upstream riparians. In 1978 Egypt threatened air strikes against a planned scheme for Ethiopia to take water from the Blue Nile; again in 1995 Egyptian President Mubarak threatened a 'response beyond anything they can imagine' when Sudan suggested it might seek to amend the 1959 Nile Waters Agreement. Michael T. Klare, *Resource Wars: The New Landscape of Global Conflict* (Metropolitan Books, New York, 2001), 158.

[75] Toffler and Toffler, 'The New Intangibles', xiv. [76] *Ibid.*

[77] World Bank, *World Development Report: Knowledge for Development*, World Bank (1998).

[78] Yogesh Malhotra, 'Measuring the Knowledge Assets of a Nation: Knowledge Systems for Development' (Paper presented at the United Nations Advisory Meeting of the Department of Economic and Social Affairs, Division for Public Administration and Development Management, Ad hoc Group of Experts Meeting – Knowledge Systems for Development, United Nations Headquarters, New York, 4–5 September 2003).

[79] *Ibid.*

of law for managing intangible property, are insufficient to deal with all intangible property dilemmas. For example, New Zealand found it necessary to amend its Crimes Act to allow electronic transfers of money to be 'things capable of being stolen' following a case in which the Court of Appeal considered that a fraudulent electronic transfer of funds was not theft.[80] In dealing with domestic offences, states have turned to a range of doctrines including theft, theft of confidential information and trespass to chattels to deal with misappropriation of digital property.[81]

Despite these difficulties, one of the defining characteristics of this age is the concept of intangible assets having hard monetary value, both as product itself and as part of the production chain. With this assignment of value our corresponding perceptions of property have also changed. Intangible property, at least for knowledge economies, has become as important as tangible property for the survival of the national economy. The movement of such intangibles to the fore is not restricted to the wealth-making sections of society; the information revolution reflects a 'civilisational shift' which can be seen in all facets of society, not least of which is the military.[82] This shift will also be reflected in our concept of what must be protected during armed conflict.

4. Military trends

Georges Abi-Saab has commented that the requirements of the principle of military necessity are defined by the evolution of military technology and strategic thought, and it is in the balance between these objective forms and the subjective requirements of humanity that the laws of armed conflict find their form and content.[83] Both technology and strategy have changed dramatically in Western militaries in recent years as the trends affecting society previously

[80] *R* v. *Wilkinson* (1999) 1 *NZLR* 403, Court of Appeal (New Zealand). The Court held that the simple electronic transfer of funds from one account to another did not amount to theft. The Court reasoned that electronic funds were not a thing 'capable of being stolen' as they were not a tangible thing, being merely an acknowledgement of a debt owed by a bank to the account holder. The problem was corrected by the Crimes Amendment Act 2003.

[81] See generally Andrew Murray, *Information Technology Law: The Law and Society* (Oxford University Press, Oxford; New York, 2010), 83–101.

[82] Toffler and Toffler, 'The New Intangibles', xiv–xv.

[83] Abi-Saab, 'Specificities', 265.

outlined have also affected the armed forces. Alvin and Heidi Toffler have long argued that the way in which a society makes war reflects the way it makes wealth; therefore as a society progresses from agricultural to industrial to a knowledge-based economy, so too do the technologies and forms of warfare available to the armed forces of that state.[84] Thus, the character of warfare is a reflection of the societal, economic and technological state of the society from which it comes.[85] The current revolution in military affairs, this sea-change in the way the military defines and thinks about carrying out its primary function, broadly reflects the transformation that is taking place in society as a whole.

Although it is evident that technology has transformed modern militaries in recent years, it is not so much the advances in weapons technology, impressive as they have been, which have had the most impact. It is the linking of those highly precise weapons to advanced sensor arrays and the joining up of multiple facets of technological advancements in command and control systems that have made the modern military so formidable. Precision munitions are made more formidable by the GPS and other sensor systems available to them. Technology has also evolved the ability to wage war to the point where the concept of a line marking the heart of the battle no longer makes sense; battlefields have become multidimensional and entire countries have become the battlespace.[86] Like the civilian sectors of society, defence forces are downsizing their operational staff and outsourcing non-essential or, in some cases, even core functions to civilian contractors. Organisational structure is becoming more decentralised with the onset of advanced command and control technology and improved infrastructure. The adoption of network-centric warfare as a framework has enabled militaries to utilise the power of networking to provide better battlespace knowledge and the related doctrine of effects-based operations allows

[84] This progression can be seen in the basic weapons of agrarian societies to the industrialised warfare of mass-produced tanks and guns, through to the high-tech weaponry seen on some of the more advanced militaries of the world. See Toffler and Toffler, *War and Anti-War*, 57–80.

[85] Gray, *Another Bloody Century*.

[86] Michael N. Schmitt, 'Asymmetrical Warfare and International Humanitarian Law' in W. Heintschel von Heinegg and V. Epping (eds.), *International Humanitarian Law Facing New Challenges: Symposium in Honour of Knut Ipsen* (Springer, Berlin; New York, 2007), 11–48, 16; Michael N. Schmitt, 'Targeting and Humanitarian Law: Current Issues' (2004) 34 *Israel Y.B. Hum. Rts* 59, 59.

the conduct of faster and more effective operations.[87] Increased situational awareness has also led to the 'pushing down' of strategic decision making so that even unit commanders in the battlespace are able and required to make strategic decisions.[88]

4.1. *A change of purpose*

General Sir Rupert Smith argues that the purpose of warfare has changed. In industrialised war, political objectives were attained by achieving strategic military objectives of such significance that the opponent conformed to the attacker's will – the intention being to decide the matter by military force.[89] Thus, Oppenheim states in his treatise on *International Law*, '[w]ar is a contention between two or more states through their armed forces, for the purpose of overpowering each other and imposing such conditions of peace as the victor pleases'.[90] Smith argues that instead we now engage in armed conflicts:[91]

> … to establish a condition or conceptual space in which the political objective can be attained by other means and in other ways. We seek to create a conceptual space for diplomacy, economic incentives, political pressure and other measures to create a desired political outcome of stability, and if possible democracy.

This move towards compellence or coercive campaigns reflects a more nuanced and complex use of military force in international relations. Michael Schmitt notes that Operation Allied Force over Kosovo serves as a classic example of a coercive or compellence campaign, as the intent was never to defeat President Slobodan Milosevic's army; rather it was to compel a return to the bargaining table and end the systematic and widespread mistreatment of the Kosovar Albanian population.[92] This change in the purpose of warfare affects the relevant doctrine to be adopted to best effect its aims.

87 An example of the 'speeding' effects of knowledge sharing in the battlespace can be seen in: Joshua Davis, 'If We Run out of Batteries, This War Is Screwed', *Wired* 6, June 2003, www.wired.com/wired/archive/11.06/battlefield.html (last accessed 12 April 2011).

88 The need for unit commanders to make strategic decisions was raised by Charles C. Krulak, 'The Strategic Corporal: Leadership in the Three Block War', *Marines Magazine* 1, January 1999, 28–34.

89 Rupert Smith, *The Utility of Force: The Art of War in the Modern World* (Penguin, London, 2005), 270.

90 Hersch Lauterpacht (ed.), *Oppenheim's International Law* (7th edn, Longmans, Green & Co., London, 1952), 202.

91 Smith, *Utility of Force*, 270.

92 Schmitt, 'Asymmetrical Warfare', 37.

4.2. *Network-centric warfare and effects-based operations*

Changes in the military are not restricted to the weaponry available to them; the organisational changes which have occurred in the commercial sector are now being implemented to yield the same benefits to the military. Modern militaries are moving towards network-centric warfare which enables them to leverage power from the effective linking or networking of their forces.[93] In traditional platform-centric warfare each component, be it a tank formation, battleship or aircraft, has its own mission and directives, albeit sometimes working in coordination. However, network-centric warfare uses the network itself to provide a combat advantage through increased situational awareness and collaboration between the components of the network, thus increasing the speed at which the forces can operate and enhancing mission effectiveness.[94] The speed of decision-making required in order to fully utilise network-centric warfare has also resulted in decision-making capabilities being pushed down the chain of command.[95] The true value of the network-centric warfare framework in the new environment however, is in the application of effects-based operations;[96] this doctrine, which has become dominant in Western military thinking, operationalises both the change in purpose of armed force and the move toward network-centric warfare.

Effects-based operations are coordinated sets of actions directed at shaping the behaviour of friends, foes and neutrals in peace, crisis and war.[97] In traditional attrition warfare, reduced to its basics, the enemy is defeated by progressively weakening its military forces.[98] This is reflected in the preambular principle set out in the St Petersburg Declaration of 1868, which states that the only legitimate object states should endeavour to accomplish during war is to weaken the military

[93] Network-centric warfare principles have been adopted by several militaries under various rubrics: network-enabled capability in the UK; network-based defence in Sweden; ubiquitous command and control in Australia.

[94] See generally David S. Alberts, John Garstka and Frederick P. Stein, *Network Centric Warfare: Developing and Leveraging Information Superiority* (2nd edn, National Defense University Press, Washington DC, 1999).

[95] Andrew M. Dorman, *Transforming to Effects-Based Operations: Lessons from the United Kingdom Experience*, Strategic Studies Institute (2008), 18.

[96] Effects-based operations are not a new concept, however their application in light of network–centric warfare is interesting. Effects-based operations are coordinated sets of actions directed at shaping the behaviour of friends, foes and neutrals in peace, crisis and war. Edward A. Smith, *Effects Based Operations: Applying Network Centric Warfare to Peace, Crisis, and War* (DOD-CCRP, Washington DC, 2002), 108.

[97] *Ibid.*, xiv. [98] Schmitt, 'Targeting', 60.

forces of the enemy.[99] However, effects-based operations utilise selective targeting and choice of means and methods of warfare to achieve a desired effect.[100] As Michael Schmitt has noted, although effects-based operations have the potential to foster international humanitarian law by systematising the search for alternative targets, they may also lead to the temptation to strike at targets which are not military in nature in order to coerce specific behaviours from opponents who may not value their military capability as highly.[101] This is discussed in more detail in Chapter 6 with respect to targeting.

4.3. *Outsourcing and civilianisation*

Another trend which has impacted the application of the laws of armed conflict is the increasing civilianisation of modern conflict. This trend is taking place through a number of processes, including the escalating prominence of internal armed conflicts in which the majority of war fighters are civilians, and the shift of the conduct of hostilities into civilian population centres. In addition, modern militaries increasingly outsource support and even core functions to contractors – some of whom, like private military or security firms, are engaged in armed tactical roles.

In the three-and-a-half centuries since the Treaty of Westphalia, the nation state has been the defining actor in international relations, and has held the monopoly on power and military force. The emergence of transnational armed groups, the increasing number of non-international armed conflicts and the expansion of the battlespace to encompass entire territories have all meant that civilians are involved in conflicts, both as participants and victims, more than ever.

Militaries are also under pressure to downsize and reduce budgets. As part of this trend, civilian contractors and employees are increasingly used to augment the defence forces as an easy and flexible way to maintain military strength according to constantly changing needs. Further, as weapons and equipment become more technologically advanced, civilians are recruited to provide essential maintenance

99 *St Petersburg Declaration*, Preamble.

100 Ultimately, the process addresses the causality between actions and their effects; concentrates on desired effects, both physical and behavioural; models the enemy as a system of systems; and considers timing because the desirability of specific effects depends on the context in which they are created. Schmitt, 'Asymmetrical Warfare', 36–7.

101 *Ibid.*, 37–8.

and support functions, sometimes from the 'factory to the foxhole'.[102] Civilians are an easy and less expensive way of maintaining access to the latest technical expertise; they can be hired when needed and discharged when the need is no longer urgent.[103] Likewise they do not require the ongoing provision of accommodation, catering, healthcare and the myriad of other services which are required to support members of the armed forces. Nowhere has the use of private military firms been more extensive and controversial than the recent conflict in Iraq. In March 2005 there were more than 20,000 foreign (non-Iraqi) private military contractors in Iraq, 6,000 of these in armed tactical roles.[104] According to the latest US Department of Defense figures, by December 2010 that figure had increased to 60,719 non-Iraqi contractor personnel (compared to approximately 47,305 uniformed personnel in country), 13,000 of whom were involved in security functions.[105] The report also notes that since December 2009, the number of contractors in Afghanistan has exceeded the number in Iraq.[106]

Civilianisation of conflict is also occurring with the growing interconnectedness of systems and the increase of dual-use objects. Cost considerations also make the military more likely to rely on civilian facilities such as airfields, ports and other communications centres.[107] For example, military communications often utilise civilian networks, particularly where they travel over satellites.[108] Enhanced interconnectedness also means that knock-on effects of attacks are more likely to affect more civilian systems than in previous conflicts. All of these factors have added to the difficulty in applying the laws of armed conflict,

[102] Michael E. Guillory, 'Civilianising the Force: Is the United States Crossing the Rubicon?' (2001) 51 *AFL Rev.* 111, 125, citing an example of Apache helicopter support technicians deployed during Desert Storm.

[103] Outsourcing allows militaries to take advantages of the competitive advantages of the contracting process.

[104] P. W. Singer, 'Outsourcing War' (2005) 84(2) *Foreign Affairs* 119.

[105] Moshe Schwartz and Joyprada Swain, *Department of Defense Contractors in Afghanistan and Iraq: Background and Analysis*, Congressional Research Service (2011), 13, www.fas.org/sgp/crs/natsec/R40764.pdf (last accessed 12 April 2011). No separate breakdown is given for the relative nationalities of those contractors involved in the provision of security services.

[106] *Ibid.*

[107] Michael N. Schmitt, *The Impact of High and Low-Tech Warfare on the Principle of Distinction*, Program on Humanitarian Policy and Conflict Research at Harvard University (2003), 8, http://ihl.ihlresearch.org/_data/n_0002/resources/live/briefing3296.pdf (last accessed 5 July 2011).

[108] Arkin puts the figure at 95 per cent of military communications travelling over civilian satellites in 1995.

in particular with respect to the fundamental principle of distinction; analyses of proportionality, precautions in attack, direct participation in hostilities have all been complicated by this increased civilian role in warfare.

5. Terminology and definitions

While an effort has been made to avoid computer jargon, a work of this kind must of necessity use technical terms and computer terminology. The author has endeavoured to provide definitions and explanations in the text; however, a glossary of computing and technical terms has been included in the appendices for the reader's convenience.

Writers in this area have also used a changing lexicon as the field has evolved. While Russian experts are still pushing for an official declaration of definitions,[109] most writers in the field have gradually adopted the US Department of Defense terms and definitions. In the beginning most legal analysts wrote in terms of 'information warfare', basing their definitions and analysis on the framework provided by Martin Libicki's seminal work.[110] Over time this term came to refer to a specific subset, namely the propaganda and misinformation aspects, of a wider field called information operations (IO), and the most recent US Joint Publication on Information Operations removes the term information warfare entirely from its lexicon.[111] The term computer network operations is now used to refer collectively to computer network attack, the subject of this book, computer network defence and computer network exploitation,[112] the intelligence-gathering aspect of such

[109] See, for example, Anatolij Streltsov, 'Threat Analysis in the Post Cold-War Order' (Paper presented at the International Expert Conference on Computer Network Attacks and the Applicability of International Humanitarian Law, Stockholm, 17–19 November 2004) 21–7, 21.

[110] Martin C. Libicki, *What Is Information Warfare?* (Center for Advanced Concepts and Technology, Institute for National Strategic Studies, Washington DC, 1995). Martin Libicki's original definition of information warfare is generally compatible with the currently accepted definition of Information Operations in that it is an umbrella term which comprises seven subcategories.

[111] US Department of Defense, *Information Operations*, Joint Chiefs of Staff, Joint Publication 3-13 (2006), iii.

[112] The US *Dictionary of Military Terms* defines computer network defence as: 'Actions taken to protect, monitor, analyze, detect, and respond to unauthorized activity within the Department of Defense information systems and computer networks'; computer network exploitation is: 'Enabling operations and intelligence collection capabilities conducted through the use of computer networks to gather data from target or adversary automated information systems or networks.'

operations. The definition of 'computer network attack' used in this work and discussed in section 1 above is that adopted by the United States joint forces and remains the standard to which most authors now subscribe.

5.1. *New laws for old?*

The primary assertion of this book is that although computer network attacks raise challenging issues for the current laws of armed conflict, for the most part, existing laws are capable of adapting to the new technology. Indeed, the Martens Clause was drafted with exactly this eventuality in mind,[113] and the International Court of Justice has noted that the clause 'has proved to be an effective means of addressing the rapid evolution of military technology'.[114] Despite some calls for there to be a new convention which addresses the issues raised by computer network attacks and other information operations,[115] the present author feels that this is unnecessary. Those who call for a new convention have generally subscribed to the idea that cyberspace represents a fundamentally different conceptual space in which to fight. However, their approach, illustrated by Brown's assertion that 'Cyberspace is nowhere', is simply not reflective of state practice in relation to other areas of internet law. There does not exist some 'matrix-like' realm of cyberspace which bears no connection to the 'real-world'. Actors still act in physical space; hardware and networks (even wireless and virtual ones) still require physical constructs. However, that is not to argue that computer network attacks fit neatly into the humanitarian law paradigm that has developed over the last century. But it is not the advent of cyberspace, *per se*, that is the problem.

[113] The Martens Clause has found expression in IHL treaties since its first inclusion in the preamble to the 1899 Hague Convention (II) with Respect to the Laws and Customs of War on Land. A modern restatement can be found in Art 1(2) of Additional Protocol I: 'In cases not covered by this Protocol or by other international agreements, civilians and combatants remain under the protection and authority of the principles of international law derived from established custom, from the principles of humanity and from dictates of public conscience.' The clause represents customary international law.

[114] *Legality of the Threat and Use of Nuclear Weapons* (1996) ICJ 226, International Court of Justice, para. 78.

[115] See, for example, Davis Brown, 'A Proposal for an International Convention to Regulate the Use of Information Systems in Armed Conflict' (2006) 47 *Harv. Int'l L.J.* 179; Duncan B. Hollis, 'Why States Need an International Law for Information Operations' (2007) 11(4) *Lewis & Clark L. Rev.* 1023.

The general principles of the laws of armed conflict are aimed at ameliorating the essential nature of conflict, which, unlike its rapidly evolving character, remains unchanged. The St Petersburg Declaration was founded on a common agreement to fix the technical limits at which the necessities of war ought to yield to the requirements of humanity. The parties were agreed:

> That the progress of civilization should have the effect of alleviating as much as possible the calamities of war; That the only legitimate object which States should endeavour to accomplish during war is to weaken the military forces of the enemy.[116]

As described above, the information revolution has fundamentally transformed society on multiple levels. Where the cultural and societal ground shifts, the underlying concepts on which our laws are based may also change; for example, the attributes of physical property that have bound formulations and interpretations of law to the tangible world are challenged by the value placed on digital property. Laws, like wars, reflect the principles and values of the societies which draft them. As Georges Abi-Saab has commented, the 'requirements of humanity' are subjective, depending on the dominant moral ideas and degree of community feeling obtained among the main contenders in society.[117] Those principles can be altered by something as momentous as the information revolution, to the point where values are fundamentally changed. Thus, when reviewing the general principles of the laws of armed conflict, this shift in values must be taken into account and the reasoning behind the statements of principle explored. Where specific laws which relate to particular means and methods of warfare sit uneasily with the concept of computer network attacks, they may need reinterpretation as discussed below.

One final point requires comment: while the move toward high-tech warfare in developed nations is increasing, warfare in many states remains for the most part brutal, physical and violently immediate. It can be easy to lose track of the purpose of these laws when one is dealing with the abstract world of bits and bytes and when targeting and indeed hostilities can be conducted from the safety of an office block half a world away from the battlespace. It is imperative that any attempt to interpret the laws of armed conflict to apply to computer

[116] *St Petersburg Declaration.* [117] Abi-Saab, 'Specificities', 265.

network attacks, must remain applicable to the traditional forms of kinetic violence for which they were first envisaged.

5.2. Methodology

It may be clear from the foregoing that this book utilises many of the tools of modern positivism;[118] however, as whole, the author has adopted a hybrid approach to international law as it applies to computer network attack. While the author has attempted to delineate carefully between *lex lata* and *lex ferenda* throughout the book, where the technology does not fit easily with existing law, a closer examination of the general principles was required. As Judge Higgins (as she then was) points out in her dissenting opinion in the *Nuclear Weapons* case:[119]

> Humanitarian law is very well developed. The fact that its principles are broadly stated and often raise further questions that require a response can be no ground for a *non liquet*. It is exactly the judicial function to take principles of general application, to elaborate their meaning and to apply them to specific situations.

This approach has been adopted when discussing the application of the general principles of humanitarian law to computer network attacks – reasoned development of authoritative starting points consistent with the object and purpose of the law of armed conflict. The author subscribes to the view that principles based on values, such as 'the dictates of humanity', are necessarily subjective and will change over time in accordance with the prevailing view of the societies from which they come,[120] but these general principles are already incorporated into the laws of armed conflict either through treaties or as obligations under customary international law. As already noted, the laws of armed conflict represent a carefully constructed balance between military necessity on the one hand and humanitarian principles on the other. To introduce a further level of humanitarian or sociological interpretation to principles of which states and, more importantly, individuals may be held in violation, is disingenuous to the carefully negotiated drafting process which often relies on minimal consensus and may

[118] For an exposition of modern positivism see the *American Journal of International Law* Symposium on Method in International Law article by Bruno Simma and Andreas L. Paulus, 'The Responsibility of Individuals for Human Rights Abuses in Internal Conflicts: A Positivist View' (1999) 93 *AJIL* 302.

[119] *Nuclear Weapons Case* (dissenting opinion of Judge Higgins), para. 32.

[120] Abi-Saab, 'Specificities', 265.

deliberately ignore the underlying interpretive or conceptual debates in order to achieve a measure of protection. Further, as Judge Cassese has commented: '[A] policy-oriented approach in the area of criminal law runs contrary to the fundamental customary principle *nullum crimen sine lege*.'[121]

Nevertheless, where this book has found the law lacking in detail, or references *lex ferenda*, a more policy-oriented approach is called for. However, it is the guiding principles of the laws of armed conflict which are to be referenced; that is, those principles which form the basis of, and are incorporated into, humanitarian law instruments, such as those prohibiting unnecessary suffering, distinction between civilian and military targets or requiring proportionality. These general principles of humanitarian law are aimed at the unchanging nature of war itself. While certain laws are aimed at the specific character of war (the technologies employed and the strategies involved), most obviously a number of the Hague Regulations and subsequent weapons conventions, when taken back to first principles and being conscious of the reasons for which they were adopted, the general principles of the laws of armed conflict will apply regardless of the technology utilised.

5.3. *A word about examples and hypotheticals*

This book deals with a method of warfare which is still in its infancy. Further, most computer network attack tools remain classified in order to protect their usefulness; once details are known by the adversary, they may be guarded against, making the tool ineffective. Likewise states (and private corporations) have historically not been open about admitting when systems have been compromised. For example, it took two years for the US Department of Defense to confirm and declassify a significant compromise of classified military networks that resulted in the loss of an undisclosed amount of information, despite reports circulating in the press at the time of the attacks.[122]

Throughout this volume, examples have been taken from domestic and peacetime computer network attacks and incursions to illustrate

[121] *Prosecutor* v. *Drazen Erdemovic* (1997) IT-96–22-A, International Criminal Tribunal for the Former Yugoslavia, Appeals Chamber (separate and dissenting judgment of President Cassesse), para. 11.

[122] Compare Noah Shachtman, 'Under Worm Assault, Military Bans Disks, USB Drives', *Wired* 19 November 2008, www.wired.com/dangerroom/2008/11/army-bans-usb-d/ (last accessed 12 April 2011); with the confirmation and information about the subsequent response in William F. Lynn, III, 'Defending a New Domain – the Pentagon's Cyberstrategy' (2010) 89(5) *Foreign Affairs* 97.

points. The aim is not to use them for their precedent value in any direct application to the laws of armed conflict, an approach which would obviously be incorrect, but rather to act as practical, real-world illustrations of the types of attacks and incursions which have taken place. Further, they serve as an indicator, where possible, of the current state of legal analysis of the concepts underlying computer network attacks and digital property.

The present author does not have security clearance and all examples used are those gleaned from publicly available sources and domestic instances of computer network attacks which indicate how such an attack could work if carried out by a party during an armed conflict. Given the covert nature of this attack form and the reluctance of victims to make details of the attacks available, the facts of each incident are often disputed in the press and remain somewhat murky. Although the facts of each example are provided in the text, a timeline and summary of each attack with its significance for the development of the technology is provided in Appendix 1 for the convenience of the reader, flagging up where disputes over the accuracy of reports has occurred. Where hypothetical examples have been used, they have been checked with computer engineers and network specialists for their general viability. In so far as is possible without compromising security, the details of specific hypothetical attacks have been checked with those with first-hand knowledge of the proposed target. The author is grateful for their assistance with these matters, and the fault for any errors remains entirely with the author. It will also be apparent that this book is fairly US-centric in its use of examples. This is for the simple reason that the United States has the most publicly available English language information and analysis of this field; where possible the author has attempted to include examples from other jurisdictions.

6. Conclusion

There are several main themes running through this book which have been set out and explained throughout this chapter; by way of summary, they are repeated here. First, despite the advent of new technology, the essential nature of armed conflict remains the same – it is the advancement of political objectives by organised violence.[123] Second,

[123] See generally Carl von Clausewitz, J. J. Graham and F. N. Maude, *On War* (new and rev. edn, Kegan Paul, Trench, Trubner & Co., London, 1940); Gray, *Another Bloody Century*.

the style and character of warfare reflects the society from which it is waged. Thus, computer network attacks are a product of, and are the greatest threat to, those societies which place a high value on information. And finally, the underlying principles of the laws of armed conflict are aimed at the fundamental nature of war. The exact content and contours of these principles are determined by the prevailing values of the society affected and by the realities of warfare. This can be seen in the way that the requirements to take precautions in attack have changed over time.

When research on this project began there was no comprehensive analysis of the laws of armed conflict as they related to computer network attack. Although some attempts had been made at uncovering the law which might apply, the studies were more descriptive than analytical.[124] During the course of writing first the thesis and now this book, computer network attack has become a fashionable topic of debate and numerous articles have been written on how computer network attacks might be governed under discrete areas of the laws of armed conflict. However, with the notable exception of a series of articles written by Michael Schmitt, none offer a systematic analysis based on a complete underlying framework. This book aims to fill that gap. Despite the attempt to adopt a comprehensive approach, a few omissions have had to be made in the interests of length. The book does not examine the laws of neutrality, nor does it look in any detail at the implications of computer network attacks for naval warfare or as an internationally wrongful act short of a use of force.

[124] See Walter G. Sharp, *Cyberspace and the Use of Force* (Aegis Research Corp., Falls Church, VA, 1999); Thomas C. Wingfield, *The Law of Information Conflict: National Security Law in Cyberspace* (Aegis Research Corp., Falls Church, VA, 2000).

PART I
Jus ad bellum

2 Computer network attacks as a use of force in international law

In June 2010 the computer security industry became aware of the Stuxnet worm, a virus aimed at industrial control systems in manufacturing and utility companies. The complexity and sophistication of the virus and its multi-pronged attack vectors quickly led researchers to the conclusion that the most likely authors of the malware were state-backed professionals.[1] The fact that Iran was the most heavily hit target of the worm, including its Bashir and Natanz nuclear facilities, also added to the speculation.[2] Iranian President Mahmoud Ahmadinejad publically accused Israel and the West of being behind the attacks, announcing that unspecified malicious software sent by Western enemies had affected Iran's centrifuges at its Natanz enrichment plant and 'succeeded in creating problems for a limited number of our centrifuges'.[3] However, at the time of writing no party has officially claimed responsibility and computer forensics have not conclusively established an author.[4] Although there are no official

[1] Stuxnet exploits four different vulnerabilities, a previously unheard of level of complexity in malware. Nicolas Falliere, Liam OMurchu and Eric Chien, *W32.Stuxnet Dossier*, Symantec (2011) version 1.4, www.symantec.com/content/en/us/enterprise/media/security_response/whitepapers/w32_stuxnet_dossier.pdf (last accessed 12 February 2011).

[2] Approximately 60 per cent of all infected PCs are located in Iran. *Ibid.*

[3] 'Iran Says Cyber Foes Caused Centrifuge Problems', Reuters, 29 November 2010; 'Iran Says Nuclear Programme Was Hit by Sabotage', *BBC News*, 29 November 2010. Iranian officials have also admitted that Stuxnet infected some computers at the Bashir nuclear power station but denied that any of the facility's control systems were affected.

[4] In a potentially embarrassing turn for Israel, the success of the Stuxnet cyber attack was celebrated in a video tribute of operational successes for the retiring head of the Israeli Defence Forces. Christopher Williams, 'Israeli Security Chief Celebrates Stuxnet Cyber Attack', *Telegraph* (London), 16 February 2011. Others claim US

reports of physical damage in the public domain, it is known that Iran decommissioned and replaced approximately one thousand IR-1 centrifuges in a short space of time, implying larger than expected levels of breakage.[5] Footage from IAEA installed surveillance cameras tends to suggest that the work was carried out 'feverishly' in response to damage and around the same time Stuxnet is believed to have attacked the plant.[6] Introduced into computers at the Iranian nuclear facilities via an infected USB device into a gateway target, the worm is designed to seek out its final target and cause damage by making quick changes in the rotational speed of motors and sabotaging the normal operation of control systems.[7] By adjusting these speeds the virus could ultimately cause centrifuges (such as those used in nuclear enrichment facilities, the purported target of the virus) to fly apart. Converters are also used to control gas pipelines, chemical plants and a number of other different machines. It should also be recalled that the 'Farewell Dossier' pipeline explosion was the result of a piece of malware operating on the same principle;[8] thus the seriousness of the threat should not be underestimated.

The cyber attacks on Estonia offer a different perspective on computer network attacks as a possible use of force. On 27 April 2007 Estonia became the victim of a prolonged series of distributed denial of service (DDoS) attacks which brought the banking system, many government services and much of the media to a halt.[9] Although no critical infrastructure was compromised, for a highly technologically dependent

involvement in testing and development, although all are unsubstantiated. See, for example, Kim Zetter, 'Did a U.S. Government Lab Help Israel Develop Stuxnet?', *Wired* 17 January 2011, www.wired.com/threatlevel/2011/01/inl-and-stuxnet/ (last accessed 10 April 2011).

[5] Albright, Brannan and Walrond, *ISIS Stuxnet Report*; Kim Zetter, 'Report Strengthens Suspicions That Stuxnet Sabotaged Iran's Nuclear Plant', *Wired* 27 December 2010, www.wired.com/threatlevel/2010/12/isis-report-on-stuxnet/# (last accessed 13 February 2011).

[6] Zetter, 'Surveillance Footage and Code Clues Indicate Stuxnet Hit Iran'.

[7] William J. Broad and David E. Sanger, 'Worm Was Perfect for Sabotaging Centrifuges', *New York Times* (New York), 18 November 2010, A1. The worm targets frequency converters, a type of power supply that changes its output frequency to control the speed of a motor.

[8] See Chapter 1 above, and Appendix 1 for a description of this event.

[9] A distributed denial of service attack (DDoS) uses many compromised computers to flood a target system with requests for information until it collapses under the strain. The compromised computers are usually ones that have been recruited to a botnet (usually without their owner's knowledge) and are controlled by a master computer.

state like Estonia that depends on the internet for everything from parking to banking to voting, the attacks caused serious disruption and caused an estimated tens of millions of euros' worth of damage.[10] Despite earlier explicit accusations that the Russian Federation was behind the offensive,[11] the Estonian government backed away from directly accusing the Kremlin of launching the attacks,[12] but requested assistance from its NATO allies under the terms of that alliance. Although no official statement regarding the cyber attacks was released by NATO, one of the clearest indications of the member states' views on computer network attacks came from Estonian defence minister Jaak Aaviksoo, who raised the matter with NATO:

> At present, NATO does not define cyber-attacks as a clear military action. This means that the provisions of Article V of the North Atlantic Treaty, or, in other words collective self-defence, will not automatically be extended to the attacked country ... Not a single NATO defence minister would define a cyber-attack as a clear military action at present. However this matter needs to be resolved in the near future.[13]

This attitude is perhaps not surprising given the general reluctance of states to consider acts of indirect aggression as armed attacks;[14] however, the incident is significant in that it is the first time that one state accused another of intentionally launching a computer network attack against it. In 2009 members of the Kremlin-backed youth group Nashi claimed that they were behind the attacks but denied that they were acting on the orders of the Russian government;[15] the claim has not been corroborated. The incident bought to the fore important issues for the law regulating the use of force in international relations. While

[10] Ian Traynor, 'Web Attackers Used a Million Computers, Says Estonia', *Guardian* (London), 18 May 2007, International 30.

[11] Tony Halpin, 'Putin Accused of Launching Cyber War', *The Times* (London), 18 May 2007, Overseas News 46.

[12] Traynor, 'Web Attackers Used a Million Computers, Says Estonia'. Russia categorically denies any involvement and subsequent reports have found no concrete evidence to substantiate the claims. While technical data shows that some of the attacks came from IP addresses allocated to the Russian government, there is no evidence that these computers were involved in initiating the attacks, or that they had not been compromised or spoofed.

[13] Ian Traynor, 'Russia Accused of Unleashing Cyberwar to Disable Estonia', *Guardian* (London), 17 May 2007, Home 1.

[14] See generally, Chapter 3 below.

[15] Charles Clover, 'Kremlin-Backed Group Behind Estonia Cyber Blitz', *Financial Times* (London), 11 March 2009, 8, www.ft.com/techblog (last accessed 27 October 2010).

controversies surrounding the rule prohibiting the use of force had tended to focus on issues such as the conditions for self-defence or the existence of a right to humanitarian intervention,[16] the advent of computer network attacks has renewed a far more fundamental question: what is the meaning of 'force' in the twenty-first century?

The prohibition of the use of force is one of the cornerstones of international law and is expressed in Article 2(4) of the UN Charter:

> All members shall refrain in their international relations from the threat or use of force against the territorial integrity or political independence of any state, or in any other manner inconsistent with the purposes of the United Nations.

However, the prohibition is not restricted to the Charter; the International Court of Justice has repeatedly noted that it represents not only a principle of customary international law but also a 'fundamental or cardinal principle of such law'.[17] The Court also recalled that the International Law Commission had considered this provision to have the character of *jus cogens*.[18] However, it must be recalled that Article 2(4) abolishes the threat or use of force only in the 'international relations' of member states; intrastate clashes and other actions by non-state actors are therefore out of reach of the Charter's provision.[19]

1. Force defined as armed force

Although the prohibition of the threat or use of force is widely acknowledged as a cornerstone of both the United Nations Charter and of customary international law, Article 2(4) and the exact meaning of the word 'force' have been a source of debate since its enactment. The drafters of the Charter did not define the term, nor has the International Court of Justice or the General Assembly done so

[16] Olivier Corten, 'The Controversies over the Customary Prohibition on the Use of Force: A Methodological Debate' (2005) 16 *EJIL* 803.

[17] *Case Concerning Military and Paramilitary Activities in and against Nicaragua (Nicaragua* v. *United States of America) (Merits)* (1986) ICJ 14, International Court of Justice, para. 190. See also *Legal Consequences of the Construction of a Wall in Occupied Palestinian Territory* (2004) ICJ 136, International Court of Justice, para. 87. *Case Concerning Armed Activities on the Territory of the Congo (Democratic Republic of the Congo* v. *Uganda)* (2005), International Court of Justice, para. 148.

[18] *Nicaragua (Merits)*, para. 190.

[19] Yoram Dinstein, *War, Aggression, and Self-Defense* (3rd edn, Cambridge University Press, New York, 2001), 85.

since. One such debate has centred on whether 'force' is limited to armed force, or includes other forms of coercion such as political and economic measures. The issue of inclusion of political and economic coercion as uses of force prohibited by Article 2(4) of the Charter has been raised repeatedly, particularly by developing and former Eastern Bloc countries, since the San Francisco conference.[20] Although no definitive conclusions have been drawn, the prevailing and commonly accepted view put forward by scholars is that the force referred to in Article 2(4) is limited to armed force.[21] As computer network attacks can result in a myriad of outcomes, and apparent state-sponsored attacks against financial infrastructure increase, these questions are likely to come to the fore again.[22] Thus, the contours of the prohibition must be fleshed out in order to ascertain when attacks will be proscribed under Article 2(4) and the corresponding customary law relating to the use of force in international law. The following discussion looks at the arguments from a textual analysis of the Charter wording, *travaux préparatoires*, historical background and academic analysis.[23]

The Charter wording

The text of the United Nations Charter refers to force as an unqualified term twice in the entirety of the document, in Article 2(4) and Article 44. The use of the term in Article 44 appears in the context of Chapter VII, and supports a restrictive definition of 'force' by placing it in close conjunction with Articles 41 and 42 using the qualified term 'armed force', clearly suggesting that the force contemplated by the unqualified term is armed. Article 44 states:

[20] Albrecht Randelzhofer, 'Article 2(4)' in B. Simma (ed.), *The Charter of the United Nations: A Commentary* (2nd edn, Oxford University Press, 2002), 118.

[21] See, for example, *ibid.*, 117; Dinstein, *War, Aggression, and Self-Defense*, 86.

[22] For example, in 2010 both the Nasdaq and London stock exchanges were hacked, although at the time of writing both attacks appear to be aimed at exfiltrating data rather than crashing the systems or altering trading algorithms.

[23] Art. 31 of the Vienna Convention on the Law of Treaties 1969 sets out the core interpretive principles that a treaty should be interpreted in 'good faith' and in accordance with 'the ordinary meaning to be given to the terms of the treaty in their context and in the light of its object and purpose'. Interpreters of a treaty may include in the context: the preamble, any treaty or instrument made in connection with the treaty, subsequent conduct and practice of the parties and any relevant rules of international law applicable in the relations between the parties. Art. 32 states that recourse may be made to supplementary means of interpretation, including the preparatory work of the treaty and the circumstances of its conclusion. The following sections consider each of these in turn.

When the Security Council has decided to use force it shall, before calling upon a Member not represented on it to provide armed forces in fulfilment of the obligations assumed under Article 43, invite that Member, if the Member so desires, to participate in the decisions of the Security Council concerning the employment of contingents of that Member's armed forces.

The decision to use force refers to the preceding Articles 41 and 42 of the Charter, which permit the Security Council to authorise actions necessary to maintain or restore international peace and security, including the use of force. Article 41 relates to 'measures not involving the use of armed force' which may include 'complete or partial interruption of economic relations and of rail, sea, air, postal, telegraphic, radio, and other means of communication, and the severance of diplomatic relations'.[24] Article 42 allows the Security Council to 'take such action by air, sea, or land forces as may be necessary to maintain or restore international peace and security' and is the article under which the Security Council mandates the use of armed force in international law under the phrase 'all necessary means'.[25] Thus the use of the term 'force' to refer to the Security Council's choice of the latter of these two options – specifically distinguished from measures not amounting to 'armed force' in the text of Article 41 – suggests that the force referred to in Article 44 and hence in Article 2(4) is also armed force.

Further support for this view can be found in paragraph 7 of the preamble of the Charter which states that 'armed force shall not be used, save in the common interest'. As Michael Schmitt has pointed out, the articles of the Charter are designed to give effect to its preambular aspirations.[26] Accordingly, if Article 2(4) were intended to extend beyond armed force, then presumably the preamble, for reasons of internal consistency, would not have included the term 'armed'. This paragraph accords with the relationship apparent between Article 2(4) and Chapter VII of the Charter, in particular Article 42 relating to Security Council authorisation of armed force, as discussed above.

[24] Note that this would tend to indicate that none of these measures would be considered breaches of Article 2(4) were they to be taken unilaterally by individual states, whether they were conducted by computer network attacks or by more conventional means.

[25] The list of actions includes 'demonstrations, blockade, and other operations by air, sea, or land forces'.

[26] Michael N. Schmitt, 'Computer Network Attack and the Use of Force in International Law: Thoughts on a Normative Framework' (1999) 37 *Col. J. Trans. L.* 885, 904.

Albrecht Randelzhofer has also put forward a teleological interpretation of the wording of Article 2(4). Randelzhofer argues that were Article 2(4) to extend to other forms of force, such as economic and political coercion, states would be left with no legal means of exerting pressure on states that violate international law.[27] Such a consequence would be unacceptable to the international community in an age where the organs of that community are unable to effectively ensure compliance with international law.

Travaux préparatoires *and historical background*

The preparatory materials of the Charter do not contain any specific discussion regarding the precise meaning of the term force in Article 2(4); however, the discussions around a number of proposed amendments to the article shed some light on the intentions of the states at the time.

The *travaux préparatoires* of Article 2(4) detail a proposal by the foreign minister of Brazil to specifically extend the prohibition to the threat or use of 'economic measures' which was firmly rejected.[28] Although Randelzhofer has cited this as evidence that military force is the only intended concern of the prohibition, David Harris states that it is unclear from the texts whether the rejection of the Brazilian amendment is proof that the Article was not intended to prohibit economic force, or that the term force in Article 2(4) was thought sufficient to cover it without specific mention.[29] The latter view appears to stem from the Belgian delegate's comments regarding Brazil's proposed amendment and the phrase 'or any other manner'.[30] Despite this, writers have generally concluded that the better view is that Western states were not prepared to admit anything other than armed force.[31]

[27] Randelzhofer, 'Article 2(4)', 118.

[28] 'Summary Report of Eleventh Meeting of Committee I/1' Doc. 215, I/1/10, 6 UNCIO (6 May 1945) 334, 559 (rejected by 2 votes in favour, 26 against).

[29] D. J. Harris, *Cases and Materials on International Law* (6th edn, Sweet & Maxwell, London, 2004), 890.

[30] The Belgian delegate suggested that the delegate of Brazil had underestimated the effect of the modifications made in the original text, calling attention particularly to the phrase 'in any other manner'; and also recalled that the subcommittee had given the point about 'economic measures' careful consideration and for good reasons decided against: 'Summary Report of Eleventh Meeting of Committee I/1' Doc. 784, I/1/27, 5 June 1945, 6 UNCIO (1945), 334.

[31] However the Western states were prepared to admit, as stated by the UK representative, that 'that was not to say that all forms of economic and political pressure which threatened the territorial integrity and political independence of another state were permissible, they might well constitute illegal intervention'.

While developing states have tried to raise the issue of prohibiting use of economic and political coercion on other occasions, each time it has received a negative vote from Western powers.

It is also of note that in the discussion following a proposed amendment to Article 2(4) requiring members to collectively resist the use of aggression (as opposed to force),[32] one of the primary objections to the amendment was the lack of definition of the term aggression. The debate caused the United Kingdom representative to comment:[33]

> Apart however from the difficulty in defining aggression and therefore of knowing what the nations were pledged to resist, the use as a standard of an inexplicit word, such as aggression, instead of something explicit such as 'force' would give an opportunity to a state to engage in an act of aggression while calling it by another name.

It is apparent from this statement that the term force was considered an explicit term, and, as Bond has pointed out, such a characterisation of the term at that time was not questionable; it was clear that force meant military or armed force, and in 1945 that meant traditional weapons employed in traditional ways.[34] Further, in the same discussion a representative of the United States pointed out that in the future there would be many kinds of aggression and that these would be covered in the Charter by the words 'threat to the peace'.[35] This indicates that at the time the discussions surrounding Article 2(4) and its proposed amendments were taking place, the perception of force was as a specific (armed) threat and all other incidents which would require flexibility of drafting would be covered by the more general phrase 'threat to the peace' under Article 39 and the collective security arrangements in Chapter VII of the Charter.

[32] 'Summary Report of Twelfth Meeting of Committee I/1' Doc. 810, I/1/30, 6 UNCIO (6 June 1945) 342. The New Zealand amendment reads as follows: 'All members of the Organisation undertake collectively to resist every act of aggression against any member'. The amendment was rejected by a vote of 26 in favour and 18 against, for failing to receive a two-thirds majority.

[33] 'Addendum to Summary Report of Twelfth Meeting of Committee I/1' Doc. 866, I/1/30(a), 8 June 1945, 6 UNCIO (1945) 356.

[34] James Bond, *Peacetime Foreign Data Manipulation as One Aspect of Offensive Information Warfare: Questions of Legality under the United Nations Charter Article 2(4)*, Naval War College (1996), 57, http://handle.dtic.mil/100.2/ADA310926 (last accessed 5 July 2011). It should also be noted that these discussions took place prior to the atomic bombing of Hiroshima and Nagasaki.

[35] 'Summary Report of Twelfth Meeting of Committee I/1' Doc. 810, I/1/30, 6 UNCIO (6 June 1945) 344.

An examination of the historical background of the Charter reveals its development against a backdrop of international efforts to eliminate the unilateral recourse to armed force, and also provides insight into the intentions of the drafters of Article 2(4). In his work on the historical context of Article 2(4), Edward Gordon has argued that the problems of interpretation arise because it is a 'legal rule located in the text of a multilateral treaty which requires adaptation to changing circumstances'; the challenge becomes one of remaining faithful to its core meaning without thereby sacrificing the flexibility required in interpreting constitution norms.[36] The historical antecedents of Article 2(4) are the Covenant of the League of Nations and the Kellogg-Briand Pact. The former (as amended), states that 'any war or threat of war' is a matter of concern for the whole League,[37] and that members of the League will preserve the 'territorial integrity and existing political independence' of members of the League against external aggression.[38] The genetic roots of the present Article 2(4) are clearly visible in the wording. In the years between the adoption of the Covenant of the League of Nations and the adoption of the Kellogg-Briand Pact in 1928, the League Assembly passed unanimous resolutions condemning wars of aggression as international crimes. Indeed, the Pan American conference in 1926 considered such wars to be crimes against the human species.[39]

Under the 1928 Kellogg-Briand Pact, states parties 'condemn the recourse to war for the solution of international controversies and renounce it as an instrument of national policy in their relations with one another'.[40] However the precise scope of the prohibition on war

[36] Edward Gordon, 'Article 2(4) in Historical Context' (1985) 10 *Yale J. Int'l L.* 271, 273.

[37] Art. 11, Covenant of the League of Nations reads as follows: 'Any war or threat of war, whether immediately affecting any of the Members of the League or not, is hereby declared a matter of concern to the whole League, and the League shall take any action that may be deemed wise and effectual to safeguard the peace of nations. In case any such emergency should arise the Secretary General shall on the request of any Member of the League forthwith summon a meeting of the Council.'

[38] Art. 10, Covenant of the League of Nations reads as follows: 'The Members of the League undertake to respect and preserve as against external aggression the territorial integrity and existing political independence of all Members of the League. In case of any such aggression or in case of any threat or danger of such aggression the Council shall advise upon the means by which this obligation shall be fulfilled.'

[39] W. Bishop Jr, *International Law: Cases and Materials 1010* (3rd edn, 1971), cited in Gordon, 'Article 2(4) in Historical Context', 274.

[40] Art. 1, General Treaty for the Renunciation of War 1928, UKTS 29 (1929), Cmnd, 3410; 94 LNTS 57.

contained in the Pact has never been whether the treaty prohibits armed force short of war as well as war.[41] Professor Ian Brownlie suggests that the best guide to the meaning of the Pact is to be found in the subsequent actions of the contracting parties.[42] He concludes that there leaves little room for doubt that it was understood to prohibit 'any substantial use of armed force'.[43]

Subsequent iterations of the rule

As can be seen from the foregoing, the rules of the Charter relating to force are brief and cannot constitute a complete code,[44] a fact acknowledged by the International Court of Justice in the *Nicaragua (Merits)* case.[45] Almost from the time that the Charter was drafted, states have tried to elaborate the prohibition on the use of force in General Assembly resolutions to provide greater clarity. However, each of the attempts has left the central issue of the essential nature of 'force' unresolved. It appears that the ambiguity of the wording has been the price of international consensus.[46]

Two other international instruments drafted around the time of the United Nations Charter, the Charter of the Organization of American States (OAS) and the North Atlantic Treaty (forming NATO), also use the term force without qualification. Perhaps unsurprisingly given their status as collective defence organisations, neither provides any support for the inclusion of economic or political coercion as a use of force. The North Atlantic Treaty uses the terminology of the UN Charter and although Article 2 separately addresses economic concerns it does not refer to economic coercion.[47]

The Charter of the OAS (as subsequently amended) refers to the prohibition on the use of force in other treaties in Article 22, however

[41] Harris, *Cases and Materials*, 861.

[42] Ian Brownlie, *International Law and the Use of Force by States* (Oxford University Press, 1963), 87.

[43] *Ibid. Cf.* D. W. Bowett, *Self-Defence in International Law* (University of Manchester Press, 1958), 136.

[44] Christine D. Gray, *International Law and the Use of Force* (2nd edn, Oxford University Press, 2004), 6.

[45] The Court stated that the UN Charter by no means covers the whole area of the regulation of force in international relations. *Nicaragua (Merits)*, para. 176.

[46] Gray, *Use of Force* (2nd edn), 8.

[47] Art. 2 states that the parties 'will seek to eliminate conflict in their international economic policies and will encourage economic collaboration between any or all of them'; North Atlantic Treaty, 4 April 1949, 34 UNTS 243 (entered into force 24 August 1949).

force is used as an unqualified term and no guidance is given as to its meaning.[48] However, when discussing such measures for the purposes of its own obligations, Articles 19 and 20 of the OAS Charter avoid using the term force without qualification by using 'armed force' and 'coercive measures of an economic or political character' as separate terms. Michael Schmitt notes that this is perhaps unsurprising given Brazil's attempt at amending Article 2(4) of the UN Charter and their membership of the OAS.[49] It is clear that the language employed by the OAS Charter is meant to be interpreted far more broadly than the 'use of force' language of Article 2(4). However, as a number of commentators have pointed out, the language is so broad as to be legally unenforceable;[50] if read literally, it outlaws diplomacy.[51]

The issue was addressed again twenty-five years after the signing of the Charter, in the 1970 General Assembly Declaration on Principles of International Law Concerning Friendly Relations and Co-operation Among States in Accordance with the Charter of the United Nations (hereinafter Declaration on Friendly Relations).[52] The section on the principle of the use of force does not clarify the matter as it merely refers to 'force'. This was the result of disagreement between the Western states, who argued that the prohibition only related to armed force, and the Soviet Bloc, European and developing states, who argued that 'all forms of pressure, including those of a political and economic character, which have the effect of threatening the territorial integrity or political independence of any state', were prohibited.[53] However the Western states were prepared to admit, presciently in the light of the

[48] Art. 22 states that 'The American States bind themselves in their international relations not to have recourse to the use of force, except in the case of self-defense in accordance with existing treaties or in fulfilment thereof'; *Charter of the Organisation of American States*, 30 April 1948, 119 UNTS 3 TIAS No. 2361.

[49] Schmitt, 'Normative Framework', 906.

[50] Richard W. Aldrich, 'How Do You Know You Are at War in the Information Age?' (2000) 22 *Hous. J. Int'l L.* 223, 254. Nicaragua attempted unsuccessfully to rely on this broad language in its case against the United States; however, the Court found that it had no jurisdiction to consider either the UN Charter wording or articles of the OAS Charter.

[51] Tom Farer, 'Political and Economic Aggression in Contemporary International Law' in A. Cassese (ed.), *The Current Legal Regulation of the Use of Force* (Martinus Nijhoff, Dordrecht, 1986) 121–32, 121. Threats, more or less subtle, have always been an important feature of the intercourse of states.

[52] Declaration on Principles of International Law concerning Friendly Relations and Co-operation among States in accordance with the Charter of the United Nations, GA Res 2625 (XXV), UN GAOR Supp. 25, 18–22, 24 October 1970.

[53] UN Doc. A/AC.125/SR.114 (1970) cited in Harris, *Cases and Materials*, 863.

Arab oil boycott of 1973 and 1974, that 'that was not to say that all forms of economic and political pressure which threatened the territorial integrity and political independence of another state were permissible; they might well constitute illegal intervention'.[54] The Declaration deals separately with political and economic coercion under the heading of the 'Principle not to Intervene', thus indicating that as far as the Declaration is concerned, force is restricted to armed coercion, and that forms of political and economic coercion were to be considered intervention.

The International Court of Justice in the *Nicaragua (Merits)* case held that the Declaration was indicative of the *opinio juris* of the international community, and showed the customary nature of the prohibition as stated in Article 2(4) of the Charter.[55] Although the judgment does not directly address the question of the status of economic or political coercion under the use of force doctrine, the Court does not include such measures in citing the acts which may be considered 'less grave' forms of the use of force. The Court quotes from the sections of the Declaration dealing specifically with the principle of the non-use of force, but also from the principle of non-intervention. In citing the latter, the Court leaves out the opening sentence of the paragraph dealing with political and economic coercion, and only quotes the second sentence relating to armed groups.[56] This is despite Nicaragua's submissions that the country had been subjected to economic coercion at the hands of the United States.[57] This omission, and the failure of the Court to consider Nicaragua's submissions in that regard when discussing Article 2(4) and the associated customary law, indicates that the Court does not include such measures in the definition of force as used in customary international law and the UN Charter.[58]

Since the *Nicaragua (Merits)* case, the international community has once again affirmed the prohibition against force in the 1987 Declaration on

[54] UK representative (Mr Sinclair), UN Doc. A/AC125/SR25 (1966), cited *ibid.*, 863–4.

[55] *Nicaragua (Merits)*, para. 188.

[56] *Ibid.*, para. 192.

[57] It should be noted, however, that Nicaragua did not attempt to argue that the economic coercion involved was sufficient to count as the threat or use of force. Although the Court stated later in the judgment that it would not rule on legal arguments not put forward by the parties (in relation to the prohibition against intervention), the fact that the submission was not even mentioned is significant.

[58] It should be noted that the definition of force as used in Art. 2(4) of the Charter was not at issue in the case as the United States has a non-binding clause to multilateral treaties.

the Non-Use of Force.[59] As with previous General Assembly resolutions, the Declaration separates the concepts of armed intervention and economic and political coercion; however, it leaves the main controversies between developed and developing states unsettled.[60]

2. Definition of armed force

The international jurisprudence indicates that while the definition of force is to be limited to armed force, the definition of armed force itself is to be interpreted widely. In particular, a direct application of armed force by a state is not necessary for the state to fall foul of Article 2(4), nor have states been able to successfully circumvent the prohibition by arguing that an incident does not affect the territorial integrity or political independence of a state.

The latter argument was raised by the United Kingdom in the *Corfu Channel Case*,[61] claiming that their mine-sweeping actions in Albanian territorial waters were not in breach of Article 2(4) as they were not directed against the territorial integrity or political independence of Albania.[62] The International Court of Justice dismissed the narrow interpretation of force, holding that the British action of sending warships to clear mines from the Corfu Channel against the express wishes of Albania amounted to a 'policy of force' such as had given rise to 'most serious abuses and such as cannot … find a place in international law'.[63] Interestingly, despite this argument being dismissed by the Court in the *Corfu Channel Case*, Belgium raised the same argument in its pleadings on provisional measures following NATO's military operations over Kosovo in 1999.[64]

[59] Arts 7 and 8, *Declaration on the Enhancement of the Effectiveness of the Principle of Refraining from the Threat or Use of Force in International Relations* GAOR 42nd sess., 73rd plen. mtg., UN Doc. A/Res/42/22 (1987).

[60] The remaining differences between developed and developing states are summarised at UN Doc. A/40/41: Gray, *Use of Force* (2nd edn), 9.

[61] *Corfu Channel Case (UK* v. *Albania) (Merits)* (1949) ICJ Reports 4, International Court of Justice, 13.

[62] The UK was attempting to collect the mines as evidence of Albania's mining operation and classified 'Operation Retail' as an act of self-help. *Ibid.*, 35.

[63] *Ibid.*

[64] Belguim's Oral Arguments, *Legality of Use of Force (Yugoslavia* v. *Belgium) (Request for Provisional Measures)*, CR99/15, 12, stating that NATO has never questioned the political independence and territorial integrity of the Federal Republic of Yugoslavia and claiming that Article 2(4) covers only intervention against political independence and the territorial integrity of a state. The Court did not address

The International Court of Justice's broad interpretation of the scope of the term force can also be seen in the *Nicaragua (Merits)* case, where the Court held that certain indirect forms of support *are* included in the prohibition against force. The Court's judgment states that acts which breach the principle of non-intervention 'will also, if they directly or indirectly involve the use of force, constitute a breach of the principle of non-use of force in international relations'.[65] In that case, the Court accepted that the provision of assistance to rebel fighters 'in the form of the provision of weapons or logistical or other support' could constitute a threat or use of force, or amount to intervention in the internal or external affairs of a state.[66] However, not all forms of indirect action or assistance were to be so considered. In particular, the Court found that 'the mere supply of funds to the contras, while undoubtedly an act of intervention in the internal affairs of Nicaragua', would not breach the prohibition of force as contemplated by Article 2(4) and the corresponding customary law.[67] In a controversial discussion, the Court also distinguishes between the gravest forms of the use of force, those that constitute an armed attack or aggression, and other uses of force, referred to by the Court as 'less grave forms'.[68] The Court used the Declaration on Friendly Relations to elucidate these lesser forms of force and determine the legal rules that may apply to them.[69] In particular, the Court emphasised the following paragraphs of the Declaration:

> Every State has the duty to refrain from the threat or use of force to violate the existing international boundaries of another State or as a means of solving international disputes, including territorial disputes and problems concerning frontiers of States.
>
> ...
>
> Every State has the duty to refrain from any forcible action which deprives peoples referred to in the elaboration of the principle of equal rights and

these arguments and refused provisional measures on the basis that it lacked prima facie jurisdiction on the merits of the case.

65 *Nicaragua (Merits)*, para. 209.

66 *Ibid.*, para. 193.

67 *Ibid.*, para. 228.

68 *Ibid.*, para. 191. Much criticism has been directed at the *Nicaragua (Merits)* decision based on this elucidation of various levels of force, particularly from American authors. See, for example, the panel discussion reported at American Society of International Law, 'The Jurisprudence of the Court in the Nicaragua Decision' (1987) 81 *ASIL Proc.* 258.

69 *Nicaragua (Merits)*, para. 191.

> self-determination of that right to self-determination and freedom and independence.
>
> Every State has the duty to refrain from organizing or encouraging the organization of irregular forces or armed bands, including mercenaries, for incursion into the territory of another State.
>
> Every State has the duty to refrain from organizing, instigating, assisting or participating in acts of civil strife or terrorist acts in another State or acquiescing in organized activities within its territory directed towards the commission of such acts, when the acts referred to in the present paragraph involve a threat or use of force.
>
> States have a duty to refrain from acts of reprisal involving the use of force.

There are two points that may be taken from this extract.[70] First, the Court considers the Declaration indicative of the international community's *opinio juris* that certain acts of indirect aggression may be a use of force prohibited by international law, albeit a lesser form.[71] This is particularly salient for computer network attacks as many of the arguments raised against such actions being considered a use of force relate to the often indirect nature of their results.[72]

Secondly, the Court included intangible assistance such as 'encouragement'. While it is reading too much into the judgment to treat this as evidence of a prohibition against intangible force, the words 'armed intervention and all other forms of interference' imply intent to reach at least some kinds of non-forcible activities. The reports of the special committee delegated the task of drafting the Declaration on Friendly Relations indicate that the participating states had little shared notion of what sort of non-forcible conduct would fall under the proscriptions

[70] Again the Court's approach in using the Declaration has been controversial. Some authors argue that the Court has used the Declaration as a source of law, a position which the states parties to the resolution would never have intended. Authors have also criticised the Court for using the text of the Declaration as indicative of *opinio juris* of the international community without also fully examining the actions of states: see generally, American Society of International Law, 'The Jurisprudence of the Court in the Nicaragua Decision'. But *cf.* Lori Fisler Damrosch, 'Politics across Borders: Nonintervention and Nonforcible Influence over Domestic Affairs' (1989) 83 *AJIL* 1, 8.

[71] This is in accord with the drafting history of Article 3g of the Definition of Aggression which rejected a proposal to include several of the broad range of activities listed as indirect aggression in the Declaration on Friendly Relations. See Pierluigi L. Zanardi, 'Indirect Military Aggression' in A. Cassese (ed.), *The Current Legal Regulation on the Use of Force* (Martinus Nijhoff, Dordrecht, 1986) 111–19, 116.

[72] The indirectness of computer network attacks is addressed in section 3.1.1 below.

in the Declaration.[73] While 'encouragement' may take many forms, some tangible and some not, the inclusion of 'organisation or encouragement' would suggest that an intangible form was envisaged as well. It is interesting in this regard to compare the Court's earlier treatment of the encouraging statements made by the Ayatollah Khomeini in the *Tehran Hostages* case.[74] Although the case did not consider the Declaration on Friendly Relations, the Court considered that the Ayatollah's statements were not sufficient to impute state responsibility for the initial actions of student militants in overrunning the US embassy in Tehran. However, the Court did hold that such statements were sufficient to turn the continued occupation of the embassy and detention of the hostages into acts of the state. The Court did not consider whether the continuing occupation (or the initial attack)[75] constituted a breach of Article 2(4) of the Charter which had been argued by the United States and dealt with the matter solely with respect to the Vienna Convention on Diplomatic Immunity.

While some commentators have argued that the *Nicaragua (Merits)* case extends the definition of force into things that begin to resemble economic and political coercion, the current author considers that the better view is that force is still restricted by the Court to military (or paramilitary) action; however, such force may be imputed through indirect means such as agency. The extension provided by the Court remains tied to armed force, whether in person or by proxy.

The Court in the *Nicaragua (Merits)* case regarded the Charter provisions as dynamic rather than fixed, and thus as capable of change over time through state practice.[76] Although the Court accepted the parties' position that the Charter provisions represented customary law, it also accepted the possibility of the development of new law on forcible intervention allowing a new exception to the prohibition on the use of

73 Damrosch, 'Politics across Borders', 10.

74 *Case Concerning the United States Diplomatic and Consular Staff in Tehran (United States v. Iran)* (1980) 74 *AJIL* 746, International Court of Justice, para. 59. In that case the Ayatollah had made several public declarations inveighing against the United States and holding the US responsible for the trouble besetting that country. On 1 November the Ayatollah had made a statement declaring it was 'up to the dear pupils, students and theological students to expand with all their might their attacks against the United States and Israel, so that they may force the United States to return the deposed and Criminal Shah, and to condemn this great plot'. Further, congratulatory statements from the Ayatollah following the attack and other statements of official approval were also made.

75 Presumably because of the lack of imputability to a state party.

76 Gray, *Use of Force* (2nd edn), 7.

force in Article 2(4). In setting out the area covered by the prohibition against force, the Court stated that Article 2(4) of the Charter represented only part of the customary international law relating to the use of force and stated that the UN Charter 'by no means covers the whole area of the regulation of the use of force in international relations'.[77] The Court pointed out that the right to self-defence stood alongside the Charter and that the Charter text did not go on to regulate all of the aspects of that rule. Bearing that in mind, we need now to turn to state practice and the theories of force propounded by other academics.

2.1. *State actions*

Given the classified and covert nature of most computer network attacks, a survey of state practice in relation to this type of incident is problematic at best. To date there have not been any computer network attacks that are conclusively attributable to a state outside of traditional conflict scenarios. However, some states have made statements regarding the use of computer network attacks and other analogous information operations. As has been stated, one of the most difficult issues surrounding computer network attacks is positive attribution of an attack to the perpetrator, as a matter of fact, as well as a matter of law. Most of the incidents detailed below and in Appendix 1 have suffered from this difficulty, thus making statements of limited use. However, although they must be used with care, the statements, actions and reactions of states also form part of the interpretive framework of Article 2(4).

While international recognition of the security threat that computer network attacks represent has been gradual, in the last few years states' recognition of the problem has solidified into action. Many states have set up cyber commands within their militaries and other institutions in reaction to reported incidents of computer network exploitation and joint exercises which have exposed vulnerabilities in their digital infrastructure.[78] For those that remained sceptical of the potential of computer network attacks to cause physical damage sufficient to amount

[77] *Nicaragua (Merits)*, para. 176.

[78] Established cyber commands at the time of writing include the United States, United Kingdom, South Korea, China as well as NATO. The establishment of the US Cyber Command, despite being lobbied for for several years, only got final approval following *Operation Buckshot Yankee* in 2008 in which US military networks were compromised by a foreign intelligence agency and sensitive data exfiltrated. See generally Lynn, 'Defending a New Domain'.

to a national security threat, the Stuxnet virus acted as evidence and a timely wake-up call.

To date the clearest indication of state practice with regard to a specific computer network attack remains the international response to the attacks against Estonia in 2007. As noted previously, the attacks did not affect critical infrastructure, damage physical property or cause human injury. It appears that states were not prepared to make definitive public statements regarding the attacks as a use of force.[79] NATO member states were certainly not prepared to accept that the attacks amounted to an armed attack which would initiate the collective self-defence provisions of the North Atlantic treaty; however, a NATO spokesperson commented that the attacks were a security issue which concerned NATO.[80] Russia denied all accusations of computer network attacks, noting that the Kremlin came under attack many hundreds of times a day.[81] Russian officials have previously stated that they will view the effects of a computer network attack as similar to that of the use of weapons of mass destruction and that:[82]

> [f]rom a military point of view, the use of information warfare means against Russia or its armed forces will categorically not be considered a non military phase of a conflict, whether there were casualties or not … Russia retains the right to use nuclear weapons first against the means and forces of information warfare, and then against the aggressor state itself.

However, this statement has not been reflected in Russia's response to computer network attacks in the 2008 conflict over South Ossetia. Small-scale attacks were occurring between the two countries for almost two months before the outbreak of hostilities;[83] however, there

[79] For example, the US called the attacks 'unacceptable' and 'pressure on an independent country' but stopped short of calling the attacks force: 'Rice Condemns Ongoing Cyber-Attacks as Estonian Embassy Siege Ends', earthtimes.org 4 May 2007.

[80] AFP, 'Cyber Attacks on Estonia Are Security Issue: NATO Chief', *The Age* (Melbourne), 26 May 2007.

[81] 'The Cyber Raiders Hitting Estonia', *BBC News*, 17 May 2007. As noted previously, responsibility for the attacks has been claimed by a member of the Russian 'Nashi' youth group, stating that the attacks were carried out without state involvement. The attribution has not been confirmed.

[82] V. I. Tsymbal, 'Kontseptsiya "Informatsionnoy Voyny" (Concept of Information Warfare)' (Paper presented at the Evolving post Cold War National Security Issues, Moscow, 12–14 September 1995), 7, cited in Timothy L. Thomas, 'Russian Views on Information Based Warfare' (1996) special edn *Airpower* 25.

[83] Kim Hart, 'Longtime Battle Lines Are Recast in Russia and Georgia's Cyberwar', *Washington Post* (Washington DC), 14 August 2008, D01.

are no reports of escalation of the military response and where reportedly Russian and South Ossetian sites have come under attacks from Georgian hackers.[84]

China has often topped the lists of states engaged in or developing computer network attack capabilities.[85] China has been accused of hacking attacks against government computers in the United States, France, Germany, the United Kingdom, India, Canada, Australia and New Zealand.[86] The Chinese government has denied any involvement in such attacks and, despite the problems of attribution, thus far it appears that the victim states have not considered the access and theft of information a use of force and are content to deal with them as cases of espionage. China has also claimed that it has sustained 'massive' and 'shocking' losses of state and military secrets via the internet.[87] Although they are non-international examples, Tibet and Taiwan have also repeatedly come under attack from alleged Chinese hackers, suspected to have the backing of the Chinese government. The Chinese government has denied all such claims, stating that 'the Chinese government always opposes the activities of hackers'.[88] In September 2002 China announced that it had developed five new information warfare

[84] *Ibid.*

[85] See, for example, McAfee, *Virtual Criminology Report*, McAfee Inc. (2007); Symantec, *Symantec Global Internet Security Threat Report*, Symantec, Vol. XIII (2008); Center for Strategic and International Studies, *Significant Cyber Incidents since 2006* (2011) http://csis.org/files/publication/110309_Significant_Cyber_Incidents_Since_2006.pdf (last accessed 12 April 2011).

[86] See, for example, Bryan Krekal, *et al.*, *Capability of the People's Republic of China to Conduct Cyber Warfare and Computer Network Exploitation Prepared for the US-China Eonomic and Security Review Commission*, Northrop Grumman (2009); Roger Boyes, 'China Accused of Hacking into Heart of Merkel Administration', *The Times* (London), 27 August 2007, www.timesonline.co.uk/tol/news/world/europe/article2332130.ece (last accessed 15 July 2011), H. Schouten, 'China Denies Role in NZ Cyber Attack', *Dominion Post* (New Zealand), 12 September 2007, B4; John Leyden, 'France Blames China for Hack Attacks', *The Register* (London), 12 September 2007, www.theregister.co.uk/2007/09/12/french_cyberattacks/ (last accessed 15 July 2011). See also Information Warfare Monitor, *Tracking Ghostnet: Investigating a Cyber Espionage Network*, Information Warfare Monitor (2009) www.infowar-monitor.net/ (last accessed 12 April 2010).

[87] Leyden, 'France Blames China for Hack Attacks'. Edward Cody, 'Chinese Official Accuses Nations of Hacking', *Washington Post* (Washington DC), 13 September 2007, A16.

[88] See, for example, 'China Denies Hacking Dalai Lama Computer', *CNN*, 25 September 2002; George V. Hulme, 'Taiwan Accuses China of Launching Cyberattack', *Information Week*, 16 June 2004, www.informationweek.com/news/22100221 (last accessed 3 January 2012).

institutes to develop information warfare patterns/weapons described as 'technological aircraft carriers' and the official news agency Xinhua reported that Chinese military leaders hoped to overcome their military weaknesses, largely outdated hardware, by attacking a technologically superior foe with electronic warfare.[89] Latest statements have confirmed a commitment to collaborate internationally on cyber security issues, as well as developed cyber strategies; however, like other governments, the Chinese government has not publicly stated what computer network attacks it would consider a use of force or armed attack.[90]

The United States has formally stated that it is their policy to respond to cyber attacks by any means appropriate, including military action.[91] In July 2002, President George W. Bush signed a secret directive ordering the government to develop, for the first time, national-level guidance for determining when and how the United States would launch cyber attacks against enemy computer networks.[92] The directive remains classified; however, US Secretary of State Clinton has remarked that 'States, terrorists, and those who would act as their proxies must know that the United States will protect our networks … Countries or individuals that engage in cyber attacks should face consequences and international condemnation.'[93] In discussing the US cyber security strategy US Deputy Secretary of Defense William Lynn has also outlined an 'active defense' strategy, which may include pre-emptive strikes against adversaries' computer networks, although they are still wrestling with how to pursue the policy within the bounds of the law.[94]

[89] 'Military Eyes Electronic Warfare', *Associated Press, South China Morning Post*, 28 September 2002.

[90] Krekal, *et al.*, *Capability of the PRC to Conduct Cyber Warfare*, 16–22.

[91] Dan Verton, 'The Prospect of Iraq Conflict Raises New Cyber Attack Fears', *Computerworld*, 30 September 2002, www.computerworld.com/s/article/74688/Prospect_of_Iraq_conflict_raises_new_cyberattack_fears (last accessed 5 July 2011).

[92] Bradley Graham, 'Bush Orders Guidelines for Cyber-Warfare', *Washington Post* (Washington DC), 7 February 2003, A1.

[93] Remarks by US Secretary of State Hillary Rodham Clinton *Remarks on Internet Freedom* at The Newseum, Washington DC, 21 January 2010, www.state.gov/secretary/rm/2010/01/135519.htm (last accessed 12 April 2011).

[94] Lynn, 'Defending a New Domain'; Ellen Nagashima, 'U.S. Eyes Preemptive Cyber-Defense Strategy', *Washington Post* (Washington DC), 29 August 2010, A05. See Chapter 3 below for a discussion of the US pre-emptive self-defence strategy.

NATO states have likewise been working within the rubric of that organisation to provide a joint approach to cyber security; however, as noted in regard to the attacks against Estonia above, they have been cautious in their conclusions. Statements by NATO officials have been careful to frame NATO's cyber defence actions within Article 4 relating to consultation in response to threats to territorial integrity, political independence or national security rather than Article 5 collective self-defence.[95]

International reaction to the Stuxnet worm has also been decidedly muted, although this attack is perhaps the clearest example to date of a computer network attack amounting to a use of force (if not an armed attack). Unsurprisingly, Iran has stated that an 'electronic war' has been launched against the country, and has stated that it will take pre-emptive cyber action against the centres which launch such attacks within the framework of international regulation.[96] While states have spoken widely about the innovative technological nature of the threat, statements concerning the legal nature of the attack have largely been absent. This may reflect a general consensus in the international community that the worm was a use of force, an unwillingness to comment while attribution has not been proved, or merely satisfaction at the outcome of the attack regardless of its legal nature. Such reactions and policy statements tend to indicate that states remain cautious about labelling computer network attacks as a use of force. Although both the United States and Russia have reserved the right to respond with force against computer network attacks they have not made any comments indicating that any computer network attack reported to date should be viewed in this manner. To date, with the exception of the 'Farewell Dossier' incident, no computer network attack conclusively attributable to another state has caused physical damage or human injury.[97]

95 Sverre Myrli, *NATO and Cyber Defence*, 173 DSCFC 09 E bis (2009), paras 58–61, www.nato-pa.int/default.asp?SHORTCUT=1782 (last accessed 12 April 2011).

96 'Iran Says Nuclear Programme Was Hit by Sabotage'; 'Iran to Take Pre-Emptive Action against Cyber Terrorism: General', *Tehran Times*, 27 February 2011, Political Desk.

97 As noted previously, at the time of writing Stuxnet has not been conclusively attributed to any actor, although it is widely believed to be the work of a state/s actors. Because of their clandestine nature, computer network attacks are perfect fodder for rumours and misinformation; for example, Chinese hackers have been inaccurately blamed for causing a massive blackout in the north-eastern United States in 2003.

2.2. *Theories of force – scholastic writings*

Although they are in agreement over the customary nature of the prohibition on the use of force, commentators remain deeply divided over the content of the rule. Theories of force and discussion of Article 2(4) of the Charter have come mainly from two different schools of legal theory, those adopting a restrictive approach and those adopting a more expansive one.[98] The first, tending to come from the positivist school, provides a definition of 'force' and determines whether a particular incident falls within the accepted definition. The extensive or contextualist approach tends to focus more on custom and the context of force.

Ian Brownlie uses a two-part definition requiring the use of a weapon which is employed for the destruction of life and property.[99] Bowett refers to the possible resort to a violent weapon which inflicts human injury.[100] The question of the need for physical means and the requirement of a weapon came to the fore with the advent of chemical and biological weapons. Ian Brownlie addressed the issue in 1963, considering whether weapons 'which do not involve any explosive effect with shock waves and heat' involved a use of force prohibited by Article 2(4) of the Charter.[101] In concluding that these weapons should fall within the prohibition, he gives two reasons. The first, is that the 'agencies concerned are commonly referred to as "weapons" and as forms of "warfare" '.[102] However, as has been seen in the preceding chapter, this may not be helpful in terms of categorising computer network attacks as force. Both the popular press and academics across disciplines have used the terms information warfare, cyber war and even computer network attack to refer to a vast range of information operations, many of which would never be considered force under the Charter paradigm.[103] Further, the terms 'war' and 'weapons' have found an increasingly political meaning in recent years. At the date of writing, the media contains articles on the present conflicts in Afghanistan and Iraq; however, they have also contained pieces on wars on terror,

[98] For an extensive analysis of the methodological debates over the prohibition on the use of force see Corten, 'Controversies'.

[99] Brownlie, *Use of Force by States*, 362.

[100] Bowett, *Self-Defence*, 184–99.

[101] Brownlie, *Use of Force by States*, 362. [102] *Ibid*.

[103] See also Todd A. Morth, 'Considering Our Position: Viewing Information Warfare as a Use of Force Prohibited by Article 2(4) of the U.N. Charter' (1998) 30 *Case W. Res. J. Int'l L.* 567, 590.

drugs, crime and poverty, all politicised uses of the lexicon of humanitarian and international law and none of them aimed at invoking its protection. Used in this manner, the term war is merely used to signify resolve,[104] a point that has been noted by legal commentators and politicians alike.[105] Thus the semantics of war and weaponry are no longer useful criteria in determining whether something is a use of force.[106] Further, as was seen in the 2001 anthrax scares in the United States and the release of sarin gas on a Tokyo subway train in 1995, the requirement of a weapons-based delivery system is not essential for a chemical or biological attack. The International Court of Justice has also noted in its advisory opinion on *Nuclear Weapons* that the provisions relating to the use of force in the Charter do not refer to specific weapons but rather apply to the use of force regardless of the weapons employed.[107]

Brownlie's second and more convincing argument was that the use of chemical and biological weapons should be viewed as force because such 'weapons are employed for the destruction of life and property'.[108] In advancing his analysis, Brownlie moves the test beyond the kinetic impact of shock waves and heat and toward a wider, results-oriented approach. This view is also representative of the requirement of a large number of commentators who argue for a physical outcome to a computer network attack. However a purely results-driven approach raises problems of its own in that it risks reopening the arguments relating to the inclusion of political and economic coercion. As Cassandra LaRae-Perez points out, once a results-based perspective is adopted, the effects of long-term, comprehensive economic sanctions of the kind

104 Frédéric Mégret, '"War"? Legal Semantics and the Move to Violence' (2002) 13(2) *EJIL* 361.

105 See, for example, Christopher Greenwood, 'International Law and the War against Terrorism' (2002) 78 *International Affairs* 301, 306. Also UK Prime Minister Tony Blair's comment: 'Whatever the technical or legal issues ... the fact is that we are at war with terrorism.' See, 'Britain at War with Terrorism' *BBC News*, 16 September 2001, http://news.bbc.co.uk/1/hi/uk_politics/1545411.stm (last accessed 5 July 2011); 'Powell Very Pleased with the Coalition-Building Results' 13 September 2001, www.usinfo.state.gov/topical/pol/terror/01091366.htm (last accessed 5 September 2008).

106 Although some of the more populist articles refer to information warfare techniques as weapons of mass disruption, obviously a play on the 'mass destruction' terminology applied to nuclear, chemical and biological weapons, such terms have not achieved widespread usage.

107 *Nuclear Weapons Case*, para. 39.

108 Brownlie, *Use of Force by States*, 362.

adopted against Iraq and Cuba, which are as severe as the use of force against those states, can be regarded as falling foul of Article 2(4).[109] However state practice shows that the majority of the international community is not ready for this to be the case. The rigorous economic embargo that the United States has enforced against Cuba since the early 1960s has not been considered to be a use of force; likewise, neither has the Arab embargo of Israel.[110]

Yoram Dinstein likewise argues for a results-based approach when discussing computer network attacks in the context of an armed attack, recognised by the International Court of Justice in the *Nicaragua (Merits)* case as a subset of the use of force.[111] He argues that violent consequences are the key to fulfilling the definition of armed attack:[112]

> From a legal perspective, there is no reason to differentiate between kinetic and electronic means of attack. A premeditated destructive [computer network attack (CNA)] can qualify as an armed attack just as much as a kinetic attack bringing about the same or similar results – the crux of the matter is not the medium at hand (a computer server in lieu of, say an artillery battery), but the violent consequences of the action taken. If there is a cause and effect chain between the CNA and these violent consequences, it is immaterial that they were produced by high and not low technology.

Dinstein's realist argument focuses solely on the consequences of an attack. As shown above, a *purely* consequence-based approach in an area that lacks the tangibility of traditional military/armed force blurs the distinction with the grey area occupied by political and economic coercion. Thus a theory of force which requires only a particular outcome is insufficient to cover the concerns raised by forms of coercion which the international community is agreed (for the most part) are not to be included as uses of force. Michael Schmitt has argued that not only is a purely consequence-based approach extraordinarily difficult to quantify or qualify,[113] it would also constitute a new normative standard altogether and as such would prove a difficult case for

[109] Cassandra LaRae-Perez, 'Economic Sanctions as a Use of Force: Re-Evaluating the Legality of Sanctions from an Effects-Based Perspective' (2002) 20 *BU Int'l L.J.* 161.

[110] Bond, *Peacetime Foreign Data Manipulation*, 59. Interestingly, in the latter case the United States was in the position of the weaker state and argued that the economic coercion by the more powerful Arab states should be considered to be a use of force, a reversal of its position in the Cuban embargo.

[111] Dinstein, 'CNA and Self-Defense', 103.

[112] *Ibid.*

[113] Schmitt, 'Normative Framework', 911.

adoption by the international community.[114] However, it is also equally clear that the requirement and current analysis with regard to weaponry cannot stand as it is and must be revisited if it is to take into account changes in technology.

The second school of legal theory writing on the subject of force is the more expansive, contextualist approach epitomised by the New Haven school writers such as Myers McDougal and Michael Reisman. In his work *Law and Minimum World Public Order,* Myers McDougal argued that force is merely a degree of major coercion and violence on a transnational scale.[115] McDougal thus places all forms of coercion on this scale and addresses the problem of the characterisation of the particular coercion as permissible or non-permissible from variables based on past actions, extrapolated forward in accordance with those variables' probable consequences upon the goal values of the kind of world order the scholar prefers:[116]

> From this perspective, the basic intellectual task is one of characterising the variable contextual factors and policies which relate to the distinction between permissible and impermissible coercion for the guidance of differing particular decision makers.

Developing the theory further, Michael Reisman sets out nine categories where he believes the use of force has achieved some form of international legal authority.[117] He argues that in determining whether a particular act of coercion is lawful or not, the question to be answered is whether a particular act (whatever its justification otherwise) enhances or undermines world order.[118] That is, the critical

[114] *Ibid.*, 917.

[115] Myres Smith McDougal and Florentino P. Feliciano, *Law and Minimum World Public Order: The Legal Regulation of International Coercion* (Yale University Press, New Haven, 1961).

[116] *Ibid.*, 153.

[117] These categories are self-defence; self-determination and decolonisation; humanitarian intervention; intervention by the military instrument to replace an elite in another state; uses of the military instrument within spheres of influence and critical defence zones; treaty-sanctioned interventions within the territory of another state; use of the military instrument for the gathering of evidence in international proceedings; use of the military instrument in enforcing international judgments; and countermeasures such as reprisals and retorsions. W. Michael Reisman, 'Criteria for the Use of Force in International Law' (1985) 10 *Yale J. Int'l L.* 279, 281.

[118] *Ibid.*, 282. Reisman's analysis is based on his concept of a complex 'operational code' for determining when the unilateral resort to force will be legitimate by states. For a summary of the factors in the operational code see Michael N. Schmitt, 'The

question is not whether coercion has been applied, but whether it has been applied (a) in support of or against community order and basic policies, and (b) in ways in which the net consequences include congruence with community goals and minimum order.[119] However, the contextualist position has fundamental difficulties. Olivier Corten points out, no 'objective law' exists that expresses social necessities or the solidarity mechanisms that characterise the international community; it is the interpreter and the interpreter alone who gives sense to what is required in a particular case by those necessities or that solidarity.[120] The idea that the fundamental goals of the community must prevail over any particular rule of law, namely the rule prohibiting armed force, would remove all certainty from international relations, leaving the outcome of any diplomatic encounter highly uncertain as to its legality. And, ultimately, that uncertainty is a greater threat to communal values and goals. This methodological schism permeates the debates on the legality of computer network attacks as a use of force and divides commentators on the correct approach to take to this emerging form of warfare.

3. Computer network attacks as a use of force

As noted above, two main theories of force have emerged out of the current writing in relation to the prohibition against force. The first is a more restrictive positivist approach which looks at the rules formulated by the international community, in this case the prohibition against force, and argues that anything falling outside that prohibition is legal. Authors adhering to this school include Ian Brownlie, Yoram Dinstein, Christine Gray and, in respect of computer network attacks, James Bond. The second is the more expansive contextualist approach which contends that all coercion falls along a continuum and the position along the continuum is the result of several factors which affect the minimum world order. Elements of the contextualist approach can be seen in the work of Michael Schmitt, and are given full expression by Michael Reisman.

As we have seen above, a results-based approach to the question as to whether a particular attack contravenes the prohibition against

Resort to Force in International Law: Reflections on Positivist and Contextualist Approaches' (1994) 37 *AFL Rev.* 105.

[119] Reisman, 'Criteria', 284.

[120] Corten, 'Controversies', 814.

'armed force' leads towards an erosion of the exclusion of economic and political coercion. Michael Schmitt has suggested that the use of force proscription is based on the desire of the international community to foster and advance the aspirational values set out in the preamble to the Charter. The prohibition on 'armed' force is thus a kind of instrumental shorthand way to restrict those acts that are most likely to endanger those objectives and aims. Schmitt argues that the international community is not concerned so much with the instrumentality of the coercion but rather the consequences of its use.[121] However, given that the range of possible consequences of any given kind of attack (i.e. denial of service, virus, direct intrusions etc.) range along a continuum, assessing the consequences of an attack can be a Sisyphean task, making the criteria for placement upon that continuum extremely difficult.

The difficulty in looking to consequences themselves as criteria for calculating lawfulness led the Charter drafters to use prescriptive shorthand to achieve their goal. Because force represents a consistently serious menace to intermediate and ultimate objectives, the prohibition of resort to it is a relatively reliable instrument-based surrogate for a ban on deleterious consequences. It eases the evaluative process by simply asking whether force has been used rather than requiring a far more difficult assessment of the consequences that have arisen.[122]

However, as seen above with reference to economic and political coercion, it cannot be the case that the only criteria to be used are the consequence-based ones. In his article setting out a normative framework for the analysis of computer network attacks, Michael Schmitt has attempted to bridge the gap between instrument-based and consequence-based interpretations. Schmitt proposes seven factors which are to be taken into account when determining whether an attack will constitute a use of force:[123]

(i) *Severity of the damage:* if people are killed or there is extensive property damage, the action is probably military; the less damage, the less likely the action is a use of force.

(ii) *Immediacy of the consequences of the attack:* When the effects are seen within seconds to minutes (such as when a bomb explodes), the

[121] Schmitt, 'Normative Framework', 911.

[122] *Ibid.* [123] *Ibid.*, 914.

operation is probably military; if the effects take weeks or months to appear, it is more likely diplomatic or economic.

(iii) *Directness:* If the action taken is the sole cause of the result, it is more likely to be viewed as a use of force; as the link between cause and effect attenuates, so does the military nature of the act.

(iv) *Invasiveness of the act into the target state:* A violated border is still an indicator of military operations; actions that are mounted from outside a target nation's borders are probably more diplomatic or economic.

(v) *Measurability of the damage:* If the effect can be quantified immediately – such as photographing a 'smoking hole' where the target used to be – the operation has a strong military character; the more subjective the process for evaluating the damage, the more diplomatic or economic.

(vi) *Presumptive legitimacy:* State actors have a monopoly on the legitimate use of kinetic force, while other non-kinetic actions – attacks through or in cyberspace – are often permissible in a wider set of circumstances; actions that have not been the sole province of nation-states are less likely to be viewed as military.

(vii) *Responsibility:* If a state takes visible responsibility for any destructive act, it is more likely to be characterized as a traditional military operation; ambiguous responsibility militates for a non-military label.

Schmitt's analysis, in particular his criteria of presumptive legitimacy, has been criticised by Jason Barkham for begging the question of legitimacy.[124] He argues that it essentially requires a victim state to ask whether an act determining the legitimacy of an attack under international law (i.e., distinguishing between acts of coercion and uses of force) by asking whether the attack is legitimate.

Many of Schmitt's criteria are undoubtedly useful in determining post-facto whether the prohibition has been breached; however, the current author considers that the criteria make insufficient allowance for the unique characteristics of computer network attacks, and the changing nature of global society, for an attack analysis. For example, while an attack is ongoing at the current state of technology, it takes weeks if not months to determine responsibility for an act. In a military age typified by remote weaponry and stand-off tactics, the criterion of a violated border is also losing its relevance. The Court in the *Nuclear Weapons* advisory case stated that in order to apply the Charter

[124] Jason Barkham, 'Information Warfare and International Law on the Use of Force' (2001) 34 *New York University Journal of International Law* 57.

law on the use of force and the law applicable in armed conflict, it was essential to take account of the unique characteristics of nuclear weapons.[125] The same approach may be adopted in respect of computer network attacks.

3.1. *Characteristics of computer network attacks*

The problem with defining computer network attacks as a use of force under current international law is an obvious one. As noted, the contemporary prohibition on the use of force (both as treaty and customary law) has its roots firmly in the text of Article 2(4) of the UN Charter, and at the time of its drafting the science of computing had not advanced to such a state where it could be considered to be any sort of threat.[126]

The problems that arise from considering a computer network attack as a use of force stem from the fundamental characteristics of such attacks. Despite the wide range of attacks which fall under the heading of computer network attack, it is possible to distil four characteristics of a computer network attack which distinguish it from its conventional counterparts: indirectness, intangibility, locus, and result. Some of these characteristics do not raise significant problems for contemporary *jus ad bellum* and are easily solved within the existing framework; however, other characteristics raise more difficult issues. This section examines each of these characteristics and the arguments that have been advanced to militate against such attacks falling within the definition of force. As will be seen, most arguments can be dealt with under existing law.

3.1.1. Indirectness

Although direct computer network attacks are certainly possible, for example the infiltration of a dam's control system to send water downstream, a large number of possible attacks will manipulate one system to achieve a knock-on effect from something else. Examples of such

[125] *Nuclear Weapons Case*, para. 36.

[126] Indeed in 1945 the breakthrough towards modern computing was just beginning. Notwithstanding the important role the Colossus machine played in the British/Allied war efforts at Bletchley Park in breaking the ciphers used by the highest level Germans for strategic commands. See www.turing.org.uk/turing/scrapbook/electronic.html for an account of the claim to invention of the computer by Alan Turing and the Virtual Bletchley Park for details of the role of electronic code breakers during World War II, www.codesandciphers.org.uk/virtualbp/ (last accessed 5 July 2011).

indirect attacks include a manipulation of GPS satellite systems to send an opposing force's missiles off target, manipulation of hospital blood type data resulting in the wrong blood type being given to enemy soldiers, or disabling air traffic control systems. These examples all involve an action which requires further action to be taken by a second actor or object to achieve the desired result. Indirectness *per se* has not been an issue for international law. As seen above, the International Court of Justice in the *Nicaragua (Merits)* case has held that indirect assistance can be a use of force contrary to international law. The Court based its reasoning as follows:

> The element of coercion, which defines, and indeed forms the very essence of, prohibited intervention, is particularly obvious in the case of an intervention which uses force, either in the direct form of military action, or in the indirect form of support for subversive or terrorist armed activities within another State. As noted above (paragraph 191), General Assembly resolution 2625 (XXV) equates assistance of this kind with the use of force by the assisting State when the acts committed in another State 'involve a threat or use of force'.[127]

However, it should be noted in these cases that the further action to be taken has involved a traditional use of armed force. Where the subsequent action would not constitute a 'threat or use of force', such as the example of the hospital records, it seems unlikely under present interpretations of the law that the action could be considered a use of force. Further, not all assistance given to the rebels was considered to be contrary to the prohibition against force; mere supply of funds to the contras did not in itself amount to a use of force.[128] Therefore the causal nexus between an act of a state and a destructive effect on the victim state will be of critical importance.

Another issue arises where the party who is being 'assisted' has no wish or intention to cause damage but is being used as the unwilling agent of the attacking actor. This is the case where computers are recruited to botnets and used to launch distributed denial of service (DDoS) attacks against a target computer or system. This was one of the main tactics of both the attacks against Estonia in 2007 and Georgia in 2008. While primarily a question of attribution, where a state-owned computer is used in this manner, the effect may appear to be a state-sponsored attack of another state's systems.[129] This was the case in

[127] *Nicaragua (Merits)*, para. 205.

[128] *Ibid.*, para. 228.

[129] The problem of attribution of attacks is examined in Chapter 3 below.

the attacks against Estonia in 2007; although computers inside the Kremlin were used in the attacks, subsequent analysis of the attacks concluded that those computers could well have been compromised by the groups of patriotic hackers generally believed to be responsible for the attacks.[130] Thus the indirectness of the attack obfuscates the true perpetrator of the attack, and increases the risk that countermeasures may be directed at innocent parties. While this may be viewed as a desirable effect by an attacking state in that it permits a high level of plausible deniability, such opacity makes the task of determining the correct legal standard to apply while the attack is ongoing that much harder.

3.1.2. Intangibility

The second characteristic of computer network attacks is the intangibility of the attack, both in terms of the method of warfare and the consequences of an attack. While it may be possible to point at a single piece of malicious code that has caused the problem, legally speaking the problem of intangibility exists on three levels. First, the target of the attack may not exist in the physical world other than as information held on a server. Secondly, the 'weapon' itself is intangible, a piece of binary coding which may cause catastrophic effect. Thirdly, the type of damage the attack causes may also be intangible. A computer network attack that does not touch the physical sphere may nevertheless cause mayhem; the oft-cited hypothetical example is an attack on the New York stock exchange that causes mayhem and panic in the United States. This last aspect of intangibility will be examined in section 3.1.4 below.

Target intangibility

Computer network attacks, by definition, target information and information systems. However, they may be divided into those which target information systems in order to affect hardware and other physical aspects, and those which target information as its own end. Where the target of an attack is ultimately a physical entity, such as manipulating software controlling the rotation speed of nuclear centrifuges, the effects of a computer network attack fit more easily with current experience. The more difficult issue arises where the target

[130] See generally Heise Security, *Estonian DDoS – a Final Analysis* http://h-online.com/-732971 (last accessed 12 April 2011); Tikk, Kaska and Vihul, *Cyber Incidents*.

of the attack is information itself. Particularly, where the effect of the attack is not to destroy the information, but to degrade the information target to the extent that it cannot be relied upon. The effects of such an attack would obviously be dependent on the system that was under attack – for instance the consequences of degrading the information resident in an air traffic control system would obviously have a far more catastrophic effect than those of a stock exchange. The former would clearly be considered a use of force, while the latter is more likely to be considered an economic harm. Under the results-based theory, the fact that the attack on the air traffic control system results in serious harm and possible loss of life places it clearly within the purview of a use of force contrary to Article 2(4). However, the test promulgated by Brownlie also talks about damage to property. In both these circumstances the property that is damaged is intangible, *vis* a database. In the event that no loss of life occurs, the only damage which has occurred is to the information or data contained in the database. Is this sufficient property damage to satisfy Brownlie's test of a use of force? Note that under New Zealand and UK intellectual property law at least, the data contained in a database is not considered intellectual property. Therefore it cannot be the case that such damage on its own without further evidence of destruction is sufficient to fall foul of Article 2(4).

Weapons intangibility

A second question that arises is whether a bitstream of malicious code is sufficiently militaristic or weapon-like to meet the required definition of 'armed' force. As seen above, Brownlie's theory of force also required a weapon, and computer network attacks do not conform to our traditional perception of weapons. However, Schmitt argues that the instrument-based distinction is a 'prescriptive shorthand' for a set of consequences which affect community values. Thus he argues:

> Armed coercion is not defined by whether or not kinetic energy is employed or released, but rather by the nature of the direct results caused, specifically physical damage and human injury. Instrumentalities that produce them are weapons. There is little debate about whether the use of chemicals or biological agents falls within the meaning of armed force, even though the means that cause the injury or death differ greatly from those produced by kinetic force.[131]

[131] Schmitt, 'Normative Framework', 913.

That is, a weapon is anything that directly causes physical damage and human injury. Jacobson has argued that 'armed' simply means equipped with weapons of war, and that weapons are tools designed to accomplish a specific mission.[132] Brownlie's approach was to fit the new technology to the definition. Given that a *purely* results-based approach does not provide the appropriate distinction between armed and other forms of coercion without reference to some other factors, those who adopt an instrument-based analysis argue the means of producing the results are still significant. That is, if a computer network attack does not meet some form of weapons criterion then it is likely that the use of such attack techniques will fall within the levels of coercion currently occupied by political and economic coercion. Where these attacks take place during an armed conflict, such acts will be seen as part of the ongoing conflict, and will be judged accordingly;[133] however, it is worth emphasising that, in this instance, these issues are raised in relation to the exercise of such force by an actor of the state before the commencement of traditional military action.

The definition of a weapon in the *Oxford English Dictionary* is a 'thing designed, used, or usable for inflicting bodily harm', or secondly, a 'means for gaining advantage in a conflict'. While the first definition fits with a traditional use of kinetic violence, it is the second definition that may provide the answer for information warfare attacks. As seen above, however, it must be used with some caution as many tactics, which may gain an advantage in a conflict situation, would not be considered weapons. James Bond gives the example of spy satellites which pass over the territory of states and yet are not considered weapons and whose use is not considered a use of force.[134] He reaches this conclusion on the grounds that satellites merely process data rather than having a direct capability of producing death or physical destruction of property. Bond thus extrapolates this to other pieces of equipment which do nothing but process data, and concludes that most, if not nearly all, instances of data manipulation would not equate to employing a weapon and would not constitute the use of force.[135]

[132] Mark Jacobson, 'War in the Information Age: International Law, Self Defence and the Problem of "Non-Armed" Attacks' (1998) 21(3) *Journal of Strategic Studies* 1.

[133] For a discussion of the law of weaponry as it pertains to computer network attacks see Chapter 8 below.

[134] Bond, *Peacetime Foreign Data Manipulation*, 83.

[135] *Ibid.*

However, the question of weapons intangibility is not helpful in determining whether new methods of warfare are uses of force prohibited under international law. An analogy with the domestic criminal law of murder is useful in this regard. Murder weapons come in all shapes and sizes, and what may be a permissible and useful tool in one regard, such as a wrench, for example, can be transformed into an instrument of death in an instant. The key factors that determine its use as a weapon is not the nature of the object itself, but rather how the object was used, against whom and why.[136]

The problem is compounded by a failure on the part of some commentators to distinguish between a particular piece of malicious code as a weapon and identifying the computer or network as a weapon. The weapon of choice for a computer network attack is a series of ones and zeros (a bitstream) which comprise a set of instructions; it may be simple (as in the case of a basic DDoS attack code), or it may be a sophisticated, multi-component piece of malware (such as the Stuxnet worm). A computer network attack is, therefore, perhaps a perfect example of the principles of domestic criminal law indicated above which interpret a weapon as something being used to injure persons or property. Thus in the example above, a wrench in the hands of a mechanic may be merely a tool of the trade or an instrument of destruction depending on his or her intent. The binary coding required for a computer network attack contains the instructions for the attack and hence represents an almost pure expression of the intent of the attacker.

3.1.3. Locus

Another argument raised against computer network attacks as a use of force concerns the problem of the locus of the attack and the locus of the target. The locus of the attack was raised early on in the literature, with some commentators claiming there was no cross-border action involved in the attack.[137] However, this argument is ill founded. While a border violation may be evidence of a use of force, international law does not require cross-border action to occur before a use of force has been found. For example, an act against a visiting foreign minister or head of state of a country has always been considered an act of force

136 See also *ibid.*, 86.

137 See, for example, Sean P. Kanuck, 'Information Warfare: New Challenges for Public International Law' (1996) 37 *Harv. Int'l L.J.* 272, 286.

by the host state.[138] Further, the actions of the United Kingdom in the *Corfu Channel* case were held to be a policy of force even though the strait concerned was of the 'class of international highways through which passage cannot be prohibited'.[139]

The main issue with the locus of the attack is that it can be very difficult to establish with any degree of certainty where the attack was actually generated. Current technology allows attackers to conceal their identity and route any attack through a number of servers based around the globe prior to hitting the target system. For example, the 1998 'Solar Sunrise' attacks, in which a number of US Department of Defense networks were compromised, appeared to be coming from multiple servers in the US as well as the United Arab Emirates, Israel, France, Germany and Taiwan. While this attack is purely a matter of transnational criminal law, the possibilities for a state-sponsored version of the same style of attack exist. The 'Titan Rain' series of intrusions routed stolen data through servers in South Korea, Hong Kong or Taiwan before sending them to computers in Guangdong province in mainland China.[140] Although most analysts believe the Chinese government to be behind the espionage attacks, China has denied all involvement and the attribution cannot be proved. From a law enforcement point of view it may be almost impossible to determine where the attack originated, and whether the attack was launched as the first blow in a conventional interstate conflict, terrorist attack, domestic crime or accident.[141] DDoS attacks may be even more difficult to trace, as attacks come from multiple computers recruited to a botnet – tracing the controlling computer can be a time-consuming and difficult job. The attacks launched against Estonia in 2007 and Georgia in 2008 are cases in point.

Other scholars have taken issue with the fact that the target of any such attack is located in the information realm, i.e. cyberspace, and have argued that this area is not regulated by the current laws of armed

[138] See, for example, the 1993 missile attacks launched by the US in response to an assassination attempt against former President George H. W. Bush.

[139] Although part of the northern Corfu Channel forms part of the territorial waters of Albania and Greece respectively, the international character of the waters and the rights of passage in international law through such waters illustrate the point at hand. *Corfu Channel Case*, 29.

[140] See generally, Nathan Thornburgh, *et al.*, 'The Invasion of the Chinese Cyberspies (and the Man Who Tried to Stop Them)' (2005) 166(10) *Time* 34.

[141] Emily Haslam, 'Information Warfare: Technological Changes and International Law' (2000) 5(2) *JC&SL* 157, 162.

conflict.[142] There are two possible answers to this argument, depending on the mode of the initial attack. The first answer to concerns over the locus of the target is a practical one. As discussed in Chapter 1 above, data does not exist in the ether of cyberspace; it must reside on a server that is actually present in the physical domain. Although the location for a target with military significance is likely to be hosted on a state's own systems, it is also possible that important functions could be located elsewhere, possibly even in a foreign state,[143] so while geographical *boundaries* are not important, the target *is* still embedded in a geographic location. This physical location has been the basis for international consensus on jurisdictional claims relating to cross-border cyber crime and electronic transactions, thus there is no reason the same approach would not be applied in respect of the international law on the use of force and humanitarian law.[144] The second answer is closely related to the following section on the results of computer network attacks which do not affect the physical sphere and is dealt with below.

3.1.4. Result

Computer network attacks incorporate many different techniques from simple denial of service to direct data or system manipulation and the possible results span the spectrum from mere inconvenience to catastrophic damage to life and property.[145] Indeed, the flexibility and wide range of possible results are two of the reasons that such weapons are attractive to the armed forces. However, the indeterminate result of a computer network attack is also the characteristic that causes the most uncertainty in applying the legal requirements of force and the laws of war.

First, while a computer network attack may result in death or damage to physical property, it need not do so. The purpose of the attack may simply be to shut off a particular service or function; for example, shutting off a particular telecommunications system to force an opponent to use a more insecure method.[146] Alternatively, the main

142 See, for example, Kanuck, 'Information Warfare: New Challenges for Public International Law', 287.

143 For example, following their virtual declaration of independence in 1998, the website of East Timor was hosted on servers in the Republic of Ireland.

144 See, for example, the UN model laws on electronic commerce and computer crime.

145 For example, the 'Farewell Dossier' incident resulted in the 'most monumental non-nuclear explosion ever seen from space', Reed, *At the Abyss*, 269.

146 Schmitt, 'Normative Framework', 888.

purpose of the computer network attack may be the denial, corruption or exploitation of the information target itself. For example, the Israeli attacks against the Syrian air defence radar system which allowed their fighters to remain undetected during their raid on a suspected nuclear facility in 2007.[147]

Given the character of these attacks, should these results be dealt with in a legal manner in the same way as conventional uses of force, or should they fall into the category of other means of coercion? As the United States Joint Chiefs of Staff have stated:[148]

> Information Warfare can make an important contribution to defusing crises; reducing the period of confrontation and enhancing the impact of informational, diplomatic and military efforts, and forestalling or eliminating the need to employ forces in a combat situation.

In terms of the prohibition against force, however, the Brownlie and Dinstein models of force require that the result of a use of force is fatality or damage to property; however, no analysis has been undertaken to determine whether such damage to property would include damage to intellectual or other intangible property. Of course, it has always been legitimate in warfare to steal or corrupt the opponent's information, supply them with disinformation and sabotage their weaponry as long as such acts do not constitute perfidy. However, this presupposes an armed conflict is already under way between the opposing states. Where such acts take place prior to the commencement of a traditional attack, the permissiveness of these acts in wartime may militate against the question of whether these acts cross the threshold of 'force' in peacetime. The current author considers that at the present time such incidents do not constitute a use of force prohibited by Article 2(4) of the Charter. Needless to say, however, they may well still constitute unlawful interference in a state and retain the character of internationally wrongful acts invoking state responsibility for reparations.

[147] See Appendix 1 for details of this attack. Fulghum and Barrie, 'Israel Used Electronic Attack in Air Strike against Syrian Mystery Target'; Fulghum, Wall and Butler, 'Israel Shows Electronic Prowess'. Although note that in that case most states would conclude that the network attack represented the start of the wider military operation and it would thus be judged in its entirety.

[148] US Joint Chiefs of Staff, *Information Warfare: A Strategy for Peace … The Decisive Edge in War* (1996) 5, http://handle.dtic.mil/100.2/ADA318379 (last accessed 5 July 2011), cited in Schmitt, 'Normative Framework', 892.

Secondly, one of the consequences of the interconnectivity of electronic resources is that military systems are often using, or are cross-wired with, civilian networks so that without extensive network reconnaissance it is very hard to determine what the effects of any particular attack may be. For example, a recent speech by the US Director of National Intelligence estimated that 98 per cent of all US government communications traffic flowed over civilian networks.[149] As well as posing a targeting dilemma, the problem becomes one of accurate mapping of the networks in order to determine what the consequences of each type of attack may be. An attack designed to electronically incapacitate a command and control centre and early warning systems may inadvertently disable critical equipment in a hospital connected via a node on a related network. The unpredictable nature of the results is one of the major problems in assessing the legality of this new form of warfare.

4. Conclusion

The present author considers that where a computer network attack, directly or indirectly, results in a physical consequence, namely destruction of physical property, injury or loss of lives, it will constitute a use of force under Article 2(4). This appears to be clear from interpretation of the relevant legal instruments despite the intangibility of the weapon used; as seen from the discussion above, the weapon criteria is losing its relevancy in today's world. However, where the result of a computer network attack does not manifest itself in the physical sphere (that is, it affects information only), or its physical results are too minimal or too removed from the chain of causation (i.e. the results were not a foreseeable consequence of the act), the attack does not fall clearly within the traditional test for Article 2(4) and will not constitute a use of force. The fact that a computer network attack does not rise to the level of a use of force does not imply that it is therefore permissible. It is likely that any computer network attack severe enough to raise this question will be considered an unlawful interference in the affairs of a state, and may in all likelihood amount to a threat to the peace.

[149] Michael McConnell, Keynote Address at the Texas Law Review Symposium: Law at the Intersection of National Security, Privacy and Technology, 4 February 2010 cited in Eric Jensen, 'Cyber Warfare and Precautions against the Effects of Attacks' (2010) 88 *Tex. L. Rev.* 1533, fn. 2.

3 Armed attack and response in the digital age

Having looked at computer network attacks as a prohibited act in the previous chapter, this chapter will look at the exceptions to the prohibition; the permissible response to the prohibited act. The UN Charter system admits only two exceptions to the prohibition on the use of force: collective measures authorised by the Security Council acting under Chapter VII of the Charter; or the inherent right of collective or individual self-defence provided for in Article 51.

The exact scope of the right of self-defence remains one of the divisive issues in international law, so it is not surprising that its application to the topic of computer network attacks is particularly complex. First, it is difficult to identify the point at which a computer network attack will rise to the level of an 'armed attack', the threshold that triggers a state's right to self-defence. As seen in the previous chapter, traditional international law focuses on personal injury, fatality and damage to physical property as measures of the seriousness of an attack. This approach is favoured by most commentators writing on the subject to date;[1] however, many catastrophically damaging computer network attacks will not cause any of these deleterious consequences. Further, it is likely that many attacks using computer network operations will not come as a single instance, but rather as a series of events which, taken separately, may not be sufficient to qualify as an armed attack. This raises the issue of the right to respond to 'pin-prick' attacks with a single use of force, a concept which has had a chequered history in international law. The likelihood that computer network attacks will

[1] See Horace B. Robertson, 'Self-Defense against Computer Network Attack under International Law' in M. N. Schmitt and B. T. O'Donnell (eds.), *Computer Network Attack and International Law* (Naval War College, Newport, RI, 2002) 122–45, 136, for examples of authors adopting the consequence-based approach.

be used in conjunction with, or as a precursor to, a conventional attack may also raise a right to respond to such attacks under the auspices of the controversial doctrine of anticipatory self-defence. For example, will intrusion into an air defence network be sufficient evidence of an imminent attack following the 2007 Israeli raid on the alleged Syrian nuclear site.

The second major difficulty with computer network attacks, as opposed to traditional kinetic attacks, is the difficulty of attributing an attack to its original perpetrator. The length of time required to trace an attack makes questions of the necessity of force to repel a computer network attack difficult to satisfy. Questions of proportionality also arise in determining whether it would *ever* be proportionate to use armed force in response to a computer network attack. This chapter examines these questions in light of current international law, the definition of armed attack and considers in what circumstances force may be used in self-defence against a computer network attack.

The final sections address the authorisation to use force under the auspices of Chapter VII Security Council resolutions, and what measures might be taken by states in response to computer network attacks which fail to meet the threshold for lawful self-defence.

1. Armed attack

The right of self-defence is a customary law right, inherent in the sovereignty of states; it has been codified for the most part by Article 51 of the UN Charter as follows:

> Nothing in the present Charter shall impair the inherent right of individual or collective self-defence if an armed attack occurs against a Member of the United Nations, until the Security Council has taken measures necessary to maintain international peace and security. Measures taken by Members in the exercise of this right of self-defence shall be immediately reported to the Security Council and shall not in any way affect the authority and responsibility of the Security Council under the present Charter to take at any time such action as it deems necessary in order to maintain or restore international peace and security.

The customary status of the right has been confirmed by the *Nicaragua (Merits)* case.[2] Article 51 of the Charter allows self-defence in response to an 'armed attack' and the International Court of Justice has

[2] *Nicaragua (Merits)*, para. 176.

confirmed in both the *Nicaragua (Merits)* and *Oil Platforms* decisions that nothing short of an armed attack (with the possible exception of an anticipated armed attack) will be sufficient to trigger the right of self-defence under international law.[3] The *Oil Platforms* case also established that it is for the state using force in self-defence to prove that it has been subjected to an armed attack.[4] However, as with the term 'force', the Charter does not provide a definition of 'armed attack'. Further, the historical predecessors to the Charter system do not offer any interpretive assistance. Neither the Kellogg-Briand Pact nor the Covenant of the League of Nations refer to the term. Both formulate self-defence as a response to aggression; hence all attempts at definition were focused on that term.[5] The drafting history of the Charter provides evidence that discussions took place on the difference between an attack and an armed attack (based on the fact that the United States draft of 11 May 1945 contained both terms) and although no definitive conclusion was reached, the term was replaced by armed attack only.[6] Professor Ian Brownlie has pointed out that it is likely that the records of the San Francisco conference contain no definition of the phrase 'armed attack' because the term was considered 'sufficiently clear' and 'self-evident'.[7] That was certainly the case in the drafting of the North Atlantic Treaty around the same time, as evidenced by the comments of the Foreign Relations Committee of the United States Senate which noted that the phrase 'armed attack' in Article 5 'is ordinarily self evident' and 'there is rarely, if ever, any doubt as to whether it has occurred or by whom it was launched'.[8] The decision of the International Court of Justice in the *Nicaragua (Merits)* case made clear that armed attacks form a subset of the term 'force' in Article 2(4); therefore those actions which have been discussed in the previous chapter as falling outside the ambit of the definition of armed force will subsequently also fail to meet the threshold of an armed attack. However, not every use of force will be an armed attack,

[3] *Ibid.*; *Case Concerning Oil Platforms (Islamic Republic of Iran* v. *United States of America)* (2003), International Court of Justice, para. 51.
[4] *Oil Platforms Case*, para. 57.
[5] Stanimir A. Alexandrov, *Self-Defense against the Use of Force in International Law* (Kluwer Law International, The Hague; London, 1996), 95.
[6] Foreign Relations of the United States 1945 (Washington, 1967) Vol. 1, 674, as cited in *ibid.*, 98.
[7] Brownlie, *Use of Force by States*, 278.
[8] Report of the Senate Committee on Foreign Relations in the North Atlantic Treaty, 6 June 1949, cited in Alexandrov, *Self-Defense*, 96.

thus resulting in a gap between those actions which constitute a use of force, and those which are an armed attack.

The *Nicaragua (Merits)* decision does not provide any clarity on the precise definition of 'armed attack', merely stating that '[t]here appears now to be general agreement on the nature of the acts which can be treated as constituting armed attacks'.[9] However, the Court fails to specify what that agreement might be, confining itself to citing the following example:[10]

> In particular, it may be considered to be agreed that an armed attack must be understood as including not merely action by regular armed forces across an international border, but also 'the sending by or on behalf of a State of armed bands, groups, irregulars or mercenaries, which carry out acts of armed force against another State of such gravity as to amount to' (inter alia) an actual armed attack conducted by regular forces, 'or its substantial involvement therein' ... The Court sees no reason to deny that, in customary law, the prohibition of armed attacks may apply to the sending by a State of armed bands to the territory of another State, if such an operation, because of its scale and effects, would have been classified as an armed attack rather than as a mere frontier incident had it been carried out by regular armed forces.

The Court goes on to consider that acts such as 'assistance to rebels in the form of the provision of weapons or logistical or other support' do not constitute an armed attack, but might nonetheless constitute a threat or use of force, or amount to intervention in the internal or external affairs of other states.[11] Despite being the subject of much criticism and discussion, the Court's decision makes clear that under current international law the definition of 'armed attack' is dependent on the 'scale and effects' of an attack which must be of sufficient gravity to elevate such actions beyond 'mere frontier incidents'.[12] The view that an armed attack must be of 'relatively large scale and with substantial effect' is reiterated by several leading scholars,[13] and is in

[9] *Nicaragua (Merits)*, para. 195. [10] *Ibid.*

[11] *Ibid.*, para. 195, 231. This position was strongly criticised by Judges Schwebel and Jennings in their dissenting opinions, Schwebel holding that the term substantial involvement in the *Definition of Aggression* meant that an armed attack could include financial and logistical support.

[12] *Ibid.*, para. 195.

[13] See, for example, Randelzhofer, 'Article 51', 796 (with accompanying citations). See also Dinstein, 'CNA and Self-Defense', 100, noting that the use of force must be of sufficient gravity before armed attack is in progress, no matter that it be of small magnitude.

agreement with the *de minimis* rule for small-scale attacks set out in Article 2 of the Definition of Aggression adopted by the UN General Assembly, an example also given by the Court in the *Nicaragua (Merits)* case.[14]

The Court's distinction between 'the most grave forms of the use of force (those constituting an armed attack)' and 'other less grave forms',[15] and the resultant gap between force and armed attack, results in the peculiar situation where an illegal use of force not tantamount to an armed attack may be launched by one state against another, leaving the victim state unable to respond in self-defence. Logically and pragmatically the gap has to be quite narrow, inasmuch as 'there is very little effective protection against states violating the prohibition on the use of force, as long as they do not resort to an armed attack'.[16] Michael Schmitt has commented that this distinction makes sense in light of the Charter's central purpose to 'maintain international peace and security' as it creates a rebuttable presumption against the resort by states to violence.[17] Thus he notes 'it is logical to interpret the prohibition on the use of force expansively, but characterize exceptions that lie outside the community decisional architecture, such as self-defense, narrowly'.[18]

Other commentators, including Judge Simma in a separate opinion in the *Oil Platforms* case, have argued, however, that it makes no logical sense to prohibit a state from forcibly defending itself or permitting its allies to come to the state's defence where it is the subject of an unlawful use of force.[19] Any fears of an unwarranted escalation of an incident can be dealt with under the existing rules pertaining to proportionality.[20] Despite such debate, following the *Nicaragua (Merits)* and

[14] The Security Council may determine whether actions falling under the examples given in Article 3 do not constitute 'acts of aggression' owing to their lack of gravity. *Definition of Aggression*, Article 2, as cited in Randelzhofer, 'Article 51', 796.

[15] *Nicaragua (Merits)*, para. 191.

[16] Dinstein, 'CNA and Self-Defense', 100; Randelzhofer, 'Article 51', 661, 664.

[17] Michael N. Schmitt, 'Bellum Americanum Revisited: US Security Strategy and the Jus Ad Bellum, Transcript of Sixteeth Waldemar A. Solf Lecture in International Law to the Judge Advocate General School, US Army, Charlottesville, Virginia, 28 February 2003' (2003) 176 *Mil. L. Rev.* 364, 419.

[18] *Ibid.*

[19] See, for example, John Hargrove, 'The Nicaragua Judgment and the Future of the Law of Force and Self-Defense' (1987) 81 *AJIL* 135, 141; *Oil Platforms Case*, Separate Opinion of Judge Simma, para. 12.

[20] See also Rosalyn Higgins, *Problems and Process: International Law and How We Use It* (Clarendon Press, Oxford, 1993), 242.

Oil Platforms cases it is now established in international law that a gap exists in the laws relating to armed force and armed attack. However, the thresholds of each of these concepts, and the responses which each allows, are yet to be established with any certainty.

This uncertainty makes the classification of any computer network attack particularly difficult. As noted previously, the possible effects of computer network attacks 'span the spectrum of consequentiality',[21] thus making classification based on the type of computer network attack impossible; each attack must be assessed on its merits. It seems certain, however, that where a computer network attack causes destruction and fatalities on a par with a conventional attack, a state will have a right to respond in self-defence. This is the conclusion reached in a report by the Office of General Counsel of the US Department of Defense, which concluded:[22]

> [I]f a coordinated computer network attack shuts down a nation's air traffic control system along with its banking and financial systems and public utilities, and opens the floodgates of several dams resulting in general flooding that causes widespread civilian deaths and property damage, it may well be that no-one would challenge a victim nation if it concluded that it was the victim of an armed attack, or of an act equivalent to an armed attack.

While undoubtedly correct, the report fails to separate and analyse the attack in its component parts, combining those components which directly cause death and destruction with less severe attacks. Schmitt focuses on the consequences of the attack rather than on the object of the attack or on the intentions of the attacker; the exception being where the intentions of the attacker are specifically to cause physical damage to tangible objects or injury to human beings, in which case Schmitt considers the resort to armed force is permitted.[23] He argues that self-defence should be limited to operations which are de facto armed attacks, or imminently preparatory thereto; the net result being a limitation on both sides to resort to computer network attack techniques which might threaten global stability and on individual responses which might themselves prove destabilising.[24]

21 Schmitt, 'Normative Framework', 912.

22 Office of General Counsel, *An Assessment on International Legal Issues in Information Operations*, United States Department of Defense (1999) www.au.af.mil/au/awc/awcgate/dod-io-legal/dod-io-legal.pdf (last accessed 5 July 2011).

23 Schmitt, 'Normative Framework', 928.

24 *Ibid.*, 886. This is in accord with his generally contextualist analysis discussed in Chapter 2.

In contrast, Walter Sharp, one of the early authors on the subject, argues that any intrusion into systems containing information which is critical to the national security of the victim state should be considered an armed attack capable of triggering the right to respond with force.[25] This idea is premised on the notion that the features of modern society and its reliance on information systems are such that non-military means can often cause more harm than military ones.[26] However Sharp's 'strict liability' approach has not been borne out by state practice, which to date has recognised such intrusions as simple espionage, an act not prohibited under international law.

The present author considers that the threshold for an armed attack conducted by computer network attack must be set in line with current international law regulating the right of self-defence. A state is therefore permitted to respond in self-defence when it is the victim of a computer network attack causing damage to property or persons of sufficient scale and effect to elevate it beyond the equivalent of a frontier incident. The denial of service attacks against Estonia in 2007, disruptive as they were, did not reach this level; neither did the attacks launched the following year against Georgia, although the analysis of that incident is complicated by the subsequent conventional attacks by Russia.[27] No loss of life, physical injury or destruction of property was suffered as a result of the computer network operations launched against either state. In contrast, the 2010 Stuxnet worm has purportedly resulted in significant destruction of property. At the time of writing, reports indicate that the worm may be responsible for Iran having to replace 1,000 of the 9,000 IR-1 centrifuges at the Natanz nuclear fuel enrichment facility. While reports of the effect of the worm on the facility and the resulting impact on Iran's nuclear programme have differed wildly, a report by the Institute for Science and International Security detailing the effect of Stuxnet has concluded that while the effects of the worm were significant, the effect on Iran's nuclear programme was only moderate:[28]

[25] Sharp, *Cyberspace and the Use of Force*, 129.

[26] *Ibid.*, 101–2.

[27] The question of the cyber attacks acting as an indication of imminent armed attacks is addressed in the following section on anticipatory self-defence.

[28] David Albright, Paul Brannan and Christina Walrond, *Stuxnet Malware and Natanz: Update of ISIS December 22, 2010 Report*, Institute for Science and International Security (2011) 3.

While it has delayed the Iranian Centrifuge program at the Natanz plant in 2010 and contributed to slowing its expansion, it did not stop it or even delay the continued build up of [low enriched uranium]

That being the case, it would appear that while the Stuxnet worm would undoubtedly amount to a use of force, the scale and effects of the attack do not appear to have sufficient gravity to amount to an armed attack.

1.1. *Anticipatory self-defence*

It is likely that computer network attacks will be used in conjunction with, or as a prelude to, a traditional armed attack. Such attacks designed to 'prepare the battle space' can come in a myriad of forms, including the disablement of intelligence-gathering sensors such as satellites and radar posts via computer network attack, disruption of military communications networks leaving units isolated and unable to be scrambled, or, on the civilian level, disablement of emergency response networks may be launched in anticipation of more traditional kinetic action to follow. It is certainly clear that states are considering how a combination of electronic and traditional attacks might be used in the future. For example, the US actively looked for computer network solutions prior to the 2003 invasion of Iraq and China is openly looking for an 'assassin's mace' to address the perceived asymmetry of the Chinese military against the US in conventional conflicts, widely believed to incorporate computer network attacks.[29] As Robertson points out, in modern warfare the electronic battlefield will play a crucial role, and any steps that a prospective attacker can take to neutralise or destroy its enemy's command and control, intelligence, communications, or weapons-control networks prior to a traditional attack would gain an enormous advantage.[30] This advantage was seen in the 2007 Israeli attack on the suspected Syrian nuclear site where the air defence radar did not detect the attacking planes until they were disappearing back over the border.[31] One might also look to the role of the distributed denial of service attacks against Georgia in 2008 which preceded the physical invasion by Russia. While such attacks may not be sufficient

[29] David A. Fulghum, 'Frustrations and Backlogs', *Aviation Week & Space Technology* 10, 10 March 2003, 33. According to reports these attempts were never put into action due to the interconnectedness of the target systems with foreign-owned networks.

[30] Robertson, 'Self-Defense against CNA', 139.

[31] See Fulghum, Wall and Butler, 'Israel Shows Electronic Prowess'.

to amount to an armed attack in and of themselves, as preparatory moves in a conventional attack they may be sufficient to justify the use of force under the doctrine of anticipatory self-defence.

1.1.1. Doctrinal debate and imminent attacks

There is no consensus in international legal doctrine over the point in time from which measures of self-defence against an armed attack may be taken.[32] In particular, states and commentators are divided over whether a right of anticipatory self-defence exists in international law. The debate centres on whether anticipatory self-defence survived the implementation of Article 51 of the UN Charter, which states that nothing in the Charter 'shall impair the inherent right of individual or collective self-defence if an armed attack occurs'. Those arguing for a restrictive interpretation of the clause interpret this to mean that the right of self-defence is now only available to member states who are the object of an actual armed attack. Gray, Kelsen and Brownlie are some of the leading proponents of this view, which is based on the premise that the Charter forbids any use of force on the part of individual members except for the right of self-defence against an armed attack.[33]

Those arguing for a more expansive approach point out that Article 51 specifically reserves the 'inherent right' of self-defence, and the customary right includes the right to respond to an imminent armed attack.[34] Although ultimately disagreeing with the position, Bothe points out that many authors acknowledge that a threat may be so direct and overwhelming that it is just not feasible to require the victim to wait to act in self-defence until the act has actually started.[35] Other authors have addressed this problem by adopting a broader notion

[32] Randelzhofer, 'Article 51', 803.

[33] Bowett, *Self-Defence*, 188. See also Brownlie, *Use of Force by States*, 275; Ian Brownlie, *International Law and the Use of Force by States: Revisited* (Europaeum, Oxford, 2001); Randelzhofer, 'Article 51', 803–4; Christine D. Gray, *International Law and the Use of Force* (3rd edn, Oxford University Press, 2008), 160; Dinstein, *War, Aggression, and Self-Defense*, 183.

[34] Authors subscribing to this expansive view include Thomas M. Franck, *Recourse to Force: State Action against Threats and Armed Attacks* (Cambridge University Press, 2002), 97; Bowett, *Self-Defence*, 187–92; Christopher Greenwood, 'International Law and the Pre-Emptive Use of Force: Afganistan, Al-Qaida, and Iraq' (2003) 4 *San Diego Int'l L.J.* 7; Robert Y. Jennings and C. Arthur H. Watts (eds.), *Oppenheim's International Law* (9th edn, Longman, Harlow, 1992), 421.

[35] Michael Bothe, 'Terrorism and the Legality of Pre-Emptive Force' (2003) 14(2) *EJIL* 227, 231.

of when an armed attack has occurred in the context of Article 51. For example, while expressly rejecting the notion of anticipatory self-defence, Yoram Dinstein nevertheless incorporates actions taken in advance of an actual attack by moving the timing of the beginning of the attack, a concept he calls 'interceptive' self-defence.[36] Dinstein argues that the beginning of an armed attack is not linked to the first shot but rather to the moment of irrevocable commitment to the attack. Once the die has been cast, the armed attack can be said to have commenced and the victim state need not 'wait impotently for the inescapable blow'.[37] This concept has echoes in the 'target locking' arguments of some states in respect of modern precision-guided missiles. For example, the US argues that an armed attack begins when the radar guiding the missile is locked on and ready to fire, and the rules of engagement of their armed forces reflect this approach.[38] Although, as Lindsay Moir points out, whether such an approach truly distinguishes between an armed attack which has begun but not yet reached the target state, and an armed attack that has not begun but is imminent, is open to question.[39]

The standard to be applied for taking action in anticipation of an attack was set out by US Secretary of State Daniel Webster in his letter regarding the *Caroline* incident of 1837. In his now classic formulation, he stated that the right of self-defence only arises where there is 'a necessity of self-defence, instant, overwhelming, leaving no choice of means, and no moment for deliberation'.[40] Christopher Greenwood

[36] Dinstein, *War, Aggression, and Self-Defense*, 187.

[37] Dinstein, 'CNA and Self-Defense', 111.

[38] Gray, *Use of Force* (2nd edn), 108, n. 48. For example, in 1998 US aircraft in the no-fly zone over Iraq fired at a missile battery when its radar had locked on to the planes patrolling the zone. Although there was controversy over whether the radar had locked on, the idea that the armed attack had started when the radar locked on was apparently accepted by Iraq and other states: *Keesings* (1998) 42368.

[39] Lindsay Moir, *Reappraising the Resort to Force: International Law, Jus Ad Bellum and the War on Terror* (Hart, Oxford, 2010), 16.

[40] In that incident, UK forces captured, fired and sent over the Niagara Falls a merchant vessel called the *Caroline* which was being used by Canadian rebels and their American forces in attacks against passing British ships. At the time of the UK attack the ship was moored in an American port and two US nationals were killed. One of the British officers, Lieutenant McLeod, was later arrested in the United States on charges of murder arising out of the incident. The British government, seeking McLeod's release, maintained that its forces had acted in self-defence, to which US Secretary of State Daniel Webster replied in what has become the accepted statement of self-defence doctrine at the time. *The Caroline Case* (1837) 29 BFSP 1137–8.

points out that the *Caroline* test was applied by the international military tribunals at Nuremburg and Tokyo, suggesting that the right to respond in self-defence against imminent attacks was part of the customary law right preserved by Article 51 of the Charter.[41] The preservation of an inherent right of self-defence and the existence of customary rights outside the Charter wording is also recognised by the Court in the *Nicaragua (Merits)* case, which noted that the Charter does not contain all the rules pertaining to self-defence, notably a definition of armed attack and the requirement of necessity and proportionality in any response.[42] The Court expressly did not comment on the lawfulness of an anticipatory response to an imminent armed attack as it was not required on the facts of the case.[43] The Court's approach in avoiding the difficult questions of self-defence and expressly stating no view on anticipatory self-defence has been followed in both the *Wall* and *Armed Activities* cases.[44] Two recent UN reports have not been so reluctant, however. The 2004 report of the UN Secretary-General's High Level Panel concluded that there was an existing right of anticipatory self-defence against imminent attacks, basing their conclusion on customary international law.[45] Likewise the Report of the Secretary-General the following year expresses the view that responses to imminent threats are fully covered by Article 51 of the Charter.[46]

1.1.2. State practice

The content of any customary right of anticipatory self-defence must be examined in light of state practice since the inception of the Charter. While state practice is far from conclusive on the matter, two instances in particular tend to indicate that the doctrine has survived. The first is the Israeli–Arab war of 1967. In the wake of escalating tensions between Syria and Israel, Egypt requested the removal of a United Nations emergency force from Egyptian territory, reinforced troops in the Sinai and dispatched troops to Jordon. Egyptian President Nasser also closed the Straits of Tiran to Israeli shipping (an act that Israel had

[41] Greenwood, 'Pre-Emptive Use of Force', 13.

[42] *Nicaragua (Merits)*, para. 176.

[43] *Ibid.*, para. 194.

[44] *The Wall Case*; *Armed Activities Case*, para. 143.

[45] United Nations, *A More Secure World: Our Shared Responsibility. Report of the Secretary-General's High-Level Panel on Threats Challenges and Change*, United Nations, UN Doc. A/59/565 (2004) 63, para. 188.

[46] UN Secretary General, *In Larger Freedom: Towards Development, Security and Human Rights for All*, United Nations, UN Doc. A/59/2005 (2005) para. 124.

previously made clear it would consider as an act of war) amid statements indicating his intention to eliminate Israel. In response to these actions Israel launched strikes against Egypt's airbases, completely destroying the Egyptian air force.[47] Israel justified its actions in the Security Council by claiming that it had been attacked first, but also by reference to a series of acts of sabotage from Syria, and stressed both the character of the Egyptian blockade as an act of war and 'perhaps more imminent than anything else, this vast and purposeful encirclement movement' caused by the transfer of large numbers of Egyptian forces to their 'very gates'.[48] Gray rejects this incident as evidence of an acceptance of anticipatory self-defence, stating that 'whatever position is taken on the facts of the outbreak of the Six Day War, the point of importance here is that Israel did not claim to be acting in anticipatory self-defence'.[49] However, Franck notes that Israel's 'words and actions clearly asserted a right of anticipatory self-defence against an imminent armed attack'.[50] In emphasising the fact that neither the Security Council nor the General Assembly condemned Israel's actions, Franck comments:[51]

> Most states, on the basis of the evidence available to them, did however conclude that such an armed attack was imminent, that Israel had reasonably surmised that it stood a better chance of survival if the attack were pre-empted, and that, therefore, in the circumstances, it had not acted unreasonably. This does not amount to an open-ended endorsement of a general right to anticipatory self-defense, but it does recognize that, in demonstrable circumstances of extreme necessity, anticipatory self-defense may be a legitimate exercise of a State's right to ensure its survival.

This accords with the advisory opinion of the International Court of Justice in the *Nuclear Weapons* case, which indirectly commented on the situation when the majority of judges were unable to conclude that the first use of nuclear weapons would invariably be unlawful if the very existence of the state were threatened.[52]

[47] For a full exposition of the facts of this incident see generally A. Mark Weisburd, *Use of Force: The Practice of States since World War II* (Pennsylvania State University Press, 1997), 135.

[48] UN SCOR, 1348th Meeting, 6 June 1967, UN Doc. S/PV.11348, paras 142–94.

[49] Gray, *Use of Force* (3rd edn), 161.

[50] Franck, *Recourse to Force*, 103.

[51] *Ibid.*, 105.

[52] *Legality of the Use by a State of Nuclear Weapons in Armed Conflicts* (1996) ICJ 26, International Court of Justice, 265, para. 105(2)E.

In comparison, when Israel attacked and destroyed the Tuwaitha Research Centre and Osarik nuclear reactor near Baghdad, Iraq in 1981, the action was 'strongly condemned' by the Security Council,[53] and in general states' reactions to the bombing were condemnatory of Israel. In most cases the reaction was based on a conclusion that Israel had failed to demonstrate that there was an imminent threat from Iraq and had thus failed to satisfy the *Caroline* requirements for anticipatory self-defence rather than a general dismissal of a right of anticipatory self-defence.[54] As Greenwood points out, the emphasis on this failure to demonstrate the existence of the imminent threat tends, if anything, to confirm the existence of a right of self-defence in cases where such an imminent threat was shown to exist.[55] He goes on to cite Rosalyn Higgins, who notes:[56]

> [I]n a nuclear age, common sense cannot require one to interpret an ambiguous provision in a text in a way that requires a state passively to accept its fate before it can defend itself. And, even in the face of conventional warfare, this would also seem the only realistic interpretation of the contemporary right of self-defence. It is the potentially devastating consequences of prohibiting self-defence unless an armed attack has already occurred that leads one to prefer this interpretation – although it has to be said that, as a matter of simple construction of the words alone, another conclusion might be reached.

In assessing what will constitute an imminent attack, Greenwood argues that there are two additional factors which must be taken into account which did not exist at the time of the *Caroline* incident:[57]

> The first is the gravity of the threat. The threat posed by a nuclear weapon, or a biological or chemical weapon, if used against a city, is so horrific that it is in a different league from the threats posed (as in the *Caroline*) by cross-border raids conducted by men armed only with rifles. Where the threat is an attack by weapons of mass destruction, the risk imposed upon a State by waiting until that attack actually takes place compounded by the impossibility for that State to afford its population any effective protection once the attack has been launched, mean that such an attack can reasonably be treated as imminent in circumstances where an attack by conventional means would not be so regarded. The second consideration is the method of delivery of the threat. It is far more difficult to determine the time scale within which a threat of

53 SC Res 487, UN SCOR, 2288th Meeting, 19 June 1981, UN Doc. S/Res/487.
54 Franck, *Recourse to Force*, 105.
55 Greenwood, 'Pre-Emptive Use of Force', 14.
56 Higgins, *Problems and Process*, 242.
57 Greenwood, 'Pre-Emptive Use of Force', 16.

attack by terrorist means would materialize than it is with threats posed by, for example, regular armed forces. These would be material considerations in assessing whether, in any particular case, an attack should be treated as imminent.

Although the existence of a right of anticipatory self-defence remains controversial, it is the view of the present author that a right of self-defence against an imminent attack is established in customary international law. The idea of computer network attacks as an indicator of an imminent attack is discussed in the following section.

1.1.3. Computer network attacks and anticipatory self-defence

There are two situations where anticipatory self-defence may be implicated in the information warfare context: first, where a computer network attack serves as an imminent threat of a conventional attack; and secondly, where electronic activity indicates a severe computer network attack (which rises to the level of an armed attack) is imminent.

As with any assessment of an imminent attack the context of a computer network attack must be taken into account. Where a computer network attack is launched as a precursor to conventional attack, the target will be important. If an attack targets early warning systems, radar posts or satellite feeds, military communications, or emergency response systems it is likely that a state will judge a traditional attack to be imminent. The Israeli attack against the Syrian air-defence network is a case in point. Had Syria become aware of the intrusion and manipulation of its air-defence radar prior to the attack, they would have been entitled to use force in response. Indeed most military forces would claim that the neutralisation of the air-defence network was the beginning of the attack itself, and would certainly be included under Dinstein's interceptive self-defence analysis.

The disruption of electrical power grids or financial systems, on the other hand, is unlikely to be sufficiently indicative of a subsequent conventional attack when viewed in isolation. Thus when Estonia was subject to distributed denial of service attacks against its banking, media and governmental sites, there were no realistic fears that this signalled the beginning of a traditional armed attack. The additional variables, such as positive attribution to a particular actor and possible motivation, are too many and too varied. However, when such attacks are viewed in conjunction with other contextual indicators, states may be more likely to conclude that a conventional attack is imminent.

The 2008 conflict between Russia and Georgia over South Ossetia illustrates the dangers associated with such an attribution, however. That conflict took place in an atmosphere of increased tensions between the states and in the face of skirmishes between separatist forces and the Georgian military. Prior to the physical invasion, a series of distributed denial of service (DDoS) attacks were launched against the Georgian government, and media targets and commentators have noted the close coordination between the targets of the computer network attacks in specific areas and the later physical strikes.[58] While it is tempting to conclude that the computer network attacks were indicative of the imminent physical invasion, analysts agree that there is no link evident between the Russian government and the attacks, and that patriotic hackers were behind the electronic disruptions.[59] Given the increasing frequency with which hacktivist groups launch such denial of service attacks to further their causes or protest political decisions, it seems unlikely that states will be able to rely on such attacks as indicators even, and perhaps especially, in times of increased political tension.

The second instance where a computer network attack may be evidence of an imminent armed attack is the use of computer network intrusions to prepare an electronic battlespace. Viruses, worms, Trojan horses and other forms of malware routinely infect unprotected computers with malicious code which may corrupt data, cause a malfunction, record keystrokes, disable virus protection and collect other information such as passwords and other access codes, feeding them back to a remote attacker. One of the most common features of these types of malware is some form of backdoor payload which allows the attacker to access and control the computer at a later date. Such malware is most often used to 'recruit' unprotected computers to vast networks of compromised computers called botnets. Usually used for spam or criminal enterprise, the controllers of the botnet can also use the command and control servers of the botnet to direct the sending of large amounts of traffic to particular IP addresses, bringing them, and

58 See, for example, US Cyber Consequences Unit, *Overview by the US-CCU of the Cyber Campaign against Georgia in August of 2008*, US Cyber Consequences Unit (2009) www.registan.net/wp-content/uploads/2009/08/US-CCU-Georgia-Cyber-Campaign-Overview.pdf (last accessed 25 February 2011); Project Grey Goose, *Phase I Report: Russia/Georgia Cyber War – Findings and Analysis* (2008).

59 US Cyber Consequences Unit, *Overview*; Project Grey Goose, *Phase I Report*; Tikk, Kaska and Vihul, *Cyber Incidents*, 66–79.

in some cases the transmission routes, to a standstill in a DDoS attack. Botnets have been utilised in most major DDoS attacks reported to date, including against Estonia in 2007 and Georgia in 2008.[60]

The inclusion of backdoor payloads in malware raises interesting questions in relation to anticipatory self-defence.[61] It is clear that a backdoor has no other purpose than to allow an intruder control over the infected computer (whether by direct intrusion or remote control) at a later date. A question which must then be addressed is whether a state has the right to respond in anticipatory self-defence against the perpetrators of a computer network attack with a backdoor payload. Although a backdoor can be used for attacks at any point until it is discovered and removed, and such attacks may have far more serious consequences, including those which would qualify as an armed attack, in most cases, the later use is to send spam or launch DDoS attacks causing inconvenience and only economic damage. Without further information about the purpose or target of any later attack, the mere creation of a backdoor by a state adversary is not indicative of the type or gravity of the attack to follow; indeed the creation of a backdoor may merely be evidence of espionage.[62] Most importantly from a legal perspective, the existence of a backdoor would not meet the 'imminent' criteria for lawful self-defence, as the timescale for any subsequent attack is variable; the attack could be launched within days or years.

Other network intrusions may more indicative, but they will be highly dependent on the circumstances. Two points deserve comment. First, the method of delivery means that there may be little warning of an impending computer network attack; this was one of Greenwood's additional factors to be considered when determining whether an attack is imminent.[63] However, Greenwood's second factor, the gravity of threat, will depend entirely on the individual threat; as pointed out in the previous chapter, one of the difficulties with assessing computer

[60] See, for example, Gadi Evron, 'Battling Botnets and Online Mobs' (2008) 9(1) *GJIA* 121, 124; Tikk, Kaska and Vihul, *Cyber Incidents*.

[61] A backdoor is a piece of code which opens a hidden or undocumented access point to the compromised computer or system, bypassing the authentication and security protocols and allowing remote access.

[62] Backdoors have been found on computers allegedly compromised by Chinese hackers.

[63] Greenwood, 'Pre-Emptive Use of Force', 16.

network attacks is that they span the 'spectrum of consequentiality'.[64] While computer network attacks against critical infrastructure such as electricity grids, dams and oil pipelines would be devastating to modern society, and may result in death and property damage, to date no computer network attack has been launched that would be comparable to the threat posed by a nuclear, biological or chemical weapon. In this regard, with the very survival of the state not in question, it is unlikely that the threat would be assessed as imminent.

1.1.4. The 'Bush doctrine' of pre-emptive self-defence[65]

In 2002 and 2006 the United States released two security strategy documents containing what appears to be a highly controversial attempt at enlarging the right to self-defence to include the use of force to pre-empt an attack which is merely threatened, but is not imminent. As Gray points out, there is some uncertainty as to the exact scope of the US claims; however, international criticism of the 'Bush doctrine' was both widespread and fierce.[66] The 2002 National Security Strategy provided that:[67]

> We must be prepared to stop rogue states and their terrorist clients before they are able to threaten or use weapons of mass destruction against the United States and our allies …
>
> … Legal scholars and international jurists often conditioned the legitimacy of pre-emption on the existence of an imminent threat – most often a visible mobilization of armies, navies, and air forces preparing to attack.
>
> We must adapt the concept of imminent threat to the capabilities and objectives of today's adversaries … The greater the threat, the greater is the risk of inaction – and the more compelling the case for taking anticipatory action to defend ourselves, even if uncertainty remains as to the time and place of the enemy's attack. To forestall or prevent such hostile acts by our adversaries, the United States will, if necessary, act pre-emptively.

[64] Schmitt, 'Normative Framework', 912.

[65] This is also known as preventative or predictive self-defence in security studies circles. There is also much confusion in the legal literature regarding what exactly is being referred to by the terms anticipatory, pre-emptive and preventative self-defence.

[66] Gray, *Use of Force* (3rd edn), 211; Moir, *Resort to Force*, 79.

[67] White House, *The National Security Strategy of the United States of America*, White House, 2002), 15, www.au.af.mil/au/awc/awcgate/nss/nss_sep2002.pdf (last accessed 5 July 2011).

The sentiments of the 2002 report are repeated in the National Security Strategy 2006.[68] While at first glance, this may appear to extrapolate logically based on the contingencies of modern weaponry, it compromises the basic premise of anticipatory self-defence laid down in the *Caroline* case and the evidence of state practice which has evolved the doctrine since. Greenwood has noted, far from removing the requirement of imminence '... practice also shows that the right of anticipatory self-defence is confined to instances where the armed attack is imminent'.[69] As noted above, while Greenwood accepts that the notion of imminence must take into account the gravity of the threat and its method of deliverance, he goes on to state: 'In so far as talk of a doctrine of "pre-emption" is intended to refer to a broader right of self-defence to respond to threats that might materialize at some time in the future, such a doctrine has no basis in law.'[70] The issue has been debated extensively in the legal literature and will not be repeated here, save to say that state practice has not indicated widespread support for the doctrine, and both UN reports on security have rejected any wider right of pre-emptive self-defence, indicating it is for the Security Council to take pre-emptive action.[71] In 2010 legal commentators watched with interest to see what the policy of the new US administration would be. However they were to be disappointed; while downplaying the idea of pre-emption, the Obama administration's National Security Strategy 2010 does not dismiss it, merely stating that states must 'adhere to standards that govern the use of force' without elaborating further on what those standards might be.[72] The relevance of this apparent

[68] White House, *National Security Strategy of the United States of America* (White House, 2006), 23. Although Gray notes the striking absence of any express reference to international law, the 2006 report does state that 'the United States will, if necessary, act pre-emptively in exercising our inherent right of self-defense', most likely an implicit reference to Art. 51 of the Charter: Christine D. Gray, 'The Bush Doctrine Revisited: The 2006 National Security Strategy of the USA' (2006) 5(3) *Chi. J. Int'l L.* 555, 561.

[69] Greenwood, 'Pre-Emptive Use of Force', 15. [70] *Ibid.*

[71] United Nations, *A More Secure World*, paras 189–92; UN Secretary General, *In Larger Freedom*, 125. See generally Christian M. Henderson, 'The 2006 National Security Strategy of the United States: The Pre-Emptive Use of Force and the Persistent Advocate' (2007) 15 *Tulsa J. Comp. & Int'l L.* 1; Gray, 'The Bush Doctrine Revisited: The 2006 National Security Strategy of the USA' for a discussion of international reaction to the 'Bush doctrine'.

[72] White House, *US NSS* 2010. For a discussion of the Bush doctrine, the US National Security Strategy 2010 and the *jus ad bellum* see generally Christian Henderson, 'The 2010 United States National Security Strategy and the Obama Doctrine of "Necessary Force"' (2010) 15 *JC&SL* 403.

continued adherence to the controversial doctrine for computer network attacks became apparent later in 2010 when it was reported that the Pentagon was considering pre-emptive strikes as part of its cyber defence strategy.[73] Although not yet released at the time of writing, the strategy was previewed by US Deputy Secretary of Defense, William Lynn III in what he describes as a 'fundamental shift in the US approach to network defense'.[74] The strategy turns on the 'active defense' of military computer systems, although there seems little consensus by officials (prior to the release of the report) as to what the term means.[75] Lynn has spoken both of 'placing scanning technology at the interface of military networks and the open internet to detect and stop malicious code before it passes into military networks', which would be fairly uncontentious, but also of responding to attacks as they happen 'or even before they arrive'.[76] If the response is to include destruction of an adversary's network before it has become an imminent threat and to a level that it would constitute a use of force, it would represent a continuation of the 'Bush doctrine'; it remains to be seen whether this is the case.

1.2. *Pinprick attacks or accumulation of events theory*

Computer network attacks falling below the armed attack threshold may still trigger a forcible response by states. The most likely strategy utilising computer network attacks is to launch a swarm of lesser attacks, rather than a single large strike qualifying as an armed attack. In a short story published in 1998, John Arquilla detailed what a sustained cyber attack might look like:[77] power blackouts, followed by weekly virus attacks of the magnitude of the Nimda, Slammer or Mydoom viruses, oil pipeline ruptures all launched within a matter of days of one another. While some of these attacks may cross the threshold of uses of force under Article 2(4), it is unlikely that any taken on their own would be considered an armed attack under current international law. An analogy may be drawn with cases of repeated cross-border incursions. States have claimed a right to act in self-defence

73 Nagashima, 'U.S. Eyes Preemptive Cyber-Defense Strategy'.
74 Lynn, 'Defending a New Domain', 103.
75 Nagashima, 'U.S. Eyes Preemptive Cyber-Defense Strategy'.
76 Lynn, 'Defending a New Domain', 103; Nagashima, 'U.S. Eyes Preemptive Cyber-Defense Strategy'.
77 John Arquilla, 'The Great Cyberwar of 2002', *Wired* 6.02, February 1998, www.wired.com/wired/archive/6.02/cyberwar.html (last accessed 12 April 2011).

against a series of small-scale incursions or attacks as collectively amounting to a single armed attack. The 'pinprick' or 'accumulation of events' theory has been unsuccessfully argued in the past by several states, including the United Kingdom,[78] the United States,[79] South Africa and Israel,[80] in an attempt to justify actions purportedly taken in self-defence, that might otherwise be disproportionate. Although the Security Council rejected the claims by these states, each time it has done so on the specific facts of the case and on grounds that such actions were disproportionate. In each instance the condemnation by the Security Council suggested that the actions were unlawful reprisals, rather than commenting on the doctrinal issue of the scope of self-defence.[81] Likewise, the International Court of Justice has avoided expressly discussing the issue, although it appears willing to contemplate the possibility of an accumulation of events amounting to an armed attack. The Court in the *Nicaragua (Merits)* case commented with regard to the incursions by Nicaragua into the territory of Honduras and Costa Rica:[82]

> Very little information is available to the Court as to the circumstances of these incursions or to their possible motivations, which renders it difficult to decide whether they may be treated for legal purposes as amounting, singly or collectively, to an 'armed attack' by Nicaragua on either or both States.

[78] In 1964 the Southern Arab Federation (SAF) which had military links with the United Kingdom, complained of an armed attack by the Yemen which consisted of a 'series of aggressions'; invoking collective self-defence at the SAF's request, the UK launched an air strike and destroyed a fort. The Security Council did not accept the reasoning and issued a statement condemning reprisals as 'incompatible with the principles and purposes of the UN'. SC Res 188, 9 April 1964, as cited in Jean Combacau, 'The Exception of Self Defence in U.N Practice' in A. Cassese (ed.), *The Current Legal Regulation of the Use of Force* (Martinus Nijhoff, Dordrecht, 1986) 9–38, 27.

[79] The US claimed it was acting in self-defence against alleged attacks by North Vietnamese naval vessels in the Gulf of Tonkin. Letter of 17 Febuary 1979, S/13094 (SCOR, 36th Year) cited in *ibid.*, 17.

[80] The Security Council has rejected claims of self-defence by Israel made to justify incursions into neighbouring states to attack Palestinian bases, when it attacked Jordan in 1966 (SC Res 228, 25 November 1966) and 1969 (SC Res 265, 1 April 1969), and Lebanon in 1969 (SC Res 270, 16 August 1969), 1970 (SC Res 279, 12 May 1970), 1972 (SC Res 313, 28 February 1972), 1973 (SC Res 332, 21 April 1973) and 1974 (SC Res 347, 24 April 1974). See *ibid.*

[81] Gray, *Use of Force* (3rd edn), 155.

[82] *Nicaragua (Merits)*, para. 231. It should be noted that the Court found that there had not been an armed attack by Nicaragua based on additional circumstances.

This statement would seem to indicate a willingness on the part of the Court to consider that a series of small-scale attacks on a target may amount to an armed attack when viewed collectively. This view is reiterated in the *Oil Platforms, Cameroon/Nigeria* and *Armed Activities (DRC v. Uganda)* cases.[83] In all three cases, while not expressly discussing the doctrine, the Court appears to permit the concept of an armed attack through cumulative attacks; however, the Court found that neither the United States, Cameroon nor Uganda respectively had sufficiently proved the facts or imputability to the other party.[84] This doctrine has obvious application for the victims of protracted campaigns involving low-intensity computer network attacks. However, as with their conventional equivalents, the devil will be in the detail, conclusively proving that the attribution of computer network attacks is a notoriously difficult and time-consuming task.[85]

One final point should be noted: not all attacks which are launched from multiple computers will require a cumulative approach to qualify as an armed attack. A DDoS attack of sufficient scale and effect to elevate it above a 'mere frontier incident' would be considered a single attack as the attack originates from a single controller.[86] That is, although the attack appears to come from large numbers of computers in disparate locations, the compromised computers all receive their attack and targeting instructions via command and control servers manipulated from a single controlling 'master' which directs the botnet.[87] This is merely the electronic equivalent of an attack using more than one soldier, or a wave of bombers in an air strike.

2. Self-defence against non-state actors

One of the most difficult issues in contemporary *jus ad bellum* is the permitted response of states to attacks by non-state actors. The difficulty

[83] *Case Concerning the Land and Maritime Boundary between Cameroon and Nigeria (Cameroon v. Nigeria: Equatorial Guinea Invervening)* (2002) ICJ Reports, International Court of Justice, para. 323; *Oil Platforms Case*, para. 64; *Armed Activities Case*, para. 146.

[84] *Oil Platforms Case*, para. 64.

[85] See section 3, below.

[86] For a detailed description of DDoS attacks, see Bennett Todd, *Distributed Denial of Service Attacks* (2000), www.linuxsecurity.com/resource_files/intrusion_detection/ddos-faq.html (last accessed 12 April 2011).

[87] In traditional botnets this is generally through an Internet Relay Chat (IRC) Server. In newer peer-to-peer based botnets, such as the Storm worm, the controller publishes commands at specific keys in the network to be found by infected machines. However, the net result is the same.

stems in part from the fact that Article 2(4) and Article 51 do not entirely correspond in their wording.[88] As has been noted above, Article 2(4) applies only in the international relations between states and thus does not apply to non-state actors; conversely, Article 51 refers to an inherent right of self-defence which is triggered 'if an armed attack occurs', with no mention of by whom the attack has been launched. Those arguing for a restricted right point to Article 51's role as an exception to the prohibition contained in Article 2(4) to argue that it should likewise be constrained to armed attacks by states. Those, including the present author, who consider there is a broader right of self-defence point to state practice, the fact that Article 51 reflects an 'inherent right' rather than establishing one, and the recognition by the International Court of Justice in the *Nicaragua (Merits)* case that the Charter by no means reflects the whole of the body of law governing force.[89]

As noted above, the International Court of Justice in its controversial decision in the *Nicaragua (Merits)* case, has addressed the issue of when a state might be held responsible for indirect military action by rebels acting 'on behalf' of a foreign state, establishing the test of 'effective control'.[90] For computer network attacks, where attribution is one of the major problems, this test will prove particularly difficult to satisfy and commentators have been quick to point out the advantages that such plausible deniability will offer to sending states. However, the question of whether a state has a right to respond with force to an armed attack by a non-state actor *absent* a degree of state complicity has proved a far more difficult question, and one on which the International Court of Justice has been reluctant to comment. The issue has been covered extensively in recent legal literature, and it is not proposed to conduct a full survey of the arguments here, other than to note that there appears to be a divergence between the limited jurisprudence of the Court and emerging state practice on the issue.[91]

It is safe to say that the Court's treatment of the matter has caused more confusion than clarification. In its advisory opinion on the *Wall*, in a statement described as both 'opaque' and 'startling in its

[88] Moir, *Resort to Force*, 22.

[89] For the latter point see *Nicaragua (Merits)*, para. 176.

[90] Although note also the alternate test established by the International Criminal Tribunal for the Former Yugoslavia in the *Tadic* case which specified an 'effective control' test.

[91] For a full discussion of the issues, see, for example, Gray, *Use of Force* (3rd edn); Moir, *Resort to Force*.

brevity',[92] the Court stated that 'Article 51 of the Charter thus recognises the existence of an inherent right of self-defence in the case of armed attack by one State against another State. However Israel does not claim that the attacks against it are imputable to a foreign State.'[93] The Court goes on to note that the threat to which Israel is responding emanates from territory which is controlled by Israel. Although some commentators have taken this to mean that there is no right of self-defence against non-state actors, Gray has argued that this is merely a statement of the basic right of self-defence and that the Court does not say this is the only time there is a right to self-defence.[94] The Court once again had an opportunity to consider the right of states to use force against attacks by non-state actors in the *Armed Activities* case between DRC and Uganda. The Court, however, chose to limit itself strictly to a discussion of the attributability of the acts complained of to a state – applying the *Nicaragua* test – and expressly declined to discuss the issue of 'whether and under what conditions contemporary international law provides for a right of self defence against large-scale attacks by irregular forces'.[95] However, as Moir has pointed out, by deciding that Uganda's actions could not amount to self-defence as the attacks complained of could not be attributed to the DRC, they effectively answered the question in the negative in any case.[96]

Surveys of state practice, on the other hand, tend to suggest a much greater willingness on the part of the international community to accept the possibility of self-defence against armed attacks by non-state actors without attribution to a state.[97] Moir notes, however, that a certain level of support and involvement on the part of the state is required, and is consistently asserted by states claiming the right.[98] Thomas Franck suggests that measures taken in self-defence against non-state actors will be accepted by the international community where victim states have identified correctly the place from where it was attacked, that the authorities in that place deliberately, knowingly

[92] Gray, *Use of Force* (3rd edn); Sean D. Murphy, 'Self-Defense and the Israeli *Wall* Advisory Opinion: An *Ipse Dixit* from the Court?' (2005) 99 *AJIL* 62, 62.

[93] *The Wall Case*, para. 139.

[94] Gray, *Use of Force* (3rd edn), 135.

[95] *Armed Activities Case*, para. 147.

[96] Moir, *Resort to Force*, 137.

[97] For a survey of the relevant practice see Gray, *Use of Force* (3rd edn), 195–202, 234–52; Moir, *Resort to Force*, 22–31, 140–7.

[98] Moir, *Resort to Force*, 149.

or recklessly permitted the attack to occur, and that the victim state's response is proportionate.[99] For computer network attacks each of these factors raise additional issues. First, as discussed in section 3 below, correctly identifying and attributing an attack to a particular perpetrator in a timely manner is one of the most difficult problems in the field of computer network attacks, both from a technical standpoint and a legal one. To be able to do so in a manner that is clear and verifiable by the international community so as to back a claim for the legitimate exercise of self-defence will be an added level of difficulty. Second, even once computer forensics have established the origins of the attacks, proving a sufficient level of support and involvement by a state will be difficult. For example, the computer network attacks against Georgia demonstrated a high level of coordination between the cyber attacks launched by 'patriotic hackers' and the military operations carried out by Russia. Russia has denied any involvement with the attacks, and no evidence has been found of state organisations guiding or directing the attacks 'be it because there was none, because the collection efforts were not far-reaching or deep enough to identify those connections, or because involvement by state organisations was conducted in a way to purposefully avoid attribution'.[100] There was certainly not enough to prove involvement of the Russian authorities in any official capacity,[101] particularly absent the context of the military operations.[102] In addition, states will find it very difficult to prevent or even be aware of computer network attacks emanating from their territory, particularly given the high prevalence of technologies such as proxy servers.

The final factor raised by Franck in determining whether the international community is likely to accept a claim by a state acting in self-defence against a non-state actor is the proportionality of the response. The question of proportionality is discussed in section 4 below; however, there are a few issues which pertain specifically to non-state actors. First, states have been quick to condemn actions

[99] Franck, *Recourse to Force*, 68.

[100] Project Grey Goose, *Phase I Report*, 3.

[101] Some investigators provided circumstantial evidence of possible involvement by individual officials within the Russian administration; however, many experts remain sceptical of these claims. See generally Tikk, Kaska and Vihul, *Cyber Incidents*, 74–6.

[102] It will be recalled that the computer network attacks began prior to the physical invasion and continued after the ceasefire.

purportedly taken in self-defence which have targeted the infrastructure of the host state.[103] Given that most computer network attacks by non-state actors utilise this infrastructure, it may prove difficult for victim states to find an acceptable target. Secondly, the dispersed nature of many computer network attacks means that the non-state actors responsible for the attack may be resident in several different host states, complicating the proportionality assessment of any forcible response. It should be noted, however, that the most obvious example of such a dispersed attack, the DDoS attack, would rarely rise to the level of gravity required for an armed attack, absent involvement in a wider programme of attacks that could be responded to under the doctrine of accumulation of events.

One final point deserves comment: computer network attacks may also provide a unique opportunity for states to take action against non-state actors without violating the territorial integrity of the state from which they are acting; however, non-state actors are less likely to have infrastructure that would provide viable targets for forcible computer network attacks.[104]

3. Attribution

One of the major problems with any computer network attack is the attribution of the attack to a particular actor. While the origin of some attacks becomes immediately evident, either because the attacker identifies themselves,[105] or because they precede traditional attacks that are easily attributed to a particular source,[106] other attacks will be launched anonymously. Given the common use of botnets and the frequency of IP spoofing that takes place in computer network attacks,[107]

[103] See, for example, the international condemnation of the Israeli attacks against Lebanon in 2006 in response to attacks by Hezbollah, and also the comment of the ICJ in the *Armed Activities* case that the taking of airports and towns many kilometres away from the site of the series of transborder attacks complained of did not seem proportionate: *Armed Activities Case*, para. 147.

[104] See, for example, the opinion of Judge Higgins in the *Wall* case, dissenting on the grounds that non-forcible measures such as the construction of a wall do not really fall within the concept of self-defence.

[105] For example, the 'I Love You' virus source code contained the 'signature' of the author.

[106] For example, the Israeli attack on the Syrian air-defences in advance of an air strike.

[107] IP spoofing essentially forges the data identifying the sending computer in the header of a data packet so that it appears to originate from a different IP address, thus any response is sent to the forged computer.

it is difficult to state with any certainty that the entity that appears to be the perpetrator of the attack is in fact the ultimate attacker.

IP spoofing can be used either to simply mask the origin of an attack, or in a deliberate attempt to place the blame for an attack on another party. An early example of the latter problem occurred in 1999 when a denial of service attack was launched against the US Department of Transport. The attack appeared to emanate from a server in Maryland run by followers of the Falun Gong movement; in fact, the attack was designed to take down both the Maryland server and the Department of Transport network server, leaving the Falun Gong bearing the blame. However, the attackers had blundered and the attack was traced to a computer located at the address of China's Ministry of Public Security. No information is publicly available regarding the US response to this attack; however, in 2002 Richard Clarke, then White House technology advisor, stated before a Senate Judiciary subcommittee hearing on cyber terrorism that the government had never been able to prove to their satisfaction that a particular government was responsible for a specific unauthorised intrusion.[108]

The proliferation of botnets also makes attribution difficult. As the 2007 denial of service attacks against Estonia proved, it can be very difficult to differentiate between attacks that originate from a particular address and those which are merely utilising a compromised computer. Although some of the IP addresses involved in the attacks against Estonia were traced to official IP addresses of Russian state authorities, Russia denied any involvement and claimed that these computers had been compromised and were being manipulated from outside the Kremlin. Subsequent reports by cyber security analysts tend to back this claim, pointing to the lack of evidence and also noting the possibility of spoofing the IP address of the attacker.[109]

Of course, problems of attribution aren't restricted to cyberspace. Armed attacks following more conventional patterns are often carried out anonymously, or responsibility is claimed by armed groups which appear unlikely to have the resources required for such an attack. Further, in the case of state-sponsored terrorism, it is unlikely that any state will step forward to take responsibility for kinetic acts, let alone electronic ones. Those factors aside, a victim state must still establish

[108] Jesse J. Holland, 'Bush Advisor Warns Cyberterrorists', *Washington Post* (Washington DC), 13 February 2002.

[109] See, for example, Evron, 'Battling Botnets'; Tikk, Kaska and Vihul, *Cyber Incidents*, 23.

a link between the attacking group and the sponsoring state, the same approach must be taken with electronic attacks.[110] This may be particularly difficult to prove where attacks are launched by groups or loose affiliations of individuals in conjunction with traditional state action. For example, when Russia moved troops into South Ossetia following the Georgian offensive in 2008, the air strikes and ground forces were accompanied by a series of computer network attacks against Georgian servers. While these attacks were certainly not serious enough to amount to armed attacks, given Georgia's limited internet connectivity, they illustrate the capacity for other actors to effectively 'join-in' a conflict with or without state authorisation. Several security analysts reported botnets allegedly controlled by the Russian Business Network, a group known to be linked to cyber crime, launching denial of service attacks at Georgian servers. However, software was also freely available for download to individual computers from Russian language websites such as 'stopgeorgia.ru' along with a list of targets, making joining the cyber offensive as simple as a few mouse clicks.[111] As noted previously, the Russian government denies any involvement with the attacks;[112] however, the issue raises the difficult question of state responsibility for non-state actors and the necessity or degree of state involvement or complicity required before force can be used in self-defence against the state. This book does not address this issue in depth; suffice it to note that there is little agreement in the international community on the issue in relation to conventional attacks, let alone computer network attacks.[113] However, it is clear that a state must not knowingly allow its territory to be used as a sanctuary for terrorists or armed bands bent on attacking military targets or civilian objects in another country.[114]

The difficulty of attribution also affects the victim state's ability to engage in forcible countermeasures in self-defence.[115] The Court in the

[110] Dinstein, 'CNA and Self-Defense', 112.

[111] John Markoff, 'Before the Gunfire, Cyberattacks', *New York Times*, 13 August 2008; Evgeny Morozov, 'An Army of Ones and Zeroes: How I Became a Soldier in the Georgia-Russia Cyberwar', *Slate*, 14 August 2008, www.slate.com/id/2197514/ (last accessed 12 April 2011).

[112] Siobhan Gorman, 'Georgia States Computers Hit by Cyberattack', *Wall Street Journal* (New York), 12 August 2008, A9.

[113] For a discussion of legal attribution and state responsibility for non-state actors' use of force see generally Gray, *Use of Force* (2nd edn).

[114] Ian Brownlie, 'International Law and the Activities of Armed Bands' (1958) 7 *International and Comparative Law Quarterly* 712, 734, cited in Dinstein, *War, Aggression, and Self-Defense*, 206.

[115] Sharp, *Cyberspace and the Use of Force*, 133, cited in Dinstein, 'CNA and Self-Defense', 111.

Oil Platforms case held that the burden of proof rests on the state invoking the right of self-defence and that the United States had failed to prove that it had been subject to an armed attack by a particular state, *vis* Iran.[116] '[A victim state] must not rush headlong into hasty action predicated on reflexive impulses and unfounded suspicions; it has no choice but to withhold forcible response until hard evidence is collated and the state of affairs is clarified, lest the innocent be endangered.'[117] Such hasty reactions could lead to the escalation of hostilities, something the ban on force was intended to prevent. However, the necessity of waiting for hard evidence of responsibility also opens the possibility that any action taken against a perpetrator once responsibility has been confirmed will be classified as an armed reprisal rather than an action taken in self-defence. Armed reprisals are prohibited under international law.[118]

4. Necessity and proportionality

All responses to attacks, whether their means of delivery are kinetic or electronic, are subject to the underlying principles of necessity and proportionality. The International Court of Justice has repeatedly confirmed that self-defence, whether individual or collective, warrants only those measures which are 'proportional to the armed attack and necessary to respond to it' as a principle well established in customary international law.[119] The principles are not stated in UN Charter but are often traced back to the Webster formulation in the *Caroline* incident.[120]

The principle of necessity requires that any measures taken avowedly in self-defence must have been necessary for that purpose; the principle is strict and objective 'leaving no room for any measure of discretion'.[121] That is, it is not sufficient that force is used after an armed attack; it must be necessary to repel that attack,[122] and non-forcible remedies must either prove futile *in limine* or have in fact been exhausted in an

[116] *Oil Platforms Case*, paras 57, 61.
[117] Dinstein, 'CNA and Self-Defense', 111.
[118] Declaration on Friendly Relations, Principle on the Use of Force, para. 6.
[119] *Nicaragua (Merits)*, paras 176 and 194; *Nuclear Weapons Case*, para. 41; *Oil Platforms Case*, para. 74; *Armed Activities Case*, para. 147.
[120] There must be 'a necessity of defence, instant, overwhelming, leaving no choice of means and no moment for deliberation'.
[121] *Oil Platforms Case*, para. 73.
[122] Greenwood, 'Pre-Emptive Use of Force', 23.

unsatisfactory manner.[123] As Roberto Ago notes, 'had [the state] been able to achieve the same result by measures not involving the use of armed force, it would have no justification for adopting conduct which contravened the general prohibition against the use of armed force'.[124] For example, in the *Oil Platforms Case* the Court noted that even if the attacks on the *Sea Isle City* and mining of the *Samuel B. Roberts* had been sufficiently imputed to Iran, the US had not complained to Iran of the military activities of the platforms, which suggested that the attacks on the platforms was not seen as a necessary act.[125]

The principle of necessity also gives rise to a related principle, namely that actions taken in self-defence must generally be taken without undue delay.[126] In the case of computer network attacks, this may make successful claims of self-defence more difficult. Where the origins of a computer network attack are not obvious, the time taken to conduct computer forensics and establish hard evidence of the identity of the perpetrator may militate against a finding that any subsequent action by the victim state is in self-defence. Similar criticisms were levelled at the United States after a delay of several weeks between the September 11 attacks on the World Trade Center and the Pentagon and subsequent action in Afghanistan.[127] Likewise, the Court in the *Nicaragua (Merits)* case also criticised the US for commencing activities purportedly in self-defence several months after the major offensive of the opposition against the government of El Salvador had been completely repulsed, although ultimately dismissing the claim of self-defence on other grounds.[128] However, Dinstein notes that this requirement must not be construed too strictly; he points to the delay of approximately five months between the invasion of Kuwait and the authorisation of all necessary means by the Security Council.[129]

The principle of proportionality requires the weighing of the response against its permitted purpose of halting and repelling the attack, or, in the case of anticipatory self-defence, preventing it from

123 Dinstein, 'CNA and Self-Defense', 109.
124 Roberto Ago, 'Addendum to the Eighth Report on State Responsibilty' (1980) II *UN Y.B. Int'l L. Comm'n* 13, 69 para. 120.
125 *Oil Platforms Case*, 73–6.
126 Dinstein, *War, Aggression, and Self-Defense*, 210; Ago, 'Addendum', 69.
127 For a discussion of the weakness of this argument, see Greenwood, 'Pre-Emptive Use of Force', 23.
128 *Nicaragua (Merits)*, 237.
129 Dinstein, *War, Aggression, and Self-Defense*, 210.

happening. Whether or not an act is proportionate will necessarily be dependent on the facts of the situation; however, the action must not be retaliatory or punitive, its lawfulness cannot be measured 'except by its capacity for achieving the desired result'.[130] This raises an interesting question with respect to computer network attacks; namely whether it would ever be proportionate to use traditional armed force against an electronic attack. The answer must surely be in the affirmative. Proportionality does not restrict the defending state to the same weapons or the same numbers of armed forces as the attacking state; nor is it necessarily limited to action on its own territory.[131] Therefore it would be open to the victim of a computer network attack to use whatever weapons it has at its disposal to repel the attack, as long as the response is proportionate to that purpose. As Ago notes:[132]

> In the case of self-defence, it was essential to avoid the error of thinking that there should be some proportionality between the action of the aggressor and the action of the state defending itself. Proportionality could be judged only in terms of the objective of the action, which was to repel an attack and prevent it from succeeding. No limitations that might prejudice the success of a response to attack could be placed on the State suffering the attack. The concept of reasonable action must of course enter into the matter, since self-defence could not justify a genuine act of aggression committed in response to an armed attack of limited proportions.

Thus, if a computer network attack is being used to prepare the battlespace for a conventional attack, the use of military force in response must be proportionate in relation to the need to repel the attack as a whole. However, it should be noted that while physically bombing the attacking computers and their owners may be legal, it is not necessarily the preferred method of response.[133] Only a few examples of state intrusion have been made public to date, and no large-scale attacks as would justify a forcible response have been reported. Further, state practice in response to electronic probes emanating from other states have not resulted in forcible responses.[134]

[130] Ago, 'Addendum', 69 para. 121.

[131] Gray, *Use of Force* (3rd edn), 150.

[132] Roberto Ago, 'State Responsibility' (1980) Vol. 1 *UN Y.B. Int'l L. Comm'n* 188, para. 2526, June 1980.

[133] Eric Talbot Jensen, 'Computer Attacks on Critical National Infrastructure: A Use of Force Invoking the Right of Self-Defense' (2002) 38 *Stan. J. Int'l L.* 207, 230.

[134] At the time of writing the only forcible response to an attack on a communications network is an outbreak of street fighting in Lebanon as a result of the cutting of a

Eric Jensen has cited technological solutions designed with a 'hack back' feature to trace an attack and reflect similar damage to the sender, or causing some other responsive action.[135] Although much of the available technology is classified, as noted above one of the major difficulties to overcome with any automated response is the correct attribution of the attack prior to launching any destructive payload. For example, the attack on the US Department of Transport that was attributed to the Falun Gong originated elsewhere, and despite initial accusations against Russia, the attacks on Estonia and Georgia are now widely believed to be the work of independent 'patriotic hackers'. As noted, routing attacks through other actors, or spoofing their IP addresses to cast blame may serve as a political end in itself.

5. Countermeasures against unlawful acts

Where a computer network attack does not rise to the level of an armed attack, a state may still respond with proportionate countermeasures. However, where forcible countermeasures are taken in response to an ordinary breach of international law, not constituting an armed attack, they are unlawful.[136]

However, an alternate line of reasoning, arising out of the controversial gap theory put forward by the majority of the International Court of Justice in the *Nicaragua (Merits)* decision, appears to have muddied the waters somewhat. The Court held that acts committed by Nicaragua against El Salvador, Honduras and Costa Rica amounted to less grave forms of 'illegal military intervention'.[137] The Court held that that 'proportionate counter-measures' were permissible on the part of the victim state; however, they did not create any entitlement by any third state to take collective countermeasures involving the use

Hezbollah telecommunications network. The response was condemned by states; however, given the non-international nature of the dispute, it does not provide any useful indication of state practice in this regard. See generally 'Hezbollah Takes over West Beirut', *BBC News*, 9 May 2008, http://news.bbc.co.uk/1/hi/world/middle_east/7391600.stm (last accessed 15 July 2011).

135 Jensen, 'Computer Attacks', 231.

136 Dinstein, *War, Aggression, and Self-Defense*, 226.

137 The Court held that Nicaragua did supply aid to rebels in the territories of El Salvador, but had insufficient evidence of the nature, scale and continuance of such aid. The Court further held that Nicaragua was responsible for certain transborder military incursions into Honduras and Costa Rica. *Nicaragua (Merits)*, paras 152, 164.

of force.[138] The Court did not venture an opinion as to what form the individual countermeasures by the victim state might take, or whether they could include the use of force.

John Hargrove has suggested that there are two possible interpretations of the Court's statement. Either the Court was saying:[139]

> (a) that there are some acts of force that nobody, not even the victim, may resist by proportionate measures of force; or it was saying (b) that the victim may resist with force provided that it does so alone. There is little to be said in explanation of the latter proposition other than it is simply a[n] … arbitrary announcement of the Court.

Hargrove argues that allowing forcible countermeasures 'would in one remarkable stroke manage both to impair the right of self-defense, and to weaken fundamentally the prohibition on the use of force by creating an open-ended and wholly new category of exceptions to Article 2(4) of the Charter, of unknown content and limit.'[140] However, Judge Simma in his separate opinion in the *Oil Platforms* case has argued that the Court in the *Nicaragua (Merits)* case cannot, in the context of that case, have understood that to mean mere pacific reprisals.[141] He argues that the Court can only have meant 'defensive military action "short of" full-scale self-defence'.[142]

> But we may encounter also a lower level of hostile military action, not reaching the threshold of an 'armed attack' within the meaning of Article 51 of the United Nations Charter. Against such hostile acts, a state may of course defend itself, but only within a more limited range and quality of responses (the main difference being that the possibility of collective self-defence does not arise, *cf. Nicaragua*) and bound to necessity, proportionality and immediacy in time in a particularly strict way.

Simma's treatment of the conditions in respect of forcible countermeasures is clearly linked to the classic self-defence analysis, rather

[138] '[A] use of force of a lesser degree of gravity cannot … produce any entitlement to take collective counter-measures involving the use of force. The acts of which Nicaragua is accused … could only have justified proportionate counter-measures on the part of the state which had been the victim of these acts, namely El Salvador, Honduras or Costa Rica. They could not justify counter-measures taken by a third state, the United States, and particularly could not justify intervention involving the use of force.' *Ibid.*, para. 249.

[139] Hargrove, 'Nicaragua Judgement', 141.

[140] *Ibid.*, 142.

[141] *Oil Platforms Case* (per Judge Simma), para. 12.

[142] *Ibid.*, per Judge Simma, paras 12–13.

than the criteria for non-forcible countermeasures set out below. This argument may prove beneficial to a state which is subjected to computer network attacks that rise to the level of force, particularly where it is unsure whether its planned proportionate response might likewise be considered an illegal use of force. It should be noted, however, that this is a minority view and that most commentators maintain that forcible countermeasures are prohibited under international law as armed reprisals.

Where a computer network attack does not amount to an armed attack (or a use of force), any countermeasures taken by the victim state will be conditional on, and limited by, the law relating to countermeasures.[143] In the *Case Concerning the Gabcikovo-Nagymaros Project*, the International Court of Justice set out a three-part test justifying proportionate countermeasures. First, the action must be taken in response to an internationally wrongful act of another state and be directed against that state. Second, the victim state must have called upon the offending state to discontinue its wrongful conduct or to make reparation for it. And finally, the effects of the countermeasure must be commensurate with the injury suffered, taking account of the rights in question.[144] In this regard the test for proportionality differs between countermeasures and self-defence, where the response must be proportional to the threat, rather than the actual harm suffered. However, the Court also stated that the purpose of the countermeasures must be to induce the wrongdoing state to comply with its obligations under international law, and that the measure must therefore be reversible.[145] In respect of a computer network attack this would seem to fit with current state practice; most attacks to date have merely resulted in states requesting the alleged perpetrator of the attack to cease their actions.[146] However, the nature of computer network attacks, including the type of attacks they make possible and, most importantly, their ability to be reversed, makes computer network attacks particularly useful as a countermeasure against a previous wrongful act of a state. For example, a series

[143] See generally United Nations, International Law Commission and James Crawford, *The International Law Commission's Articles on State Responsibility: Introduction, Text, and Commentaries* (Cambridge University Press, Cambridge, New York, 2002).

[144] *Case Concerning the Gabcikovo-Nagymoros Project* (1997) ICJ Reports 3, International Court of Justice, para. 85.

[145] *Ibid.*, para. 87.

[146] Although it should be noted that in all reported cases to date the suspected perpetrator has denied any involvement with the attack.

of blackouts in response to an internationally wrongful act may be a useful coercive measure.

The 1998 Zapatista 'Floodnet' attacks on the Pentagon's website by the Electronic Disturbance Theatre also provide a model for how electronic countermeasures might work against non-state actors. US Department of Defense specialists created a program that would recognise the Floodnet applet installed on computers trying to access the Department of Defense website.[147] Once the applet was identified, a program was sent back to the activist's computer to shut down their web browser, thus ending the attack.[148] Although criticisms were levelled at the Department for not thoroughly considering the legal ramifications of such a response,[149] the United States' decision to respond electronically, in this instance, against an attack which they had been expecting, must be considered a proportionate countermeasure, particularly against a demonstrably non-violent protest group. There was some speculation that this approach was being renewed by the US Department of Defense in its 'active defence' policy released in 2011;[150] statements made by Lynn prior to release of the policy indicated that the US intended to stop threats *before* they reached Department of Defense networks. However, the unclassified version of the Pentagon's cyber-security strategy appears to focus the concept of active defence on network speed hunting for malicious activity *within* Department of Defense networks and systems, rather than pushing back beyond the network boundaries.[151] The danger with adopting the push-back approach against states is the risk of escalation of such countermeasures into forcible responses, a situation which the Security Council would be likely to determine a threat to international peace and security, as seen in the following section.

[147] An applet is a small piece of code designed to perform one specific task within another application.

[148] Brian Friel, 'DoD Launches Internet Counterattack', *Government Executive*, 18 September 1998, http://govexec.com/dailyfed/0998/091898b1.htm (last accessed 12 April 2011).

[149] George I. Seffers, 'Legalities Cloud Pentagon's Cyber Defence', *Defence News*, 25 January 1999, 3.

[150] Based on statements made by Deputy Defense Secretary William J. Lynn III. Lynn, 'Defending a New Domain', 101; Nagashima, 'U.S. Eyes Preemptive Cyber-Defense Strategy'.

[151] US Department of Defense, *Strategy for Operating in Cyberspace*, July 2011, US Department of Defense (2011), www.defense.gov/news/d20110714cyber.pdf (last accessed 18 January 2012), 8.

6. Security Council authorisation and collective measures

Other than self-defence under Article 51, the only other exception to the prohibition on the use of force under the UN Charter is the authorisation of collective security measures under Chapter VII. The Security Council has 'primary responsibility for the maintenance of international peace and security' and may recommend or authorise member states to engage in measures, which may include the use of force, in order to pursue this aim. Article 39 provides:

> The Security Council shall determine the existence of any threat to the peace, breach of the peace, or act of aggression and shall make recommendations, or decide what measures shall be taken in accordance with Articles 41 and 42, to maintain or restore international peace and security.

This raises two separate issues for the use of computer network attacks. First, when will the Security Council consider computer network attacks to be a 'threat to the peace, breach of the peace, or act of aggression'; and second, under what circumstances might measures authorised by the Security Council under Chapter VII include computer network attacks?

As noted, prior to the Security Council recommending action, it must determine that a 'threat to the peace, breach of the peace, or act of aggression' exists under Article 39. In practice, the Council has almost exclusively exercised its powers by finding a 'threat to international peace and security', even in situations where a breach of the peace or act of aggression is obvious;[152] thus the analysis will concentrate on this broader standard. Whether or not there are any limits on the Security Council's ability to determine a threat to the peace is the subject of debate amongst legal scholars;[153] however, it is generally accepted that the Council possesses considerable discretion in this respect. The resulting range of situations in which the Security Council has found a threat to the peace is increasingly broad and includes the danger of violent countermeasures by states to violations of international law, regardless of their admissibility.[154] The Council has also

[152] Jochen Frowein, 'Article 39' in B. Simma (ed.), *The Charter of the United Nations: A Commentary* (2nd edn, Oxford University Press, 2002) 717, 722; Gray, *Use of Force* (3rd edn), 256. All instances in which the Security Council has found an act of aggression or breach of the peace have involved inter-state conflicts.

[153] See generally Erika De Wet, *The Chapter VII Powers of the United Nations Security Council* (Hart, Oxford, 2004), 133–4.

[154] Frowein, 'Article 39', 722.

found threats to the peace in internal conflicts such as those in Liberia, Rwanda, Sierra Leone and East Timor;[155] violations of human rights and humanitarian law in Somalia, Rwanda and Eastern Zaire;[156] violations of democratic principles in Haiti and Sierra Leone;[157] terrorism;[158] nuclear proliferation and even failure to cooperate with international prosecutions.[159] These examples illustrate the broad discretion that the Security Council has to determine that a threat to the peace exists; however, Frowein has argued that this does not mean that the notion of a threat to the peace has become limitless. He argues that a threat to the peace exists when, in a particular situation, a danger of the use of force on a considerable scale arises.[160] Although the Security Council has not considered any computer network attacks to date,[161] based on this assessment it would certainly appear that a computer network attack would constitute a threat to the peace where it is of sufficient gravity that a state is likely to respond to it with force, regardless of whether it is categorised as an armed attack, or where the type of attack indicates further violence to follow, whether electronically or by kinetic means.

Once the Security Council has determined a threat to the peace, the Council may make recommendations, or require states to take action under Articles 40, 41 and 42 of the Charter, for the restoration or maintenance of international peace and security. Interestingly, once the Security Council has deemed a situation a threat to the peace, it 'is free to take measures against any entity which it considers to be an obstructive factor in the restoration of peace'.[162] Thus, having determined the

155 See SC Res 788, 19 November 1993 on Liberia; SC Res 918, 17 May 1994, SC Res 929, 22 June 1994 on Rwanda; SC Res 1132, 8 Oct 1997, SC Res 1289, 7 Feb 2000, SC Res 1306, 5 July 2000 on Sierra Leone; SC Res 1264, 15 September 1999 on East Timor.

156 SC Res 794, 3 December 1992 on Somalia; SC Res 929, 22 June 1994 on Rwanda; SC Res 1078, 9 November 1996 on Zaire.

157 For Haiti: SC Res 841, 16 June 1993; SC Res 917, 6 May 1994; SC Res 940, 31 July 1994 and most recently SC Res 1529, 29 February 2004. For Sierra Leone: SC Res 1132, 8 October 1997; SC Res 1270, 22 October 1999; SC Res 1289, 7 February 2000; SC Res 1306, 5 July 2000.

158 For example, SC Res 1526, 30 January 2004.

159 SC Res 1172, 6 June 1998 on nuclear proliferation; SC Res 748, 31 March 1992 on Libya's failure to cooperate with prosecution of the Lockerbie bombers.

160 Frowein, 'Article 39', 726.

161 Georgia raised the issue of alleged Russian cyber attacks in the context of the 2008 conflict with Russia over South Ossetia; however, the point was not taken up by the other members and no resolution was forthcoming, S/PV.5961, 19 August 2008. As noted previously, attribution of the attacks is far from certain.

162 Dinstein, *War, Aggression, and Self-Defense*, 287.

situation in Angola in 1997 to be a threat to international peace and security, the Council then imposed sanctions on UNITA (Union for the Total Independence of Angola), a non-state entity.[163] This will be of particular significance in respect of computer network attacks which are launched by disaffected groups 'joining in' conflicts, such as happened in Georgia in 2008, although as outlined above, positive attribution will always be a factor.

Article 40 allows the Security Council to establish any provisional measures it deems necessary or desirable to prevent an aggravation of the situation, before deciding what recommendations to make under Article 39. Provisional measures are made without prejudice to the rights, claims or position of the parties, but failure to comply with them will be taken into account by the Security Council.

Article 41 allows the Security Council to decide:

> [W]hat measures not involving the use of armed force are to be employed to give effect to its decisions, and it may call upon the Members of the United Nations to apply such measures. These may include complete or partial interruption of economic relations and of rail, sea, air, postal, telegraphic, radio, *and other means of communication*, and the severance of diplomatic relations. (emphasis added)

This would also allow for the disruption of individual internet access, such as that demonstrated by the United States in response to the 1998 attack by the Electronic Disturbance Theatre, and may also allow for the total disconnection of a country from the internet by a third party, in the same manner as the self-imposed internet blackouts attempted internally by regimes in Egypt, Myanmar and Libya.[164] Some attempts have been made to argue that such a measure would be the electronic equivalent of a blockade (which would fall within the forceful measures provided for in Article 42);[165] however, the present author does not consider that this is an appropriate classification, despite the

[163] SC Res 1127, 1997, cited in *ibid*.

[164] James Cowie, 'What Libya Learned from Egypt', *Renesys Blog*, 5 March 2011, www.renesys.com/blog/2011/03/what-libya-learned-from-egypt.shtml (last accessed 12 April 2011); Jordan Robertson, 'The Day Part of the Internet Died: Egypt Goes Dark', *The Washington Times* 28 January 2011, www.washingtontimes.com/news/2011/jan/28/day-part-internet-died-egypt-goes-dark/ (last accessed 12 April 2011); OpenNet Initiative, *Pulling the Plug: A Technical Review of the Internet Shutdown in Burma*, OpenNet Initiative (2007), http://opennet.net/sites/opennet.net/files/ONI_Bulletin_Burma_2007.pdf (last accessed 12 April 2011).

[165] See media reports following both the Estonian and Georgian denial of service attacks.

severe economic effects that this would undoubtedly have on highly technologically dependent states.[166] Electronic measures under Article 41 could arguably also encompass denial of service attacks launched against the media, banking and telecommunications infrastructure of a state, although the shaping of the response to the threat may prove difficult in practice.

Where non-forceful measures have been unsuccessful, or if the Security Council determines that such measures would be inadequate, the Council may authorise action under Article 42 by such air, sea or land forces as may be necessary to maintain or restore international peace and security; examples include demonstrations, blockade and other operations by air, sea or land forces of Members of the United Nations.[167] It appears unlikely that the Security Council would consider it necessary to authorise force against a computer network attack; however, where an attack or ongoing series of attacks cannot be prevented or stopped by electronic means, it is within their purview to do so. As noted earlier, force is not necessarily the preferred means of countering a cyber attack because of the distributed nature of many methods of attack.

A more interesting scenario is the possibility of the Security Council authorising computer network attacks of sufficient gravity to amount to a use of force in response to a threat to the peace. For example, in the past few years the Security Council has passed a series of resolutions expressing its concern over the proliferation risks presented by the Iranian nuclear programme.[168] Although the Council has not explicitly stated that it considers that the programme represents a threat to international peace and security, this is far from unusual; it is clear from the practice of the Council that no formal pronouncement with an express reference to Article 39 is necessary; the use of the language is apparently sufficient.[169] The Council does, however, explicitly state that it is acting under Article 41 of Chapter VII of the Charter, and each resolution calls for member states to implement a series of specific non-forceful measures against Iran. The author has previously concluded that the Stuxnet worm amounts to a use of force as it is

[166] For example, the OECD estimates the economic impact of shutting down internet and mobile phone services in Egypt for five days in 2011 at US$90 million in direct costs alone.

[167] Art. 42, UN Charter.

[168] SC Res 1929 (2010); SC Res 1803 (2008); SC Res 1747 (2007); SC Res 1737 (2006).

[169] Gray, *Use of Force* (3rd edn), 256.

generally believed to have caused physical damage to approximately 1,000 nuclear centrifuges at the Natanz enrichment facility. Although Resolution 1929 (2010) specifically stresses that it does not authorise any measures exceeding the scope of the resolution,[170] or the threat or use of force, had the measures decided on been taken under Article 42 with its usual phraseology 'all necessary means', the Stuxnet worm would have proved an efficient way to achieve one of the aims of the resolution, namely the suspension of enrichment-related activities.

7. Conclusion

As with the discussion in the previous chapter relating to the use of force, the classification of computer network attacks as armed attacks sits on top of a deep doctrinal divide between those who would argue for a wide interpretation of the right to self-defence and those who would restrict it. The present author believes that in regard to computer network attacks, a restrictive view of the armed attack threshold and the subsequent right to self-defence is preferable. Classification of computer network attacks which would amount to armed attacks should therefore be restricted to those attacks which cause physical damage to property or persons of a sufficient scale and effect. Given the particular difficulties and length of time involved in accurately attributing attacks and the fact that states acting in self-defence are not restricted to responding in kind, the danger of computer network attacks escalating into traditional conflict is apparent. That is not to say that states cannot respond to computer network attacks that do not rise to the level of an armed attack; a state may still respond with proportionate countermeasures or appeal to the Security Council for a finding that the attacks amount to a threat to the peace.

[170] Thus even if Stuxnet had not resulted in physical damage but merely stopped the enrichment facility from working it could not have fallen within the terms of the Security Council authorisations.

PART II
Jus in bello

4 The applicability of the laws of armed conflict to computer network attacks

The laws of armed conflict apply to all situations of armed conflict, whether or not war is declared, and regardless of whether the parties involved recognise the state of armed conflict or, indeed, the opposing force. The determination is deliberately a factual rather than a legal one. None of the instruments relating to the laws of armed conflict deal with computer network attacks explicitly, therefore the question must be asked whether the laws of armed conflict should apply to computer network attacks at all, and, if so, under what circumstances would a computer network attack be sufficient to trigger the application of those laws? As with much of the application of the law to computer network attacks, the advancement of technology into a qualitatively different type of weaponry (rather than merely a difference in scale), requires a re-examination of the terminology. The question of the applicability of the laws of armed conflict to computer network attacks arises in three distinct circumstances: first, where computer network attacks are used with traditional weapons in an ongoing conventional armed conflict; second, where computer network attacks are launched on their own; and finally, where the use of conventional weapons is insufficient in and of itself to qualify as an armed conflict, but it is accompanied by extensive computer network attacks. In some circumstances, a computer network attack may also represent an opening salvo in a wider conflict, and might therefore indicate the beginning of the application of the laws of armed conflict. The existence or not of an armed conflict is of particular relevance, as the outbreak of an armed conflict between two states will lead to many of the rules of the ordinary law of peace being superseded, as between the parties to the conflict, by the rules of humanitarian law.[1] For example, the right to seize property, use force and detain nationals will become materially different.

1. Armed conflict

Different thresholds have been established by both customary and treaty law for the application of the laws of armed conflict in both international and non-international armed conflicts. In the case of international armed conflicts, Article 2 common to the Geneva Conventions provides that they apply in full to 'all cases of declared war or of any other armed conflict which may arise between two or more of the High Contracting Parties, even if the state of war is not recognised by one of them' or in 'all cases of partial or total occupation of the territory of a High Contracting Party, even if the said occupation meets with no armed resistance'.[2] The Conventions also apply to and in respect of any non-contracting party, where that party accepts and applies the provisions of the Conventions itself. Additional Protocol I provides that the provisions of that protocol will apply in the same circumstances as Common Article 2 of the Conventions, and states that they also apply to 'armed conflicts in which peoples are fighting against colonial domination and alien occupation and against racist regimes in the exercise of their right of self-determination'.[3] This provision has proved highly controversial and is one of the reasons that several states, including the United States, have not ratified Additional Protocol I.[4]

In relation to conflicts not of an international character, that is, internal armed conflicts, Common Article 3 of the Conventions and Article 1 of Additional Protocol II both employ the term 'armed conflict' as the trigger for the application of humanitarian law principles.[5] While Common Article 3 merely refers to armed conflicts occurring

[1] Christopher Greenwood, 'Scope of Application of Humanitarian Law' in D. Fleck (ed.), *The Handbook of Humanitarian Law in Armed Conflict* (2nd edn, Oxford University Press, 2008) 45–78, 40 (per Chua J).

[2] Common Art. 2, Geneva Conventions 1949. The conventions also apply where a state declares war but doesn't engage in actual hostilities: *ibid.*, 47.

[3] Art. 1, Additional Protocol I.

[4] Greenwood, 'Scope of Application of Humanitarian Law', 48. Greenwood notes that it is unlikely that states consider this to be declaratory of customary international law, although there is evidence that it has had an influence on state practice.

[5] For a full discussion of the application of the laws of armed conflict to internal armed conflicts see Lindsay Moir, *The Law of Internal Armed Conflict* (Cambridge University Press, New York, 2002).

in the territory of a high contracting power, Additional Protocol II generally contains a higher threshold of application and thus applies to a smaller range of conflicts. Additional Protocol II includes additional criteria, such as control of territory by an organised armed group under a responsible command, so as to enable them to carry out sustained and concerted military operations and be capable of implementing the Protocol. The effect of the application of these additional criteria with respect to computer network attacks is discussed in section 3 below.

The term 'armed conflict' is not defined anywhere in the Conventions or Additional Protocols. This was a deliberate attempt by the drafters of the Conventions to avoid the political and legal wrangling that had occurred over the legal definition of war, and the ensuing distinctions between a state of war, a police action, or any other form of hostile action.[6] Thus, the determination is intended to be factual rather than legal. In respect of international armed conflicts, Pictet's commentary to the Conventions takes a broad view, stating:[7]

> Any difference arising between two states and leading to the intervention of armed forces is an armed conflict within the meaning of Article 2, even if one of the parties denies the existence of a state of war. It makes no difference how long the conflict lasts, or how much slaughter takes place.

Likewise the International Committee of the Red Cross (ICRC) takes an expansive view of armed conflict, maintaining that in the case of cross-border operations, the first shot suffices to trigger an international armed conflict:[8]

> By using the words 'from the outset' the authors of the Convention wished to show that it became applicable as soon as the first acts of violence were committed, even if the armed struggle did not continue. Nor is it necessary for there to have been many victims. Mere frontier incidents may make the Convention applicable, for they may be the beginning of a more widespread conflict.

Given this apparent denial of a *de minimis* level of intervention, it would appear that computer network attacks could well come within the ambit of armed conflict. However, this view is not universally held and

[6] 'Article 2' in Jean S. Pictet, *The Geneva Conventions of 12 August 1949: Commentary* (International Committee of the Red Cross, Geneva, 1952), 32.

[7] *Ibid.*

[8] ICRC, 'Article 6', Commentary to Geneva Convention IV, 59.

the statement regarding the length and intensity of the conflict is not necessarily borne out by state practice.[9] Christopher Greenwood notes that it is by no means clear that most states would regard an isolated incident or exchange of fire as an armed conflict, however serious the consequences, bringing into operation the full panoply of the Geneva Conventions.[10] While there are examples of relatively minor incidents where a state has claimed protection of the laws of armed conflict, there have been a number of border clashes and naval incidents which have not been treated as armed conflicts.[11] For example, during the *Dogger Bank Incident* of 1904 the Russian Navy's North Sea fleet opened fire on British fishing trawlers, believing them to be Japanese warships. The incident was closed by payment of compensation to the British government for the lives of the two men lost, the sinking of one trawler and injury and damage to other trawlers and crew.[12] On 8 June 1967, Israeli fighter jets and torpedo boats attacked the *USS Liberty* in the eastern Mediterranean, killing 34 crew members and wounding 171 more. The officially accepted explanation for the attack has been that it was a tragic mistake, and the US accepted an apology and compensation for the losses despite the controversy over the official findings.[13] Similarly, in 1987 when the *USS Stark* was struck by missiles launched from an Iraqi fighter jet under a misapprehension that it was an Iranian tanker, the United States was prepared to accept an apology and compensation for the 37 lives lost and damage to the frigate.[14]

[9] This is also supported by the series of decisions of the Ethiopia/Eritrea Claims Commission. Though controversial and criticised for several reasons, the decision of the Commission in declaring 12 May 1998 as the start of the armed conflict that was the subject of their awards indicates their unwillingness to apply the laws of armed conflict to the border skirmishes that took place in the preceding weeks.

[10] Christopher Greenwood, 'The Law of War (International Humanitarian Law)' in M. D. Evans (ed.), *International Law* (Oxford University Press, 2003) 789–821, 791.

[11] Greenwood, 'Scope of Application of Humanitarian Law', 48.

[12] *Finding of the International Commission of Inquiry Organized under Article 9 of the Convention for the Pacific Settlement of International Disputes, of July 29, 1899 (the Dogger Bank Incident)* (1905) 2 *AJIL* 931–6, The International Commission of Inquiry between Great Britain and Russia arising out of the North Sea incident.

[13] Several crew members and intelligence officials dispute the findings of the official inquiry, stating that the attacks were deliberate. See generally, William D. Gerhard and Henry W. Millington, *Attack on a Sigint Collector, the U.S.S. Liberty* (National Security Agency, 1981). For an article concluding the attack was deliberate see Walter L. Jacobsen, 'A Juridical Examination of the Israeli Attack on the USS Liberty' (1986) 36 (Winter) *Naval Law Review* 69.

[14] Jim Hoagland, 'U.S., Iraq to Confer on Air War', *Washington Post* (Washington DC), 25 May 1987, 1.

In contrast, when a US Navy pilot was shot down and captured by Syrian forces over Lebanon in 1983, the United States maintained that this incident amounted to an armed conflict and the pilot was thus entitled to prisoner of war status.[15] Reports from Syria also appeared to assume that this was the case.[16] Similarly, when US helicopters fired on the Iranian vessel the *Iran Ajr* during a mine-laying operation and forced its crew to abandon ship, the rescued sailors and the bodies of their less fortunate compatriots were swiftly repatriated. Although the status of those particular sailors was never publicly discussed between the United States and Iran, the ICRC delivered a note to the United States stating that 'such situations and their consequences fell within the scope of the Geneva Conventions'.[17]

A survey of these incidents, some of which led to extremely strained diplomatic relations, tends to suggest that states' willingness to classify events as an armed conflict appears to be based on the perceived intentions of the other party, an assessment which is often largely influenced by *realpolitik*. Thus, Nils Melzer comments 'in the absence of a formal declaration of war, an international armed conflict requires a minimal transgression, which expresses the belligerent intent of the acting state against another'.[18] However, where prisoners have been taken, a willingness to extend the protections of the Conventions to captured personnel appears to be a major driver behind the classification of such incidents as armed conflicts.

Non-international armed conflicts, on the other hand, are generally accepted to be subject to a requirement of a certain level of intensity in order to distinguish them from internal disturbances and tensions. Although the principle that the laws of armed conflict do not apply to situations of 'internal disturbances and tensions, such as riots, isolated and sporadic acts of violence and other acts of a similar nature' is first

[15] Greenwood, 'Scope of Application of Humanitarian Law', 48. Although President Reagan later stated, 'I don't know how you have a prisoner of war when there is no declared war between nations. I don't think that makes you eligible for the Geneva Accords [sic]', it appears that this was simply an error on the President's part. 'President's News Conference on Foreign and Domestic Issues', *New York Times*, 21 December 1983, A22.

[16] Thomas L. Friedman, 'Widened Cabinet Sought in Beirut', *New York Times*, 8 December 1983, A18; Thomas L. Friedman, 'Syria Says Airman Seized in U.S. Raid Will Not Be Freed', *New York Times*, 6 December 1983, A1.

[17] ICRC, 'External Activities: September–October 1987' (1987) 27(261) *IRRC* 650. Note, however, that the ICRC did not make clear whether this determination was made by reference to the laws relating to neutral shipping or armed conflict.

[18] Nils Melzer, *Targeted Killing in International Law* (Oxford University Press, 2008), 250.

set out in Article 1 paragraph 2 of Additional Protocol II, the principle is generally applied to all non-international armed conflicts.[19] In 1995 the Appeals Chamber of the International Criminal Tribunal for the Former Yugoslavia in the *Tadic* case considered the term 'armed conflict', holding that:[20]

> an armed conflict exists wherever there is resort to armed force between states or protracted armed violence between government authorities and organised armed groups or between such groups with a state.

The Appeals Chamber thus took a more inclusive approach to the question, asserting that the threshold for the application of Common Article 3 is actually relatively low, although it does not apply as widely as Pictet's commentary would suggest.[21]

1.1. *The requirement of armed force*

Although Pictet's definition of armed conflict refers only to a 'difference' between states, the test in *Tadic* shows that armed force or armed violence is the requirement for both international and non-international armed conflicts.[22] As noted above, the refinement of the test by the Tribunal also separates the level of violence required for international and internal armed conflict by requiring a level of protraction of the violence in conflicts not of an international nature. This is in keeping with the requirement that the laws of armed conflict are not to apply to internal disturbances and tensions, such as riots, isolated and sporadic acts of violence and other acts of a similar nature.[23] Both internal and international armed conflicts, however, require the use of armed force. As with the terms 'force' and 'armed attack', armed force is not defined in international law (its definition undoubtedly being considered self-evident). However, as discussed in Chapter 2 above, armed force is to be construed somewhat broadly; in particular

[19] Dieter Fleck, 'The Law of Non-International Armed Conflicts' in D. Fleck (ed.), *The Handbook of International Humanitarian Law* (2nd edn, Oxford University Press, 2008) 605–33, 616.

[20] *Prosecutor* v. *Dusko Tadic (Interlocutory Appeal on Jurisdiction)* (1995) Case No. IT-94-1-AR, International Criminal Tribunal for the Former Yugoslavia, Appeals Chamber, para. 70.

[21] Moir, *The Law of Internal Armed Conflict*, 43.

[22] It should be noted that there was no question in the *Tadic* case of whether there had been such force or violence used against the people of the former Yugoslavia, the case addressed the question of the international or internal nature of the armed conflict that took place in the Balkans.

[23] See Art. 1(2) Additional Protocol II.

the term includes indirect forms of support for the application of force. Thus where a computer network attack, directly or indirectly, results in injury or death, or destruction of physical property, it will constitute a use of armed force. Whether a computer network attack can amount to a use of armed force will be a factual determination, likely to be established over time by state practice; however, a survey of the current thinking shows that it is likely that a certain level of physical damage will also be required.

1.2. *Intervention of the armed forces*

In respect of international armed conflicts, Pictet's commentary on Common Article 2 of the Conventions requires the intervention of the armed forces of a state as a precondition for armed conflict. This approach raises two problematic issues in respect of contemporary conflicts. First, in modern armed conflict, particularly in an age characterised by the civilianisation of the military and outsourcing of key defence functions, the armed forces of a state may not be the only actors engaged in its armed conflicts. The use of unmanned Predator drones by the United States Central Intelligence Agency in the ongoing war in Afghanistan is a case in point. For example, on 13 January 2006 the CIA ordered an air strike by a Predator drone that fired air-to-ground missiles at the Pakistani village of Damadola, close to the border with Afghanistan.[24] The air strike was targeting a high level Al Qaeda leader, but failed to eliminate him; eighteen other people were killed in the attack. The United States military denied any involvement in the strike.[25] Although this attack took place in the context of an established and wider armed conflict, an attack such as this launched in peacetime would not fall within Pictet's definition, and illustrates the changing nature of the participants involved in contemporary armed conflicts. However, while it may not always be the armed forces of a state who conduct such activities, it is clear that some nexus with governmental authority will be required to instigate an international armed conflict.

Second, military forces are often used against other states and groups for tasks other than an armed conflict – for example, for aerial

[24] Dafna Linzer and Griff Witte, 'U.S. Airstrike Targets Al Qaeda's Zawahiri', *Washington Post* (Washington DC), 14 January 2006, A09, www.washingtonpost.com/wp-dyn/content/article/2006/01/13/AR2006011302260_pf.html (last accessed 15 July 2011).

[25] *Ibid.*

surveillance and reconnaissance.[26] Michael Schmitt thus contends that a dispute resulting in the commitment of armed forces cannot be the sole criterion for establishing an armed conflict. He argues that the reference to the armed forces is more logically understood as a form of prescriptive shorthand for activity of a particular nature and intensity.[27] That is, when a dispute reaches the level that a state deems it necessary to involve the armed forces, it has reached a sufficient level to be considered an armed conflict.

Internal armed conflicts are even more problematic. Involvement of the armed forces of a state are not required for an internal armed conflict; however, the identity of the parties involved will determine which legal regime applies. Although nothing in Common Article 3 defines internal armed conflicts in terms of the parties involved, Additional Protocol II is more selective. The Protocol applies to internal armed conflicts between the armed forces of a state and dissident armed forces or organised armed groups;[28] thus conflicts between other government agencies and such groups do not qualify.[29] The definition provided by the Appeals Chamber in *Tadic* reflects the breadth of the parties covered by Common Article 3 and applies to protracted armed violence by any government authority. Also, unlike Common Article 3 and the definition in *Tadic*, Additional Protocol II does not apply to conflicts between armed groups within the territory of a high contracting party. Under Additional Protocol II, organised armed groups (or dissident armed forces) must be under a responsible command. According to the Trial Chamber in *Akayesu* this entails 'a degree of organisation within the armed group … such as to enable the armed group to plan and carry out concerted military operations, and to impose discipline in the name of a *de facto* authority'.[30] It is clear from

[26] Michael N. Schmitt, 'Wired Warfare: Computer Network Attack and the *Jus in Bello*' in M. N. Schmitt and B. T. O'Donnell (eds.), *Computer Network Attack and International Law* (US Naval War College, Newport, RI, 2002) 187–218, 191.

[27] *Ibid.*, 372.

[28] Art. 1, Additional Protocol II.

[29] An explanatory note inserted into the Report of Committee I describes 'armed forces' as: 'All the armed forces … According to the views expressed by a number of delegations, the expression would not include other government agencies the members of which may be armed; examples of such agencies are the police, customs and other similar organisations.' According to Moir this leaves grey areas in the protocol. For a discussion of armed forces in internal conflict, see Moir, *The Law of Internal Armed Conflict*, 38–40, 104–5.

[30] *Prosecutor* v. *Jean-Paul Akayesu* (1998) Case No. ICTR-96-4-T, International Criminal Tribunal for Rwanda, para. 626.

the highly dispersed nature of the participants in the computer network attacks launched against Estonia, Georgia and the participants in other low-level computer network attacks such as those launched by supporters of the website helpisraelwin.com and its variants, that a fairly minimal level of organisation is needed to be able to launch sustained and debilitating attacks against a country's infrastructure. It is also likely that such organisations may well be networked in nature rather than the hierarchical structure traditionally associated with organised armed groups. However, the requirement that such groups be capable of imposing discipline may prove a more difficult criterion to fulfil. Louise Doswald-Beck has commented that any computer network attack launched by a group, however well-organised, is likely to be seen solely as criminal behaviour to be dealt with by agencies other than the military, even though the potential for damage could be enormous.[31] However, given the gradual move to recognition of armed attacks by non-state actors, the present author considers that this position can no longer be supported. The qualification of terrorist attacks as acts which may bring into operation the laws of armed conflict has proved particularly problematic since the attacks against the United States in September 2001; likewise the application of those laws to counter-terrorist operations. However, not all terrorist (or counter-terrorist) operations, even those severe enough to amount to armed attacks, will necessarily give rise to, or constitute an armed conflict. Internationally, terrorist acts that are attributable to a state may give rise to an armed conflict between that state and the target of the attacks.[32] Where no state is involved, terrorist operations may constitute non-international armed conflicts either under Common Article 3 or Additional Protocol II if the terrorist group is sufficiently organised and their acts are sufficiently violent and protracted.[33] However, where terrorist groups do not meet these criteria, their actions will simply

[31] Louise Doswald-Beck, 'Some Thoughts on Computer Network Attack and the International Law of Armed Conflict' in M. N. Schmitt and B. T. O'Donnell (eds.), *Computer Network Attack and International Law* (Naval War College, Newport, RI, 2002) 163–86, 165.

[32] Greenwood, 'Scope of Application of Humanitarian Law', 58, para. 215. Greenwood also notes the possibility of counter-terrorist activities becoming an international armed conflict where the victim state takes military action on the territory of another state, and brings about fighting between its forces and the forces of that state.

[33] Gary D. Solis, *The Law of Armed Conflict: International Humanitarian Law in War* (Cambridge University Press, 2010), 157.

constitute criminal acts and be dealt with under the domestic laws in place in the relevant jurisdictions.

2. Application to computer network attacks

A number of authors have discussed the applicability of the laws of armed conflict to computer network attacks. Mark Shulman has no difficulty in finding that '[a]s with other armed conflict, defensive [information warfare] operations are subject to the restraints of LOAC and its principle of proportionality', despite observing that 'information warfare is neither "armed" in the traditional sense, nor does it necessarily involve "conflict"'.[34] Other writers, however, are less sure of the application of the laws of armed conflict. Richard Aldrich claims that a physical manifestation such as an explosion is required:[35]

> 'Armed conflict', as presently understood, seems far less likely to be applied to the simple manipulation of bits inside a computer, although this may soon change since the nefarious manipulation of bits could, in some cases, already cause significantly more harm than could a bomb.

Emily Haslam has analysed the approaches of Shulman and Aldrich and concludes that, while it is welcome that the authors do not treat computer network attacks and other information operations homogeneously, they fail to establish a test which either works within the framework of the laws of armed conflict, or sets out the appropriate components of information warfare which should be taken into account (means and results respectively).[36] Other authors have addressed the issue in different ways; for example, after a flawed analysis equating armed conflict to the definition of aggression and using the terms armed force and armed attack synonymously, Hanseman concludes that the laws of armed conflict will apply to computer network attacks

[34] Mark R. Shulman, *Legal Constraints on Information Warfare*, Center for Strategy and Technology, Air War Center, Occasional Paper No. 7 (1999). Note that Shulman's use of the term information warfare rather than computer network attack relates, in part, to his broader definition but also the date of the paper. As discussed in Chapter 1, earlier literature tends to use the term information warfare rather than specifying computer network attacks. Shulman's paper concentrates on information attacks that seek to alter 'information without visibly changing the physical entity within which it arises'.

[35] Richard W. Aldrich, 'The International Legal Implications of Information Warfare' (1996) *Airpower* 99, 102.

[36] Haslam, 'Information Warfare', 167.

where the 'consequences of the attack are equivalent to the damage done by traditional weapons'.[37] Scott, writing on disruption of telecommunications, argues that the laws of armed conflict readily apply to computer network attacks: 'In determining the constraints imposed on computer network attack by the law of war the focus of analysis must be the intent and likely results of an attack, not the novel method of attack.'[38]

As noted above, the question of the applicability of the laws of armed conflict to computer network attacks arises in three distinct circumstances:

(i) where computer network attacks are utilised as part of an ongoing conventional armed conflict;
(ii) where computer network attacks are launched on their own; and
(iii) where the use of conventional weapons is insufficient in and of itself to qualify as an armed conflict, but it is accompanied by extensive computer network attacks.

2.1. *Application during conventional armed conflict*

The question of whether the laws of armed conflict apply to computer network attacks launched during a conventional conflict can be dealt with fairly briefly. The 2008 conflict between Russia and Georgia is generally accepted to be the first instance of an armed conflict accompanied by computer network attacks. To the extent that the computer network attacks can be positively attributed to one of the parties to the conflict (which, as noted previously, has not been proved in this instance), those attacks which are directly related to military operations will be covered by the laws of armed conflict as soon as it comes into force.

One possible argument that has been raised is that the law should not be applied, as the Geneva Conventions were drafted significantly before the technology to launch such attacks was available.[39] This argument can be dismissed on several grounds. First, the inclusion of the Martens

[37] Robert G. Hanseman, 'The Realities and Legalities of Information Warfare' (1997) 42 *AFL Rev.* 173, 184.

[38] Roger D. Scott, 'Legal Aspects of Information Warfare: Military Disruption of Telecommunications' (1998) 45 *Naval Law Review* 57, 59.

[39] Schmitt, 'Wired Warfare'. Schmitt also raises and dismisses a further possible argument that the LOAC do not apply to computer network attack because they are not specifically mentioned in the Conventions. However, an examination of the Martens clause shows that new methods and innovations are clearly anticipated by the laws.

clause in the Geneva Conventions and their Additional Protocols,[40] and the specific inclusion of Article 36 of Additional Protocol I relating to new weapons, indicates that the drafters of the Conventions anticipated the development and use of new weapons, means and methods of warfare.[41] The fact that the drafters require states to determine the legality of new methods of war by reference to the Protocol in and of itself indicates their acknowledgement of the applicability of the laws to newer technology. Secondly, the issue was addressed and dismissed by the International Court of Justice in relation to nuclear weapons in the *Nuclear Weapons* case in 1996.[42] The Court held:[43]

> Indeed, nuclear weapons were invented after most of the principles and rules of humanitarian law applicable in armed conflict had already come into existence; the Conferences of 1949 and 1974–1977 left these weapons aside, and there is a qualitative as well as quantitative difference between nuclear weapons and all conventional arms. However, it cannot be concluded from this that the established principles and rules of humanitarian law applicable in armed conflict did not apply to nuclear weapons. Such a conclusion would be incompatible with the intrinsically humanitarian character of the legal principles in question which permeates the entire law of armed conflict and applies to all forms of warfare and to all kinds of weapons, those of the past, those of the present and those of the future.

Such a clear statement by the Court indicates that the Conventions and other general legal principles of the laws of armed conflict are applicable to computer network attacks, despite the fact that the technology is new, or the fact that such attacks are qualitatively different to weapons systems which have come before. This has led some commentators to state categorically that 'there is no doubt that an armed conflict exists and the law of armed conflict applies, once traditional

[40] Art. 63 GCI; Art. 62 GCII; Art. 142 GCIII; Art. 158 GCIV; Art. 1(2) API; preambular para. 4 APII.

[41] Art. 36 of Additional Protocol I states: 'In the study, development, acquisition or adoption of a new weapon, means or method of warfare, a High Contracting Party is under an obligation to determine whether its employment would, in some or all circumstances, be prohibited by this Protocol or by any other rule of international law applicable to the High Contracting Party.'

[42] *Nuclear Weapons Case*; Schmitt, 'Wired Warfare', 189.

[43] *Nuclear Weapons Case*, para. 86. The Court also cited with approval the written statement of New Zealand: 'International humanitarian law has evolved to meet contemporary circumstances, and is not limited in its application to weaponry of an earlier time. The fundamental principles of this law endure: to mitigate and circumscribe the cruelty of war for humanitarian reasons.' (New Zealand, Written Statement, 15, paras 63–64)

kinetic weapons are used in combination with new methods of computer network attack'.[44]

However, it should be noted that computer network attacks may be distinguished from both conventional weapons and nuclear weapons in one significant respect that was not at issue in the *Nuclear Weapons* case and therefore not anticipated by the Court. The extent and type of damage inflicted by a computer network attack depends entirely on the objective and design of the attack itself. Conventional weapons and their nuclear counterparts have a single effect when employed against a target – *vis* physical destruction, injury and loss of life – and thus may be regulated as a category. The difficulty comes when dealing with a form of attack that may or may not cause physical destruction and may only indirectly cause loss of life or injury to individuals. In that regard it should be noted that the concept of hostilities also covers certain non-violent acts, provided that they are directly connected to military operations against the enemy (e.g. logistics or the gathering of intelligence).[45]

2.2. *Computer network attack on its own*

The flexibility of the attack medium and the diversity of possible consequences of computer network attacks have raised a further argument against the application of the laws of armed conflict to such attacks. Although it is clear that one cannot apply a blanket rule against all computer network attacks, the question must be asked, can a computer network attack, or series of such attacks, trigger the application of the laws of armed conflict? For example, the Stuxnet virus was introduced into computers at two Iranian nuclear facilities in 2009–10. In November 2010 President Mahmoud Ahmadinejad announced that unspecified malicious software sent by 'western enemies' had affected Iran's centrifuges at its Natanz enrichment plant and had 'succeeded in creating problems for a limited number of our centrifuges'.[46] Although he did not mention Stuxnet by name and there are no reported incidents of physical damage in the public domain, it is

[44] Knut Dörmann, 'Applicability of the Additional Protocols to Computer Network Attacks' (Paper presented at the International Expert Conference on Computer Network Attacks and the Applicability of International Humanitarian Law, Stockholm, 17–19 November 2004) 139–54, 141. See also, Doswald-Beck, 'Thoughts on Computer Network Attack', 165.

[45] Dinstein, *Conduct of Hostilities* (2nd edn), 2.

[46] 'Iran Says Cyber Foes Caused Centrifuge Problems', Reuters, 29 November 2010.

known that Iran decommissioned and replaced approximately 1,000 IR-1 centrifuges at about the time of the reported cyber attacks, implying larger than expected levels of failure.[47] Questions of attribution of the attack aside,[48] would this be sufficient to bring into operation the laws of armed conflict? Forensic analysis of the Stuxnet code indicates that the attackers sought to manipulate the frequency converters, controlling the rotors of the centrifuges used for uranium enrichment.[49] A report by ISIS into Stuxnet notes that the malware's commands do not guarantee destruction of the centrifuges, however the code increases the speed of the rotor to very close to the maximum speed the rotor can withstand mechanically, and well above the frequency at which the Natanz plant is said to normally operate.[50] The report concludes that if the maximum tangential speed were to reach the frequency of the rotor as set by Stuxnet, the rotor would be likely to fly apart. Stuxnet leaves the frequency at the higher level for fifteen minutes (while masking the effects by shutting off warning signals) and then returns the frequency converter to its normal operating frequency.[51] A second phase of the attack lowers the frequency for a period of time before returning it to normal operations. The report postulates that the intention of the attackers may simply be to disrupt normal operation of the plant rather than to destroy it completely.[52]

The criteria established by the ICTY in *Tadic*, the ICRC commentary by Pictet and subsequent state practice, indicate that a computer network attack will be considered the start of an armed conflict where the attacker is a state organ or armed group that launches a computer network attack which is intended to cause, or which actually causes, physical damage to life and/or property. It is the perceived intention and consequences of the attack that must be addressed and this is where Pictet's definition departs from state practice. Further, where

[47] Albright, Brannan and Walrond, *ISIS Stuxnet Report*; Zetter, 'Report Strengthens Suspicions that Stuxnet Sabotaged Iran's Nuclear Plant'.

[48] At the time of writing the origin of the Stuxnet malware is unknown, although speculation is rife that the attacks were either by a state or states (the United States and Israel being the most common suspects), or state sponsored, given the complexity and sophistication of the attacks.

[49] Falliere, OMurchu and Chien, *Stuxnet Dossier.*

[50] Albright, Brannan and Walrond, *ISIS Stuxnet Report*, 4.

[51] Note that at the time of writing, although progress had been made on deciphering the Stuxnet code, some attack sequences are still being analysed by Symantec, and there remained significicant questions about its operation.

[52] Albright, Brannan and Walrond, *ISIS Stuxnet Report*.

the attack is launched by an armed group, the attack must be part of a protracted series of attacks (whether or not such attacks are computer network attacks) in order to establish that they are not isolated or sporadic acts of violence. Assuming for the purposes of analysis that Stuxnet can be attributed to a state actor (or organised group whose actions can be attributed to a state) is it sufficient to bring into operation the laws of armed conflict? Under the ICRC commentary model, the answer must be yes. Most reports on Stuxnet conclude that the malware has caused physical damage to Iran's nuclear centrifuges. Without a requirement of a *de minimus* level of intensity, this would constitute the beginnings of an armed conflict. However, as noted above, it is by no means certain that states would consider this to be the case. It seems likely that this incident would be considered to be an isolated incident, despite the continuing nature of the attack,[53] more akin to a border skirmish, particularly as very little is known about the fuel enrichment plant's countermeasures to the attack and Iran has not responded in public (although that may well be because of a lack of definitive attribution).

2.2.1. Armed force

This author has argued that computer network attacks can constitute a use of force under Article 2(4) of the UN Charter,[54] but a separate question must be asked whether such attacks are to be considered 'armed force' in such a way as to initiate the application of the laws of armed conflict? Given the broad scope of possible information operations, it is not clear where on the spectrum the line will be drawn regarding what amounts to an armed attack and what may be a use of armed force, or whether a further demarcation will occur between force generally and armed force. Following the discussion regarding armed attack in Chapter 3 above,[55] it seems clear that any computer network attack launched by the armed forces, or some other organ of a state and causing large scale physical damage or personal injury equivalent to that caused by a conventional attack will be considered both an armed attack and the start of an armed conflict. However, an attack designed to merely neutralise the air-defence network of a country by switching it off, would not; although if the attacking state's air force were to take advantage of that window of opportunity to launch a conventional

[53] Stuxnet is programmed to launch alternating attack sequences every 27 days.

[54] See Chapter 2 above. [55] See Chapter 3 above.

attack, that would then trigger the application of the laws of armed conflict, the start of which would be backdated to the computer network attack as the beginning of the attack as a whole. Admittedly this may only be a matter of minutes or hours. For example, had the 2007 raid by Israel against the alleged Syrian nuclear site escalated further, the laws of armed conflict would have applied from the start of the engagement with the single Syrian radar site at Tall al-Abuad near the Turkish border.[56]

Michael Schmitt has argued that the term armed conflict occurs when a group takes measures that injure, kill, damage or destroy, and includes actions 'intended to cause such results or which are the foreseeable consequences thereof'.[57] He goes on to argue that in the case of computer network attacks:[58]

> ... humanitarian law principles apply whenever computer network attacks can be ascribed to a State are more than merely sporadic and isolated incidents and are either intended to cause injury, death, damage or destruction (and analogous effects), or such consequences are foreseeable. This is so even though classic *armed* force is not being employed. By this standard, a computer network attack on a large airport's air traffic control system by agents of another State would implicate humanitarian law. So too would an attack intended to destroy oil pipelines by surging oil through them after taking control of computers governing flow, causing the meltdown of a nuclear reactor by manipulation of its computerized nerve centre, or using computers to trigger a release of toxic chemicals from production and storage facilities. On the other hand, humanitarian law would not pertain to disrupting a university intranet, downloading financial records, shutting down Internet access temporarily or conducting cyber espionage, because, even if part of a regular campaign of similar acts, the foreseeable consequences would not include injury, death, damage or destruction.

This analysis appears convincing in most respects. However, Schmitt's extension of the term to incorporate the foreseeable consequences of an attack (which he uses to cover such actions as the shutting down of air traffic control systems) should be balanced particularly against the perceived intention of the attacking party. Where an attack is launched against a target which is not so obviously linked to the violent

[56] The radar site was attacked with a combination of electronic attacks, computer network attacks and precision bombing. This would have been the case even without the precison bombing. See generally: Fulghum, Wall and Butler, 'Israel Shows Electronic Prowess'.

[57] Schmitt, 'Wired Warfare', 373.

[58] *Ibid.*, 374 (footnotes omitted).

consequences (that is, the consequences are a knock-on effect from the initial attack), a blanket prohibition would: (a) beg the question of what is foreseeable to the attacker; and (b) preclude any assertions of mistaken identity such as those promulgated in the case of the *Dogger Bank*, *USS Liberty* and *USS Stark* incidents. With respect to the forseeability question, it is also worth considering that in the Stuxnet attacks, the code increases the rotational frequency of the rotors to a level that appears to be within (although very close to) the maximum speed the rotor can withstand mechanically.[59] However, the code only raises the frequency for fifteen minutes, during which time the centrifuge may not reach the maximum frequency that would guarantee its destruction.[60] Thus it may be that the intention of the attackers was not to damage or destroy the centrifuges but rather to disrupt the operation of the plant. If that were the case, is it then foreseeable that the malware would cause physical destruction? Probably. Does it make a difference that the normal operational frequency of the centrifuges at Iran's Natanz plant was lower than the usual nominal frequency for that type of centrifuge in an effort to reduce the number of breakages because of a particular sensitivity to frequency of rotation?[61] It seems unlikely that states would consider such an event to be the start of an armed conflict, although undoubtedly an internationally unlawful act.

2.2.2. Humanitarian principles

The motivation of the application of the laws of armed conflict is to limit the damage caused by hostilities and provide care for the casualties.[62] As Louise Doswald-Beck points out, this would militate in favour of an expansive interpretation of when the laws of armed conflict should begin to apply.[63] Greenburg *et al.*, while not discussing the definition of armed conflict, address the issue of whether information warfare is war, and point out that international law draws 'a strong distinction between traditional, kinetic force and the infliction of hardship or suffering on a government or population'.[64] Computer network attacks

59 Albright, Brannan and Walrond, *ISIS Stuxnet Report*, 4.

60 *Ibid*.

61 *Ibid*; Zetter, 'Report Strengthens Suspicions that Stuxnet Sabotaged Iran's Nuclear Plant'.

62 Dörmann, 'Additional Protocols'; Doswald-Beck, 'Thoughts on Computer Network Attack', 164.

63 Doswald-Beck, 'Thoughts on Computer Network Attack', 164.

64 Lawrence T. Greenberg, Seymour E. Goodman and Kevin J. Soo Hoo, *Information Warfare and International Law* (CCRP, Washington DC, 1998), 19.

which merely cause discomfort, inconvenience or even a certain level of suffering are not sufficient to equate to an armed conflict. Michael Schmitt has argued that the purposes of humanitarian law are such that it aims to protect those that do not directly participate in hostilities and their property; the protection itself is framed in terms of injury or death or, in the case of property, damage or destruction.[65] Thus it must be reasoned that armed conflict occurs when a group takes measures that injure, kill, damage or destroy.[66]

2.3. Computer network attacks in support of conventional attacks

The third situation in which the applicability of the laws of armed conflict is raised occurs where a conventional attack is launched, which would not by itself qualify as an armed conflict, but which is supported by extensive computer network attacks. In that situation the accompanying computer network attacks would serve as an indicator of the intentions of the opposing party. For example, it would prove difficult for a state to claim a case of mistaken identity in the bombing of a ship, if it were later discovered that the radar system had been tampered with so that the ship's commander believed that the attacking aircraft were in fact allied military planes, and incoming missiles were not detected because the onboard defence system had been remotely turned off. There has been much speculation that this combined tactic will be used by armed groups to multiply the impact of any conventional attack. For example, the consequences of a small conventional attack in a metropolitan city would increase several-fold if at the same time the city experienced a power-cut, including power to traffic signals and hospitals, the emergency response telephone number was disconnected or jammed, and the water supply was cut off. It is likely that such a combination of attacks would also be considered sufficient to qualify as an armed conflict under the accumulation of events theory set out in Chapter 3 above.

3. Territory

Internal armed conflicts conducted through the means of computer network attack also raise an additional issue in respect of territory. Common Article 3 of the Geneva Conventions applies to armed conflicts not of an international character 'occurring in the territory of

[65] Schmitt, 'Wired Warfare', 373. [66] *Ibid.*

one of the High Contracting Parties'. Likewise, Article 1 of Additional Protocol II also requires that the armed conflict in question must take place in the 'territory of the high contracting party'. Despite some commentators' arguments that conflicts involving computer network attacks take place in the somewhat ethereal plane of cyberspace, in general such discussion has been replaced by an understanding that it is the effects of these attacks on tangible objects and individuals which creates obligations and responsibilities under the laws of armed conflict. This is so even in the few discrete cases where communications equipment controlling assets in a particular state are located offshore, or even where the state's entire online presence is hosted in a foreign state.[67] Although not relating to computer network attack, the decision of the Appeals Chamber of the International Tribunal for the Former Yugoslavia in *Kunarac et al.* is instructive:[68]

> There is no necessary correlation between the area where the actual fighting is taking place and the geographical reach of the laws of war. The laws of war apply in the whole territory of the warring states or, in the case of internal armed conflicts, the whole territory under the control of a party to the conflict, whether or not actual combat takes place there, and continue to apply until a general conclusion of peace or, in the case of internal armed conflicts, until a peaceful settlement is achieved. A violation of the laws or customs of war may therefore occur at a time when and in a place where no fighting is actually taking place. As indicated by the Trial Chamber, the requirement that the acts of the accused must be closely related to the armed conflict would not be negated if the crimes were temporally and geographically remote from the actual fighting. It would be sufficient, for instance, for the purpose of this requirement, that the alleged crimes were closely related to hostilities occurring in other parts of the territories controlled by the parties to the conflict.

Thus any state or armed group finding itself in an armed conflict involving the use of computer network attack will be required to apply the laws of armed conflict to the whole of the territory of the state; or alternatively, in the case of the armed group, to any territory under its control. This latter requirement raises the question of in what manner an armed group can control territory.

[67] See, for example, Charles Arthur, 'The Day East Timor Was Deleted', *Independent* (London), 28 February 1999, Features 8; Chris Nuttall, 'Virtual Country "Nuked" on Net', *BBC News*, 26 January 1999, http://news.bbc.co.uk/1/hi/sci/tech/263169.stm (last accessed 14 December 2011).

[68] *Prosecutor* v. *Dragoljub Kunarac et al.* (2002) Case No. IT-96-23 and 23/1, International Criminal Tribunal for the Former Yugoslavia, para. 57 (footnotes omitted).

Article 1 of Additional Protocol II requires armed groups to 'exercise such control over a part of its territory as to enable them to carry out sustained and concerted military operations and to implement this Protocol'. The Diplomatic Conference considered several proposals to clarify the amount of territory which should be controlled;[69] however, they chose not to adopt any of these proposals and instead linked the control of territory to the ability first to launch sustained and concerted military operations, and secondly to apply the Protocol. While the restrictiveness of this provision has been roundly criticised by scholars in the context of conventional internal conflicts,[70] when applied to computer network attacks it may in fact, not act so restrictively. In the information age, physical control over territory is largely irrelevant. It is possible for armed groups to launch sustained and concerted computer network attacks against a state without ever capturing any significant territory. Of course, the desirability of undertaking such a strategy would be dependent on the purposes of the rebellion – although the current trend towards coercive campaigns may lend itself more to this form of action than others. It should also be noted that wars of national liberation (in which control of territory would be key) are incorporated into the scope of Additional Protocol I, which does not contain any requirement for territorial control.[71] In contrast to their approach to Common Article 3, the Diplomatic Conference decided that some cut-off point was required to show that conflicts must have reached a critical point before Additional Protocol II should apply.[72] The criterion of 'sustained and concerted' military operations was arrived at in an effort to find criteria for the critical point, implying duration and intensity, but on a more objective assessment.[73] This selection of criteria, however, may well have enabled conflicts utilising computer network attacks to be covered by Additional Protocol II in situations where their more conventional counterparts remain under the auspices of Common Article 3.

[69] Proposals considered by the committee included a requirement that it should be a 'non-negligible part of the territory' or a 'substantial part of the territory': see Claude Pilloud, *et al.*, *Commentary on the Additional Protocols of 8 June 1977* (Martinus Nijhoff, Geneva, 1987), para. 4465.

[70] Moir, *The Law of Internal Armed Conflict*, 105–6. The provision has been particularly criticised in respect of those internal conflicts involving guerrilla warfare.

[71] As noted above, this provision has proved highly controversial and is one of the reasons why states such as the United States have refused to ratify the Protocol.

[72] Moir, *The Law of Internal Armed Conflict*, 106.

[73] Pilloud, *et al.*, *Commentary*, para. 4465; Moir, *The Law of Internal Armed Conflict*, 106.

The second criterion for application of the Additional Protocol II is that the armed group's control of territory must be sufficient to enable it to apply the Protocol.[74] Waldemar Solf has argued in respect of the civil conflict in El Salvador:[75]

> I doubt that a movement that does not control a single town and whose political arm is situated in another country, with only loose links to the movement's organised armed groups, has the capability of implementing the Protocol. I question whether it can implement the judicial standards of article 6, the standards established for the treatment of detained persons under article 5, and the standards established under articles 7–12 for the protection of wounded, sick, shipwrecked, and medical personnel units.

While the same issue is raised in conflicts involving computer network attacks, if the conflict is waged *only* by these means, it may be that the necessity for the above standards set out by the Protocol and highlighted by Solf is negligible. Concerted computer network attacks which obey the principles of distinction and do not result in severe physical damage, may not result in wounded or detained persons. That will be highly dependent on the purpose of the conflict. Of course it may be argued in counterpoint that if there is no need for the protection of wounded etc., there is no need for the application of the Protocol in the first place.

4. Conclusion

As with the previous chapters, application of the laws of armed conflict to computer network attack requires us to revisit first principles in order to interpret humanitarian norms for the information age. It appears likely that the laws of armed conflict will apply to most computer network attacks launched by states (or, in the case of internal armed conflicts, organised armed groups) where there is a physical manifestation resulting in damage to property and, more importantly, injury or death to individuals. While it may not be necessary for the level of damage or injury caused to rise to the level of an armed attack as discussed in Chapter 3, any attack will need to be of significant

[74] For a criticism of this rather circular argument see Moir, *The Law of Internal Armed Conflict*, 108; Michael Bothe, *et al.*, *New Rules for Victims of Armed Conflicts* (Martinus Nijhoff Publishers, Leiden, 1982), 625.

[75] Waldemar A. Solf, 'Comment: Non-International Armed Conflicts' (1981–1982) 31 *Am. U. L. Rev.* 927, 932. These comments were made with respect to the 12-year civil conflict in El Salvador which ended in 1992.

seriousness to raise it above the *de minimis* level indicated by current state practice. The attacks must be more than isolated incidents and, in the case of internal armed conflict, the online hostilities must be protracted and of a nature to raise them above the level of riots and other internal disturbances.

5 Participants in conflict: combatant status, direct participation and computer network attack

The beginning of the twenty-first century has seen huge changes in the war-fighting capacities of the modern military. Regardless of one's opinion about the existence of a revolution in military affairs, one thing is certain: the range of people involved and the technologies available to them have changed significantly. Information operations, and in particular computer network attacks, have raised many challenging questions for the laws of armed conflict. This chapter focuses on the problems that computer network attacks raise in regard to the participants in armed conflicts. Indeed, one of the most pressing problems facing the laws of armed conflict may not be how to deal with combatants or civilians who carry rifles on the front lines, but rather in determining the status of personnel armed with CPUs and keyboards sitting at a desk a continent away.[1] The reason for this is two-fold: first, it is not obvious how the requirements for lawful combatancy will translate onto a medium where anonymity is the norm and distance and proximity are largely irrelevant. Secondly, the specialist nature of new technologies and the downsizing of military forces have resulted in increased civilianisation of state armed forces. Care must be taken in deciding what roles can be outsourced to civilian contractors without jeopardising their legal protections under international conventions.

The law of armed conflict makes a fundamental distinction between combatants and civilians.[2] The former have the right to participate in hostilities and may attack, kill and wound enemy combatants and destroy military objectives. Conversely, civilians are not permitted to

[1] Kenneth Watkin, *Combatants, Unprivileged Belligerents and Conflicts in the 21st Century* (HPCR, Cambridge, MA, 2003).

[2] See *Nuclear Weapons Case*, 257.

directly participate in hostilities. Their status as civilians enables them to enjoy protection from the dangers arising from military operations and they must not be made the object of a direct attack. Where civilians do take a direct part in hostilities, they lose their protection for the duration of their involvement and may be liable for punishment for their actions either through domestic or international criminal processes. This chapter addresses both the question of which legal requirements must be met by combatants involved in computer network attacks in order to maintain their combatant privileges, and also what level of involvement in computer network attacks will constitute direct participation in hostilities, such that civilian employees or participants would lose their civilian protection. This chapter also examines states' obligations in relation to young would-be cyber soldiers and assesses whether the increasing numbers of embedded civilian contractors are at risk of falling foul of mercenary provisions.

1. Combatant status

The problem of combatant status is one which has caused difficulty for the international legal community. While the principle of distinction is firmly established in customary international law, actually establishing who is a combatant, who is a civilian, and what rights and duties are imposed on them both on the battlefield and following capture has proved more problematic.

Additional Protocol I defines combatants as '[m]embers of the armed forces of a Party to the conflict (other than medical personnel or chaplains ...)'; the preceding section establishes that:[3]

> [t]he armed forces of a party consist of all organized armed forces, groups and units which are under a command responsible to that Party for the conduct of its subordinates, even if that Party is represented by a government or an authority not recognized by an adverse Party. Such armed forces shall be subject to an internal disciplinary system which, '*inter alia*', shall enforce compliance with the rules of international law applicable in armed conflict.

However, this definition has proved highly controversial and is one of the reasons why some states have refused to ratify the Protocol. The reason for the controversy is the inclusion of armed groups within the definition. The more widely accepted definition under Geneva Convention III falls within the definition of prisoners of war.[4] As Watts

[3] Art. 43(1) and (2) Additional Protocol I.
[4] Art. 4(A)(1), (2) (3) and (6) Geneva Convention III.

notes, four of the groups enumerated in Article 4(A) of the Convention are acknowledged as distinct from the civilian class, and generally acknowledged to take active or direct part in hostilities as combatants: (1) members of the armed forces of a party; (2) militia, volunteer corps and organised resistance movements belonging to a party; (3) armed forces of parties to the Conventions not diplomatically recognised by their enemy; (4) members of a *levée en mass*.[5]

Civilians are defined negatively by Article 50 of Additional Protocol I as those who do not fall within the above categories. One category of civilians does merit special mention, however: persons accompanying the armed forces without actually being members of it are entitled to prisoner of war status provided that they are authorised by a party to the conflict, but are not considered combatants.[6]

The term 'combatant' has been used to describe both those persons with a right to take direct part in hostilities, but also to describe any person who actually engages in hostile acts in an armed conflict on behalf of a party to the conflict, whether or not they are permitted to do so; in those circumstances the term has been qualified with the adjective lawful or unlawful.[7] Thus, in international armed conflicts, combatants may be further distinguished into two categories: first, those people who are members of the armed forces of a belligerent party (with the exception of medical and religious personnel), even if their specific tasks are not related to active hostilities; and second, any other person who takes an active part in hostilities.[8] This second group are unlawful combatants.[9]

Yoram Dinstein argues that unlawful or unprivileged combatant status can be achieved in one of two ways: either the individual's primary status is that of a combatant and they lose their privileged status through lack of compliance with the requirements of lawful combatancy; or they are civilians who directly participate in hostilities.[10] Unlawful or unprivileged combatants may be targeted in the

5 Sean Watts, 'Combatant Status and Computer Network Attack' (2010) 50(2) *Virginia Journal of International Law* 391, 417.

6 Art. 4(A)(4) Geneva Convention III.

7 UK Ministry of Defence, *The Manual of the Law of Armed Conflict* (Oxford University Press, Oxford; New York, 2004), 38.

8 Dinstein, *Conduct of Hostilities* (2nd edn), 33.

9 Also known as unprivileged belligerents, for the purposes of this work they will be referred to as unlawful combatants. See generally Richard R. Baxter, 'So-Called "Unprivileged Belligerency": Spies, Guerrillas, and Saboteurs' (1951) 28 *BYBIL* 323.

10 Dinstein, *Conduct of Hostilities* (2nd edn), 36.

same manner as a combatant but they do not enjoy any of the privileges of lawful combatancy, nor those of civilian protection from direct attack or the dangers arising from military operations.[11] The most important privileges of lawful combatancy are the legal shield that it provides for acts which would otherwise be illegal (for example, murder), and the entitlement to prisoner of war status in the event of capture by the enemy. As Michael Schmitt has indicated, reduced to basics, the detention issue, which surfaces only in international armed conflict, is whether civilians who take up arms qualify for treatment as: (1) prisoners of war under the 1949 Third Geneva Convention; (2) civilians under the 1949 Fourth Geneva Convention; or (3) 'unlawful combatants' who enjoy only basic protection, such as that set forth in Common Article 3 to the 1949 Geneva Conventions and Article 75 of Additional Protocol I.[12]

The concept of combatant status is one which sits more securely in international armed conflicts; indeed, many scholars would argue that it has no place in discussions involving internal armed conflicts. However, as discussed in Chapter 1, traditional interstate conflicts account for only a small minority of the world's current conflicts.[13] The majority of conflicts are fought by non-state actors, whether they are insurgent groups fighting against the state,[14] hybrid conflicts with both internal and international elements,[15] multiple armed groups fighting each other on the territory of a state,[16] or a transnational group fighting internal or state armed forces.[17] The notion of a lawful combatant does not fit any of these groups. While Article 13(2) of Geneva Conventions

[11] Yoram Dinstein, 'Unlawful Combatancy' in F. Borch and P. Wilson (eds.), *International Law and the War on Terror* (Naval War College, Newport, RI, 2003) 151–74.

[12] Michael N. Schmitt, 'The Interpretive Guidance on the Notion of Direct Participation in Hostilities: A Critical Analysis' (2010) 1 *Harvard National Security Journal* 55, summarising Dinstein, 'Unlawful Combatancy'.

[13] Of the 118 conflicts recorded by the Uppsala Conflict Database between 1989 and 2006, only 5.8 per cent were traditional interstate conflicts. However, a further 21.3 per cent constituted internationalised internal conflicts which would include international elements. Lotta Harbom and Peter Wallensteen, 'Armed Conflict, 1989–2006' (2007) 44(5) *Journal of Peace Research* 623, 624.

[14] See, for example, the conflicts between the Tamil Tigers in Sri Lanka, Maoist rebels in Nepal, Chechen separatists fighting Russia.

[15] The conflict that took place in the territory of the former Yugoslavia is an example of this.

[16] There are currently five separate armed groups on the territory of the Democratic Republic of the Congo fighting against each other. Somalia is another example.

[17] For example, Al Qaeda against the Northern Alliance and the United States.

I and II and Article 4 of Geneva Convention IV provide protection for volunteer groups and militia who are not incorporated into the armed forces, they are generally accepted as applying only to situations of occupied territory or wars of national liberation.[18] The position under Article 43(2) of Additional Protocol I is much clearer – only members of the armed forces of a party to the conflict (with the exception of medical and religious personnel) have the right to directly participate in hostilities. It should be noted, however, that the use of the term 'armed forces' in Additional Protocol I refers both to regular and irregular troops. It is clear that the states participating in the Geneva Diplomatic Conference which resulted in the Additional Protocols did not intend to go so far as to upgrade rebels to the status of 'lawful combatants', a move which, in their view, would have entailed legitimising the rebels' struggle.[19] While states remain the primary focus of international relations and allegiance, they are unwilling to accept the principle that insurgent fighters are anything other than criminals. However, as the global conflict paradigm shifts from a model where states maintain the monopoly on politically motivated violence, to a model where substate, trans-state and in some cases supra-state actors are encroaching on traditionally state-controlled areas, the changing scope of conflict is being reflected in our interpretation of the laws that govern hostilities.

In 2009 the International Committee of the Red Cross released their interpretive guidelines on the notion of direct participation in hostilities.[20] The guidance was the result of six years of consultation with experts in the field and represents the ICRC's view of the current state of the law with respect to direct participation in hostilities as applied during the conduct of hostilities. Importantly, the ICRC concluded that for the purposes of non-international armed conflicts, members of organised armed groups who perform a continuous combat function for that group are not civilians, and may be targeted for the duration

[18] This work will not discuss the categorisation of the Chechen conflict as a war of national liberation.

[19] Antonio Cassese, 'The Status of Rebels under the 1977 Geneva Protocol on Non-International Armed Conflicts' (1981) 30 *International and Comparative Law Quarterly* 416, fn. 2. In the last two weeks of the final session of the conference, any provisions which could imply recognition of insurgent parties were deleted; Adam Roberts and Richard Guelff, *Documents on the Laws of War* (3rd edn, Oxford University Press, 1999), 482.

[20] ICRC, *Interpretive Guidance on the Notion of Direct Participation in Hostilities* (International Committee of the Red Cross, Geneva, 2009).

of their membership (as opposed to the duration of their participation). Although the guidance specifically does not grant any change in status to members of such groups, the move to no longer consider these participants to be civilians is significant. As Michael Schmitt points out, despite the cautionary caveat that this only applies to direct participation and not to status during detention,[21] treating direct participants differently from civilians proper seems to support the proposition that they are a separate category and, thus, not entitled to the protections civilians enjoy during detention under the Fourth Geneva Convention.[22]

1.1. *Requirements of combatant status*

Yoram Dinstein has usefully identified seven cumulative conditions for lawful combatancy.[23] The first four are cumulative conditions set out by the Hague Regulations and Geneva Conventions for the applicability of prisoner of war and lawful combatant status: (i) being under the command of a person responsible for his or her subordinates; (ii) having a fixed distinctive sign recognisable at a distance; (iii) carrying arms openly; and (iv) conducting operations in accordance with the laws and customs of war.[24] An additional two may be implied from Article 4(A)(2) of Geneva Convention III, that of (v) organisation, and (vi) belonging to a party to the conflict. Finally, a seventh condition may be inferred from case law, which denies prisoner of war status to any person owing a duty of allegiance to a detaining power.[25] Members of the armed forces of a party, militia and volunteer forces must comply with these conditions to be accorded the status of a prisoner of war or lawful combatant.[26] Several of these conditions raise particular

[21] Note that the Guidance specifically states that the conclusions of the report are not intended to serve as a basis for any determination outside the conduct of hostilities – for example those held in detention. *Ibid.*, 11.

[22] Schmitt, 'Interpretive Guidance', 14.

[23] Yoram Dinstein, *The Conduct of Hostilities under the Law of International Armed Conflict* (Cambridge University Press, 2004), 33–7.

[24] Art. 13(2) Geneva Conventions I and II and Art. 4(2) Geneva Convention IV.

[25] *Public Prosecutor* v. *Koi et al.* (1968) AC 829, Privy Council (per Lord Hodson). The Privy Council considered that the principle was one of customary international law; *cf.* Rogers, who argues that this decision has probably not survived the introduction of Additional Protocol I, A. P. V. Rogers, *Law on the Battlefield* (2nd edn, Manchester University Press, 2004), 32.

[26] Pictet, *Commentary*, 48. See also *Osman Bin Haji Mohamed Ali and Another* v. *Public Prosecutor* (1969) 1 AC 430, Privy Council (Malaysia), 449.

issues with respect to computer network attack; others simply require reinterpretation for the digital age.

1.1.1. Responsible command

The first condition, that of being commanded by a person responsible for his/her subordinates, merely excludes individuals, or groups of individuals, from independently waging war on the enemy. Warnings against this kind of behaviour have been seen in the press by US officials attempting to dissuade US-based hackers from 'joining-in' the conflicts against Afghanistan and Iraq.[27] Such actions are clearly not permitted, and no individual engaged in such attacks would be entitled to claim combatant immunity for their part. As Watts notes, not only is the requirement of a responsible command designed to keep rogue actors from the rubric of war, it also ensures that states can trace unlawful warlike acts to responsible leaders from whom reparations could be extracted.[28]

1.1.2. Distinction

The second and third conditions, that of having a fixed distinctive sign recognisable at a distance and that of carrying arms openly, may be dealt with together as they raise similar problems for computer network attack. The problem stems primarily from the anonymity that is characteristic of the internet, namely that it is impossible to tell who is sitting at any particular computer. The intention of the two requirements is to eliminate confusion in the distinction between civilians and combatants, and to prevent deception.[29] However, the rules were drafted in an era when warfare involved a certain amount of physical proximity between opposing forces. For the most part, combatants could see one another and hence distinguish between combatant and non-combatant, friend and foe. In the instance of a computer network attack, where the adversaries are plainly not in sight of each other (and may be half a world away), the usefulness of these conditions has diminished. The principle of distinction on which they are based, however, remains fundamental.

Although problematical, the issue is not without precedent. A parallel may be drawn with troops who are serving aboard or travelling

[27] David F. Gallagher, 'Hackers; Government Tells Vigilantes Their "Help" Isn't Necessary', *New York Times,* 20 February 2003, G1 5.

[28] Watts, 'Combatant Status and CNA', 437.

[29] Dinstein, *Conduct of Hostilities* (1st edn), 37.

upon vehicles, engines of war, aircraft, tanks and boats etc. All are required to be marked with the distinctive sign of the belligerent party whenever partisans are on board.[30] Given the impossibility of determining the user of a particular computer at any given moment, the requirement to display a distinctive sign may also be applied to the computer from which the attack is launched. One method of achieving such markings would be to require any computer network attack to emanate from a designated military IP address.[31] A form of electronic marking is already in use for medical transports appearing on radar or IFF technology, albeit with the opposite intention of marking a protected object.[32] Such an approach would also address the issue of the obligation of an individual to wear uniform while carrying out such an attack. Members of armed forces not wearing uniform aboard properly marked warships or military aircraft and taking part in hostilities are and remain combatants regardless of this circumstance.[33]

Attractive as this suggestion may initially seem, it is not without problems of its own. In the age of computer network attack where range and visibility are no longer requirements for targeting, requiring a computer to be marked as a military computer is tantamount to painting a bulls-eye on any system to which it is connected. At any one time the internet is being searched or 'crawled' by millions of software bots intent on finding connected computers.[34] A bot searching for military designated IP addresses would be able to find them

[30] This is in line with the long-established regulations in international law regarding the flag in the case of war at sea. Pictet, *Commentary*, 60. Although note that this does not absolve the combatants on board from wearing their personal distinctive emblems once they are separated from the aircraft or other vehicle: Dinstein, *Conduct of Hostilities* (2nd edn), 45.

[31] Every device (computer, server etc.) that communicates over the internet is assigned a four-number numerical address (e.g. 168.212.226.204) that uniquely identifies the device and distinguishes it from other computers on the internet. Each address is registered with one of three registry bodies to avoid duplicates. Creating a class of military addresses, or another form of military network designator, would be a relatively simple matter.

[32] Additional Protocol 1, Annex 1, Art. 8. IFF stands for Identification Friend or Foe, a secondary radar system that transmits an identification code when the transponder is triggered by detection of the target by the primary radar.

[33] Knut Ipsen, 'Combatants and Non-Combatants' in D. Fleck (ed.), *The Handbook of Humanitarian Law in Armed Conflicts* (Oxford University Press, 1999) 65–104, 101.

[34] Bots (also called spiders) are used legitimately to create search engines, mailing lists, indexes etc., and less legitimately to trawl for undefended computers that might provide access to systems or recruitment possibilities as a zombie or slave.

in a matter of minutes.[35] Once found, there are no lines to retreat behind and no way to move the computer out of range other than to disconnect it, a solution which is likely to disrupt the normal running and/or usefulness of the system. Any computer remaining connected to the internet in any way would be solely reliant on its electronic defences to prevent intrusion. In addition, it is not only potential enemy forces that will attempt to access military computers. the US Department of Defense is an attractive target for regular hackers, and the number of attacks on its systems has grown steadily. In 1992 US Department of Defense and military computers came under attack from intruders an estimated 53 times.[36] By 1997 the annual number of attacks had risen to 780; that number had risen again to almost 40,000 times in 2002, despite a brief dip in attacks following the 11 September 2001 terrorist attacks, when many US military networks were disconnected from the internet.[37] US governmental and military systems remain high cachet targets for independent hackers, not to mention those hackers deliberately attempting to access classified military information.

Additionally, most attacks do not proceed directly from the originating computer to the target. Attacks are likely to be routed through several intermediary servers (each with its own IP address) in various locales before the attack reaches the target computer. Tracing an attack back to its origin takes time and at the present state of technology it is not always possible to ensure that the apparent source of the attack is in fact the end of the trail. While the legitimacy of this tactic is perhaps more pertinent to discussions of perfidy and camouflage, it illustrates a problem with the solution proposed.

On the other hand, it may be argued that in the high-tech battlespace there is no practical need for such distinguishers. During a computer network attack against military assets, the originator is either a lawful

[35] In fact lists of military IP addresses have been circulating the internet for several years; however, the IP ranges specified are for fixed installations which may contain multiple dynamic IP addresses within the range. See, for example, 'U.S. Gov IP Addresses You Should Not Scan', *Hellbound Hackers* 21 June 2007, www.hellboundhackers.org/articles/721-US-GOV-IP-ADDRESSES-YOU-SHOULD-NOT-SCAN.html (last accessed 5 July 2011).

[36] Schmitt, 'Normative Framework', 885 fn. 25.

[37] James F. Dunnigan, *The Next War Zone: Confronting the Global Threat of Cyber Terrorism* (Citadel Press Books, New York, 2003), 85. Note that Dunnigan's figure for attacks in 1999 was 22,144. Latest figures from the Pentagon show the number of attempted intrusions from all sources in 2005 totalled 79,000.

combatant or a civilian directly participating in hostilities; in either case, he or she may be legitimately targeted. While this holds true for targeting judgments made in the heat of battle, a more sophisticated determination of an individual's status is required in the event that the originator is captured to ensure protection of the rights of prisoners of war. Obviously this situation is far more likely to apply to the combatant who is not sitting a continent away but is physically present in the battlespace.

What, then, of the requirement to distinguish the individual combatant from the civilian population? Although the technological revolution in military affairs means that warfare is moving away from a situation where there is a clear set of enemy lines, it is not always the case. A common sense approach to the problem should suffice. Where a combatant engages in a computer network attack in circumstances where they are in physical proximity to opposing forces such that there is a risk that they may be mistaken for a civilian, the requirement to wear a uniform or other distinctive mark would remain. Where there is no danger of deception or of the combatant being mistaken for a civilian, the need for an individual to wear a distinguishing emblem is irrelevant.[38]

The problems involved in applying the requirement to wear distinguishing marks and carry arms openly have arisen in other cases of non-traditional conflict, namely the situation of guerrilla fighters in occupied territory. The requirement that combatants wear a fixed distinctive emblem visible from a distance has been relaxed somewhat as a result of Article 44(3) of Additional Protocol I, which recognises that there are some situations in which the nature of hostilities makes it impossible (or suicidal) for a combatant to distinguish him or herself at all times.[39] In those cases the requirement is restricted to the

[38] Mark R. Shulman, 'Discrimination in the Laws of Information Warfare' (1999) 37 *Col. J. Trans. L.* 939, 956.

[39] The article provides: '[I]n order to promote the protection of the civilian population from the effects of hostilities, combatants are obliged to distinguish themselves from the civilian population while they are engaged in an attack or in a military operation preparatory to an attack. Recognising, however, that there are situations in armed conflicts where, owing to the nature of the hostilities an armed combatant cannot so distinguish himself, he shall retain his status as a combatant, provided that, in such situations, he carries his arms openly: (a) during each engagement, and (b) during such time as he is visible to the adversary while he is engaged in a military deployment preceding the launching of an attack in which he is to participate.'

engagement and such times as the individual is visible to the adversary in the preceding military deployment. The controversial provision is aimed primarily at guerrilla fighters, whose use of covert tactics are designed to address inequality between the military and logistical means of the parties.[40] However, an argument may be made that computer network attacks are an example of a type of warfare, the nature of which is anticipated by this provision. Computer network attacks are by their very nature a covert method of warfare and many authors have cited its possible use as a force multiplier for militarily weaker opponents.[41] If this is the case, it raises the possibility that preparatory moves for computer network attacks may be attempted from non-military computers (for example electronic probing and reconnaissance, sending a virus with a backdoor payload to enable access to vulnerable systems or recruiting zombie computers to a botnet for a distributed denial of service attack),[42] but that once the attack proper starts, it would need to emanate from a designated 'combatant' computer or system.[43]

1.1.3. Compliance, organisation, belonging and allegiance

The remaining requirements for lawful combatancy do not change markedly with the advent of computer network technology. The obligation to conduct hostilities in accordance with the laws of armed conflict will be the same, regardless of the technology employed by the combatant.

[40] Pilloud, *et al.*, *Commentary*, 527. Some states have argued that this provision is mainly restricted to resistance movements in occupied territories, and indeed some states (e.g. the United Kingdom) have stated in their reservations to the Convention that their acceptance of this clause is limited to such territories and wars of self-determination.

[41] See, for example, Schmitt, 'Normative Framework', 897; Michael J. Robbat, 'Resolving the Legal Issues Concerning the Use of Information Warfare in the International Forum: The Reach of the Existing Legal Framework, and the Creation of a New Paradigm' (2000) 6 *BU J. Sci. & Tech. L.* 10.

[42] Note that New Zealand has specifically included a declaration interpreting the term 'visible' to include visible any form of surveillance, electronic or otherwise. This would appear to be broad enough to encompass sweeps of all activities against military IP addresses, a situation which may require all preparatory manoeuvres against NZ to emanate from a designated computer.

[43] This may raise further issues about whether a distributed denial of service (DDoS) attack could ever be legal as it may amount to hiding in the demography, one military computer amongst thousands of civilian zombies. However, such an attack would tend to be considered a nuisance attack rather than one of the main threats that could cause damage.

The level and type of organisation required to satisfy the fifth requirement is affected by the changing structures of parties to conflicts generally, but the nature of their weaponry does not raise any particular issues.[44] Certain computer network attack techniques, for example distributed denial of service attacks, allow for a more dispersed structure of the armed group, allowing group members who are geographically dispersed to play a more active role in coordinated actions. However, this is a factual issue rather than a legal one. If the group does not have the requisite organisation (whether in network or hierarchical form) and does not maintain discipline and supervision, its members cannot be lawful combatants. The Trial Chamber of the ICTR in *Akeyesu* stated:[45]

> The armed forces opposing the government must be under responsible command, which entails a degree of organization within the armed group or dissident armed forces. This degree of organization should be such so as to enable the armed group or dissident forces to plan and carry out concerted military operations, and to impose discipline in the name of a *de facto* authority.

Likewise, the sixth condition, namely that a combatant must 'belong to a party to the conflict', will deny protection to vigilante groups of hackers from 'joining in' the confrontation in the same way that protection is denied to independent guerrilla groups fighting for a cause without a relationship to a belligerent party.[46] This may raise difficult factual issues with proving a connection between specific groups launching computer network attacks and a state; however, the legal test will remain the same. For example, the 2008 conflict between Russia and Georgia was accompanied by extensive computer network attacks. While there appeared to be a significant level of coordination between the computer network attacks and the actions taken by the Russian troops, the Kremlin has disavowed all connection with the online attacks.[47] While it is clear that any organised armed group for whose actions a state is responsible under the law of state responsibility will 'belong' to that state, it is far from settled what level of control

[44] See generally, John Arquilla, David F. Ronfeldt and United States Dept. of Defense, Office of the Secretary of Defense, *Networks and Netwars: The Future of Terror, Crime, and Militancy* (RAND, Santa Monica, 2001), for a description of the evolving nature of parties from military style hierarchies to networks.

[45] *Akayesu*, para. 626.

[46] *Public Prosecutor* v. *Koi et al.*

[47] See generally US Cyber Consequences Unit, *Overview*; Siobhan Gorman, 'Hackers Stole IDs for Attacks', *Wall Street Journal*, 17 August 2009, A5.

over state actions will be required.[48] The Appeals Chamber of the ICTY in *Tadic* has held that:[49]

> states have in practice accepted that belligerents may use paramilitary units and other irregulars in the conduct of hostilities only on the condition that those belligerents are prepared to take responsibility for any infringements committed by such forces. In order for irregulars to qualify as lawful combatants, it appears that international rules and State practice therefore require control over them by a Party to an international armed conflict and, by the same token, a relationship of dependence and allegiance of these irregulars vis-à-vis that Party to the conflict. These then may be regarded as the ingredients of the term 'belonging to a Party to the conflict'.

In a controversial section of its *Interpretive Guidance* on direct participation in hostilities the ICRC also adds:[50]

> The concept of 'belonging to' requires at least a de facto relationship between an organized armed group and a party to the conflict. This relationship may be officially declared, but may also be expressed through tacit agreement or conclusive behaviour that makes clear for which party the group is fighting.

While computer network attacks lend themselves to the use of 'cyber militia' and offer the attractiveness of 'plausible deniability' for a state, unless a relationship can be established between the group and the state the participants will not be considered lawful combatants.

The seventh requirement that the person does not owe a duty of allegiance to the capturing power will apply equally in the case of electronic attackers as it does to traditional combatants.

1.2. Saboteurs and spies

The laws of armed conflict do not prohibit sabotage or espionage;[51] however, engaging in either act will affect the status of the perpetrator in the event that they are captured. The clandestine nature of both acts means that a combatant will lose his or her combatant immunity and become an unlawful combatant; as such, spies and saboteurs are not entitled to treatment as a prisoner of war if captured. Civilians are not entitled to prisoner of war status in any event, unless they

48 *Cf.* the tests set down by the ICTY in *Tadic (Interlocutory Appeal)* of 'overall control' and the more stringent 'effective control' test set by the ICJ in *Nicaragua (Merits)*.

49 *Tadic*, para. 94.

50 ICRC, *Interpretive Guidance*, 23. For a discussion of the issues this raises see Schmitt, 'Interpretive Guidance', 17.

51 So long as acts of sabotage are directed at a legitimate military target.

are accompanying the armed forces under the auspices of the third Geneva Convention.[52] Sabotage and espionage raise two specific issues for the use of computer network attacks in armed conflicts. First, both involve clandestine behaviour and acting under false pretences, the personnel that carry them out generally do not wear uniform (the element that separates lawful intelligence gathering from espionage). Second, both sabotage and espionage traditionally require the combatant to be operating behind enemy lines, that is, in territory controlled or occupied by an adverse party. However, advanced technology now means that sensitive information may be retrieved from, and damage may be caused in, the territory of the adverse party without ever setting foot there. Wartime electronic eavesdropping and aerial surveillance are well-accepted methods of gathering information provided that the operative remains outside the territory of the adverse state (or wears a uniform or distinctive emblem while engaged in such activities).[53] However, the proactive and clandestine nature of the intrusion involved in some types of computer network attacks, and the actor's ability to manipulate data and information inside the territory, while remaining physically outside, raise interesting issues for those seeking to utilise this form of warfare.

1.2.1. Sabotage

Acts of sabotage are not prohibited under the laws of armed conflict where they are carried out by uniformed soldiers and target legitimate military objectives. However, while sabotage is not in and of itself internationally culpable (unless committed by a civilian resident in occupied territory), where it is carried out by non-uniformed members of the armed forces it will result in the loss of combatant privilege and prisoner of war status on capture.

The cases of *Ex p Quirin*, in which eight German saboteurs were arrested and tried for unsuccessful sabotage missions in the United States,[54] and *Mohamed Ali*, which involved the sabotage of a building in

[52] Art. 4(A)(3); Ipsen, 'Combatants and Non-Combatants', 111.

[53] Some questions arise when the espionage takes place in the exclusive economic zone or territorial waters of the target country; see the 1968 case of the *USS Pueblo* in North Korean waters, or the more recent 2001 case of the US EP-3 surveillance plane which crashed after colliding with a Chinese F8 fighter jet. It should be noted that both were peacetime incidents.

[54] *Ex p Quirin et al.* (1942) 317 US 1, Supreme Court of the United States. In *Ex p Quirin*, eight Germans landed secretly on the shores of the United States in German uniform with explosives for the purposes of sabotage. While they landed in

Singapore,[55] make it clear that those entering into the territory of the opposing party with the intent to cause destruction to life or property, without uniform, were not entitled to prisoner of war status and were liable to trial and punishment. While the court in *Quirin* convicted the saboteurs for 'unlawful combatancy', it is not clear that states would consider saboteurs belonging to the armed forces liable for an offence, separate from the specific acts of sabotage today. In *Quirin* the Supreme Court of the United States held:[56]

> The spy who secretly and without uniform passes the military lines of a belligerent in time of war, seeking to gather military information and communicate it to the enemy, or an enemy combatant who without uniform comes secretly through the lines for the purpose of waging war by destruction of life or property, are familiar examples of belligerents who are generally deemed not to be entitled to the status of prisoners of war, but to be offenders against the law of war subject to trial and punishment by military tribunals.

The Court went on to hold:[57]

> By passing our boundaries for such purposes without uniform or other emblem signifying their belligerent status, or by discarding that means of identification after entry, such enemies become unlawful belligerents subject to trial and punishment.

Similarly, in the *Ali* case, the Privy Council confirmed the lower court's decision that:[58]

> ... members of enemy armed forces who are combatants and who come here with the assumption of the semblance of peaceful pursuits divesting themselves of the character or appearance of soldiers and are captured, such persons are not entitled to the privileges of prisoners of war.

Thus in both *Quirin* and *Ali*, the act of sabotage itself and the basis of the defendants' classification as unlawful combatants was perpetrated by entering onto the territory of the victim state in civilian dress and

German uniforms, on landing they changed into civilian clothing and proceeded to travel to their destinations, where they were captured.

55 *Mohamed Ali*: During a time of confrontation between Singapore and Malaysia (then part of Indonesia), Ali and another member of the Indonesian army entered a bank in Singapore in civilian clothing and deposited a bag containing nitroglycerine in the stairwell. The bag exploded, killing three civilians. Three days later the perpetrators were rescued from the sea and arrested, still wearing civilian clothes and carrying no identity documents.

56 *Ex p Quirin*, 31. 57 *Ibid.*, 37.

58 *Mohamed Ali*, 445 (citing Chua J).

committing (or attempting to commit) acts of destruction. Computer network attacks therefore may amount to acts of sabotage where a state engages in acts which disrupt, disable or damage resources in a clandestine manner, in the territory controlled by the adversary (in other than occupied territory). This is precisely the type of attack for which computer network attacks are likely to be used. The difficulty that technology now brings is that such acts of sabotage are now capable of being committed in the territory of the belligerent party without the perpetrator physically setting foot in that territory. It is now possible to commit acts of sabotage by entering into an adversary's computer systems in a clandestine manner (i.e. through a backdoor), or to introduce malicious code into a system and cause significant damage to state interests. Given that the physical act of crossing a border or passing into enemy occupied territory is no longer necessary to cause covert damage, does that mean that any computer network attack committed by members of the armed forces will amount to sabotage, such that they would lose their combatant privilege status? Like most issues involving computer network attack, the legal status will depend on the type of attack being envisaged.

The simplest examples of sabotage by computer network attack are those utilised on a daily basis by civilian virus writers around the world. An attacker sends an email to the recipient which incorporates a virus or other malicious code; the code activates upon opening the email or email attachment and damages information resident on the recipient's computer networks.[59] As long as the email does not purport to be from a person or organisation with protected status or claim to offer terms of surrender or some other perfidious simulation, this would amount to a lawful ruse of war (see Chapter 8 below) and the combatant would retain their status in the event of capture. This is the electronic equivalent of sending dangerous items though the mail, for example a letter or parcel bomb. The computer system is merely being used as a delivery device.

Direct intrusion into a system or network, however, may be more akin to slipping across borders to directly cause damage; thus a question may be raised as to whether the actor being physically present in

[59] Damage estimates from computer network attacks such as viruses are notoriously difficult to quantify as there are no agreed standard measures. However, it is estimated that viruses cost businesses billions of dollars every year.

the territory is a fundamental element of sabotage. Yoram Dinstein has argued in relation to espionage that the combatant must be physically located in an area controlled by the enemy for the offence to crystallise.[60] An alternative reading would require only that the effects of the act take place in the territory, in much the same way that traditional 'shot across the border' cases and more recent domestic cases of computer intrusion are prosecuted in the state where the damage occurs.[61] The current author argues that it is the act of deception for the purposes of destruction which negates combatant status. As Baxter points out:[62]

> International law deliberately neglects to protect unprivileged belligerents because of the danger their acts present to their opponent. The peril to the enemy inherent in attempts to obtain secret information or to sabotage his facilities and in attacks by persons whom he often cannot distinguish from the peaceful population is sufficient to require the recognition of wide retaliatory powers.

This would also fit with the reasoning set out in the *Hostages Trial* where 'guerrillas were actually said, in legal intendment, to resemble spies in that the enemy punished such activities not because of their illegality in an international sense but because of the danger they presented to him'.[63] The fact that acts of covert damage can now be performed from outside the territory controlled by the enemy does not eliminate the danger and, in fact, makes it more difficult to detect. Under this analysis, a covert intrusion into a system resident in the territory of a victim state, with the intention of causing damage, while disguised as something other than a combatant, is likely to be considered sabotage with the resultant loss of combatant status for any operative thus caught. An intrusion attempt directly from a military computer would remain legitimate, as there is no deception involved.

60 See section 2.2 below; Dinstein, *Conduct of Hostilities* (1st edn), 209.

61 See generally Antonio Cassese, *International Criminal Law* (Oxford University Press, 2003), 278, citing *Rivard* v. *United States* (1967) 375 F. 2d 882, US Ct. App., 5th Cir.: 'All the nations of the world recognize the principle that a man who outside of a country wilfully puts in motion a force to take effect in it is answerable at the place where evil is done.' This is the fundamental territorial principle set out in the *Lotus Case: The Lotus* (1927) Series A No. 10, Permanent Court of International Justice.

62 Baxter, 'Unprivileged Belligerency', 343.

63 *Ibid.*, 336, citing *United States* v. *List et al.* (1949) Trials of War Criminals, XI (1950), 1245; War Crimes Reports, VIII (1949) 56.

1.2.2. Espionage

In the case of electronic espionage, multiple peacetime instances of which have been made publicly available,[64] it is doubtful that this will raise any new difficulties.

Strictly speaking, most acts of cyber espionage would fall under the heading of computer network exploitation rather than computer network attack, as they are aimed at the collection of information rather than its degradation or destruction.[65] However, as these acts have an effect on both the conduct of computer network attacks themselves and the status of the personnel involved it will be discussed here. Gathering intelligence about the enemy is an accepted and necessary part of any armed conflict, particularly for the fulfilment of a party's obligation to take precautions in attack. Thus the Hague Regulations expressly note that 'the employment of measures necessary for obtaining information about the enemy are considered permissible'.[66] However Article 29 of the Regulations provides the definition of espionage:

> A person can only be considered a spy when, acting clandestinely or on false pretences, he obtains or endeavours to obtain information in the zone of operations of a belligerent, with the intention of communicating it to the hostile party.
>
> Thus, soldiers not wearing a disguise who have penetrated into the zone of operations of the hostile army, for the purpose of obtaining information, are not considered spies.

[64] One example, code-named 'Titan Rain', consists of a series of coordinated attacks launched against US computer systems since 2003. The attacks are highly sophisticated intrusions against unclassified networks, in which the attackers have gained access, copied as many files as possible from the computer, transmitted them via way-stations to China and then made a near clean exit. Systems compromised include NASA, the World Bank, military sites such as Redstone Arsenal military base, and defence contractors such as Lockheed Martin. Unusually for cyber intrusions, the origin of the attacks was quickly and clearly traced to three routers operating in Guangdong province China; however, it is unclear whether the attacks are emanating from military, corporate or individual operators. The Chinese government denies any involvement. Bradley Graham, 'Hackers Attack Via Chinese Web Sites', *Washington Post* (Washington DC), 25 August 2005, A1; Thornburgh, *et al.*, 'The Invasion of the Chinese Cyberspies (and the Man Who Tried to Stop Them)'.

[65] The US Department of Defense defines computer network exploitation as 'Enabling operations and intelligence collection capabilities conducted through the use of computer networks to gather data from target or adversary automated information systems or networks.'

[66] Art. 24, Hague Regulations.

A spy may be either a combatant or a civilian. Although espionage in time of war is not a violation of the laws of armed conflict (as evidenced by Article 24 of the Hague Regulations above),[67] or indeed of any other international law, it is generally prohibited and will be prosecuted under the domestic criminal laws of the state. Being captured as a spy results in the loss of prisoner of war status and, as Dinstein points out, it follows from the clandestine nature of their activities that spies are unlawful combatants. Additional Protocol I confirms that any member of the armed forces taken by the adverse party while engaged in espionage is not entitled to prisoner of war status.[68]

As noted above, two factors raise interesting issues for electronic espionage during armed conflict. The first, that espionage must be carried out 'clandestinely and under false pretences', relates to the wearing (or lack thereof) of uniforms or other distinguishing markers. The impact of this for computer network attacks has been examined in the preceding sections on distinction and sabotage. In the same manner that sabotage is distinguished from ordinary military operations in the section above, it is the clandestine character of the activity and the spy's intention to deceive that distinguishes espionage from the reconnaissance, scouting or surveillance performed by military forces and by individual members of the armed forces.[69] It is this fact that prevents the operators of unmanned aerial vehicles (UAVs), which have been operating in war zones since the 1950s, from being classed as spies. An intrusion attempt launched directly from a military computer would remain legitimate, as there is no deception involved.

The second issue relates to where the act takes place. Under the Hague Regulations, a key factor in the offence of espionage is the attempt to gain information in the 'zone of operations of a belligerent'; Additional Protocol I extends that area to all territory controlled by the enemy.[70]

[67] See also the *Flesche* case (Holland, Special Court of Cassation, 1949), [1949] ADIL 266, 272: 'espionage ... is a recognised means of warfare and therefore is neither an international delinquency on the part of the state employing the spy nor a war crime proper on the part of the individual concerned', cited in Dinstein, *Conduct of Hostilities* (2nd edn), 243.

[68] Art. 46, Additional Protocol I.

[69] Erik Castrén, *The Present Law of War and Neutrality* (Suomalaisen Tiedeakatemian Toimituksia, Helsinki, 1954), 152, cited in Dinstein, *Conduct of Hostilities* (1st edn), 111.

[70] The Commentary to Additional Protocol I states that the Additional Protocol did not intend to change the substance of the traditional rules of espionage adopted in The Hague, but merely sought to supplement and elaborate them. This conclusion was confirmed by the wording of Art. 39(3) (Emblems of nationality), which refers to the 'existing generally recognized rules of international law applicable to espionage'.

Yoram Dinstein has stated that this means that a spy must be physically located in an area controlled by the enemy: 'A person stationed on his own state's side of the front line – say, clandestinely monitoring or deciphering enemy radio signals – is not a spy.'[71] While this statement is undoubtedly correct, it does not follow that the combatant must necessarily be physically located in enemy-controlled territory. A distinction can be made between the passive collection of radio signals from outside a specified zone of operations or control, and the active intrusion into a system either controlled by an adverse party or resident on enemy territory for the purposes of information collection. Passive collection of information (even when assisted by the strategic placement of a listening post) does not require an act which would allow the opposing state to claim any grievance or assert jurisdiction over that act. A similar computer-based equivalent would be use of a listening post system such as Echelon to intercept emails, telephone calls and other data traffic off satellite communications, or taking advantage of an unsecured wireless network to intercept traffic travelling across that network.[72] However, the present author argues that it is the *act* of collection of the information which must occur in the territory controlled by the enemy. Active penetration into a network controlled by the enemy in order to collect information (by breaching firewalls and other electronic defences) would be enough to situate the collection of information from that system in the territory of the adverse party. As noted above, this approach to jurisdiction has been adopted in many cross-border computer crime cases at the domestic level as well as the traditional 'shot across the border' cases.[73] In this regard it must be recalled that espionage is prosecuted under the domestic laws of a state.

There is one additional factor which must be taken into account with computer-based forms of espionage. Spies are excused for any liability for their actions under national law once they rejoin the army to which they belong.[74] However, the rule is explicitly limited to combatants; no

71 Dinstein, *Conduct of Hostilities* (1st edn), 209.

72 Echelon is the signals intelligence collection system operated between the US, UK, Australia, Canada and New Zealand.

73 See generally Cassese, *International Criminal Law*, 278, citing *Rivard* v. *United States*: 'All the nations of the world recognize the principle that a man who outside of a country wilfully puts in motion a force to take effect in it is answerable at the place where evil is done.'

74 Art. 31 Hague Regulations; Art. 46(4) Additional Protocol I.

similar dispensation exists for civilian spies.[75] Thus where a member of the armed forces who is engaged in computer network espionage never leaves his or her lines, it seems unlikely that they will ever be at risk of capture before regaining their combatant privileges. However, a combatant stationed inside the territory of the opposing forces, or in any other territory, would be exposed to the same risks as any other spy. Civilian spies remain liable for their actions, thus any contractors involved in computer network espionage may be captured and punished as spies at any time during the conflict and will remain liable for their actions long after the conflict is over.

2. Direct participation by civilians

Civilians are entitled to protection from the dangers arising from military operations and may not be targeted until, and for such time as, they take an active or direct part in hostilities.[76] Where civilians do take a direct part in the hostilities, they lose their protected status for the period of their involvement and may be liable for punishment either through domestic or international criminal processes for their actions. However, the question of what actions will amount to direct participation in hostilities is one which continues to raise difficult issues for the laws of armed conflict. Civilians have played a vital supporting role in warfare throughout history, and modern warfare is no exception. Civilians are widely employed by the armed forces, both as contractors and as full-time employees, or they may accompany the armed forces for a variety of other reasons. The use of contractors, in particular, has increased exponentially in recent years as the combined effects of the technological revolution and 'privatisation through outsourcing' have been used to ensure continuing military might, while reducing costs.[77]

[75] Dinstein, *Conduct of Hostilities* (2nd edn), 243. This was clearly stated by the Special Court of Cassation in the case of Alfred Flesche, a German national convicted of spying in the Netherlands prior to the Nazi invasion, holding: 'Though Article 29 covers civilians also, Article 31 applies only to those in military service, as clearly appears from its text and as is expressly stated by several writers,' *Re Flesche* (1949) 16 *International Law Reports* 266, Holland, Special Court of Cassation, 272.

[76] Common Art. 3 Geneva Conventions; Art. 51(3) Additional Protocol I. Common Article 3 of the Geneva Conventions employs the term 'active' rather than 'direct' as used in the Additional Protocols. The distinction between active and direct participation was discussed by the International Criminal Tribunal for Rwanda in the *Akayesu* case which held that the terms are so similar that they should be treated as synonymous: *Akayesu*, para. 629.

[77] Guillory, 'Civilianising the Force: Is the United States Crossing the Rubicon?', 111.

For example, in 2006 over 38,000 contractors were serving with coalition forces in Iraq in support functions from cleaners and cooks, with an additional 30,000 providing security for both the military and other contractors and guarding convoys and military installations.[78]

The use of civilians has increased particularly with regard to technologically advanced methods of warfare. Civilians maintain complex weapons systems such as the F-117 Nighthawk fighter, B-2 Spirit bomber, M1 Abrams tank, and TOW missile system, and have both maintained and operated the Global Hawk and Predator UAVs.[79] This high level of outsourcing takes place for a number of reasons. First, it is far more cost-effective to hire civilian contractors to maintain and operate the systems which run the military than to train military personnel to do so. Despite their (generally) higher salaries, contractors do not require the training and infrastructure costs of military personnel. Second, the systems being used are seldom standard inventory; for the most part they are too specialised and often still in the throes of research and development.[80] It is clear that military personnel are being trained in computer network attack capabilities;[81] however, it is also apparent that civilian contractors are used by the military to run their information systems. As these information systems become both targets and weapons, such contractors find themselves involved in hostilities on an unprecedented scale. It is therefore essential to carefully delineate which tasks are permissible non-combatant support and which will constitute unlawfully participating in hostilities.

Sean Watts notes an additional consideration for states using civilian contractors: in addition to potential individual criminal consequences, civilian participation in hostilities raises concerns of state

[78] As at 1 May 2006, the number of civilian logistics personnel was reportedly 38,305, Brookings Institution, *Iraq Index*, Brookings Institute (2006) 15, www.brookings.edu/iraqindex (last accessed 5 July 2011). It is unclear whether this data relates only to the US Central Command Area of Responsibility. Figures on the numbers of security contractors are harder to come by, but a recent figure in a Frontline report put this figure at an additional 35,000 both Iraqi and non-Iraqi contractors. www.pbs.org/wgbh/pages/frontline/shows/warriors/faqs/ (last accessed 5 July 2011).

[79] Michael N. Schmitt, 'Humanitarian Law and Direct Participation by Private Contractors or Civilian Employees' (2004) 5(2) *Chi. J. Int'l L.* 511, 512.

[80] Michael N. Schmitt, 'Direct Participation in Hostilities and 21st Century Armed Conflict' in H. Fischer, *et al.* (eds.), *Crisis Management and Humanitarian Protection* (Berliner WissenschaftsVerlag, Berlin, 2004) 505–29, 523.

[81] See, for example, C. Todd Lopez, 'Military Students Get Lesson in Cyberwarfare', *Air Force Print News*, 3 May 2006, http://searchsecurity.techtarget.com/originalContent/0,289142,sid14_gci1186049,00.html (last accessed 15 July 2011).

responsibility.[82] He argues that the existing treaty-based definitions of those classed as combatants could be interpreted to restrain not only individual conduct, but also the state's composition of their fighting forces through a self-imposed limit. States that employ civilians to perform tasks amounting to direct participation in hostilities would be in breach of those limits.[83]

2.1. *Requirements of direct participation*

Despite much academic commentary and a six-year research project by the ICRC, there remains significant debate over the definition and requirements of 'active' or 'direct' participation in hostilities.[84] The commentary to Article 43 of Additional Protocol I indicates that the phrase includes 'acts which are intended by their nature or their purpose to hit specifically the personnel and the *matériel* of the armed forces of the adverse party'.[85] The commentary goes on to state that 'direct participation in hostilities implies a direct causal relationship between the activity engaged in and the harm done to the enemy at the time and place where the activity takes place'.[86] A similar assessment is seen in the commentary to Additional Protocol II relating to Article 13(3) of the Protocol; it requires a 'significant causal relationship' between act and immediate consequence.[87] As a result, any determination of status will need to be done on a case-by-case basis, an approach confirmed by the Trial Chamber of the International Criminal Tribunal for the Former Yugoslavia in *Tadic*:[88]

> It is unnecessary to define exactly the line dividing those taking an active part in hostilities and those who are not so involved. It is sufficient to examine the relevant facts of each victim and to ascertain whether, in each individual's circumstances, that person was actively involved in hostilities at the relevant time.

The case-by-case approach was likewise adopted by the Israeli Supreme Court in its 2006 *Targeted Killings* decision.[89]

82 Watts, 'Combatant Status and CNA', 423. 83 *Ibid.*

84 See generally ICRC, *Interpretive Guidance*; Schmitt, 'Interpretive Guidance', as well as the forum discussion in the 2010 volume of *New York University Journal of International Law and Politics* 637–916.

85 Pilloud, *et al.*, *Commentary*, para. 1679.

86 *Ibid.* 87 *Ibid.*, para. 4787.

88 *Prosecutor* v. *Dusko Tadic* (1997) 36 ILM 908, International Criminal Tribunal for the Former Yugoslavia, para. 616.

89 *The Public Committee against Torture in Israel et al.* v. *The Government of Israel et al.* (2007) 46 ILM 375, Supreme Court of Israel, paras 34–7.

2.1.1. Earlier scholarship

Hays Parks has defined civilian participation in warfare as a spectrum between total non-participation at one end (what has sometimes been termed the 'pure' civilian) through war effort, military effort and finally to direct participation in military operations at the other end.[90] Such categories were based on the 1971 ICRC Conference of Government Experts which resulted in the Additional Protocols. It is widely recognised that acts that help the general war effort do not constitute direct participation;[91] however, where the line is to be drawn is a subject of intense debate between commentators which has still not been settled. A. P. V. Rogers has argued for a narrow construction of a 'direct part in hostilities', stating that actions such as arms production, military engineering work or military transport (including the oft-cited example of the ammunition truck driver) would not be deemed direct participation.[92] He argues that to hold otherwise places civilian protection severely at risk. However, the same argument has been used by both Hays Parks and Michael Schmitt to advocate a wider construction. Parks has stated that civilians working toward the military effort are far more combatant than civilian, and has argued for an expansive approach, which would include intelligence gathering and logistical support for combatant forces as direct participation.[93] Using the traditional example of the status of the civilian driver of an ammunition supply truck, Parks thus finds that the truck driver should be a lawful target of attack.[94] Rogers has dismissed this view as creating a class of quasi-combatants based on job description, a situation that has been expressly rejected by the commentary to the Protocol.[95]

In an attempt to find a more general rule, Michael Schmitt has argued that the best approach is to assess the criticality of the act to the direct application of violence against the enemy.[96] Although direct cause and effect is unnecessary, there must be sufficient causal

[90] See generally, W. Hays Parks' categories of civilian participation as considered by the 1971 ICRC Conference of Government Experts: W. Hays Parks, 'Air War and the Law of War' (1990) 32 *AFL Rev.* 1, 132.

[91] Pilloud, *et al.*, *Commentary*, para. 1679.

[92] Rogers, *Law on the Battlefield*, 9.

[93] See Parks, 'Air War and the Law of War', 132.

[94] *Ibid.*, 132.

[95] Rogers, *Law on the Battlefield*, 9; Pilloud, *et al.*, *Commentary*, para. 1679.

[96] Schmitt, 'Direct Participation', 505; Schmitt, 'Humanitarian Law and Direct Participation', 534.

proximity to a foreseeable consequence of harm or other disadvantage to the enemy:[97]

> [T]he civilian must have engaged in an action that he or she knew would harm (or otherwise disadvantage) the enemy in a relatively direct and immediate way. The participation must have been part of the process by which a particular use of force was rendered possible, either through preparation or execution. It is not necessary that the individual foresaw the eventual result of the operation, but only that he or she knew their participation was indispensable to a discrete hostile act or series of related acts.

In a passage later quoted with approval by the Israeli Supreme Court in the 2006 *Targeted Killings* case,[98] Schmitt further argues that any grey areas should be interpreted in favour of finding direct participation to protect the law and provide an incentive for civilians to remain as distant from conflict as possible. Thus under Schmitt's analysis, the civilian driver of an ammunition supply truck is not taking a direct part in hostilities when driving from the factory to the ammunition depot, but would be when driving from the depot to the front.[99] Both Parks and Schmitt maintain that to grant immunity to civilians who are intimately involved in the conflict is to risk engendering disrespect for the law by combatants who are put at risk by their actions.[100] In other words, the actions of the relatively small number of civilians supporting the military should not endanger the lives of civilians who have no part in the conflict.

Some scholars have argued that direct participation includes not only activities involving the delivery of violence, but also acts aimed at protecting personnel, infrastructure or material.[101] These problems have arisen in Iraq in relation to private security contractors engaged to defend military installations.[102] US military doctrine (and

[97] Schmitt, 'Humanitarian Law and Direct Participation', 533.

[98] *Targeted Killings*, para. 34.

[99] Schmitt, 'Direct Participation', 504.

[100] *Ibid.*, 505; Parks, 'Air War and the Law of War', 132.

[101] US Air Force, Pam. 110–34, Judge Advocate General: Commander's Handbook on the Law of Armed Conflict (1980, §2–8), commenting that the rescue of downed airmen would constitute taking direct part in hostilities, cited in Jean-François Quéguiner, *Direct Participation in Hostilities under International Humanitarian Law*, Program on Humanitarian Policy and Conflict Research at Harvard University (2003), http://ihl.ihlresearch.org/_data/n_0002/resources/live/briefing3297.pdf (last accessed 5 July 2011). See also, UK Ministry of Defence, *UK Manual*, §12.69.

[102] See PBS Frontline, *Interview with Stephen Schooner – Private Warriors* www.pbs.org/wgbh/pages/frontline/shows/warriors/interviews/schooner.html (last accessed

international law) states that contractors are allowed to carry arms for the purposes of defence only;[103] however, Peter Singer has noted that, despite this, contractors are in fact being used for roles which clash with that doctrine, and that the very function for which they are being hired mandates their use in combat roles.[104] For example, if states are hiring a private military company to guard a key installation in a combat zone, or to escort a convoy through insurgent territory which is renowned for attacks on convoys, it is then disingenuous to argue that the contractors are armed solely for self-defence.[105] Michael Schmitt has stated that a civilian government employee or private contractor defending military personnel or military objectives from enemy attacks directly participates in hostilities.[106] These arguments accord with Article 49 of Additional Protocol I, which states that the term 'attacks' incorporates acts of violence against the adversary, whether in offence or defence. This definition would appear to extend to those technicians who engage in active defence of military computer networks such that harm is caused to the attacking adversary or their equipment.

One further point deserves brief comment. Traditionally, geographic proximity to the battle lines has also been used as a rough guide to ascertaining the status of the civilian concerned; however, this measure is no longer decisive in twenty-first century combat. Not only have traditional battle lines been replaced by amorphous battlespaces, but physical proximity to that space is no longer required. Even before computer networks are considered, missiles and other weapons may now be loaded onto aircraft or otherwise launched from continents away and strike any point on the globe. Likewise, a civilian operator of an unmanned aerial vehicle who launches a missile at a target in a conflict zone is directly participating in hostilities.

5 July 2011). Also, for example, the use of the private military firm Blackwater to defend the Coalition Government Headquaters in Najaf, Iraq which subsequently came under attack. Dana Priest, 'Private Guards Repel Attack on US Headquarters', *Washington Post*, 6 April 2004, A01. See section 2.4 below.

103 Letter from Donald H. Rumsfeld, Secretary of Defense, to The Honorable Ike Skelton (4 May 2004), cited in Schmitt, 'Humanitarian Law and Direct Participation', fn. 11.

104 PBS Frontline, *Interview with Peter Singer – Private Warriors* www.pbs.org/wgbh/pages/frontline/shows/warriors/interviews/singer.html (last accessed 5 July 2011).

105 *Ibid.*

106 Schmitt, 'Humanitarian Law and Direct Participation', 538.

2.1.2. The ICRC interpretive guidance

These earlier debates formed the background to the expert process in the ICRC project to provide guidance on the interpretation of direct participation in hostilities. In 2009, the ICRC published its long-awaited report of the six-year project; however, despite extensive consultation with experts in a series of meetings during the course of project, the final document represents the views of the ICRC alone rather than the consensus document that was hoped for.[107] Needless to say, the interpretive guidance is not a binding legal document, and remains contentious in some of its views.[108] However, given the comprehensive nature of the project,[109] the breadth of expert views drawn on and the unique standing of the ICRC in matters of international humanitarian law, it is a timely and useful step in the interpretation of customary and treaty law on the notion of direct participation. The ICRC Interpretive Guidance defines direct participation in hostilities by reference to three cumulative criteria:[110]

1. the act must be likely to adversely affect the military operations or military capacity of a party to an armed conflict or, alternatively, to inflict death, injury, or destruction on persons or objects protected against direct attack (*threshold of harm*); and
2. there must be a direct causal link between the act and the harm likely to result either from that act, or from a coordinated military operation of which that act constitutes an integral part (*direct causation*); and

[107] The resulting Interpretative Guidance has proved somewhat contentious and several members of the expert meetings withdrew their names from the report lest their inclusion be taken as support for the propositions. In the end the ICRC opted to publish the guidance without listing participants.

[108] A full discussion of the Interpretive Guidance and its critics is beyond the scope of this work. For criticism and responses, see, for example, the forum discussion in the 2010 volume of *New York University Journal of International Law and Politics* 637–916. Summaries of the Guidance and proceedings of the expert meetings are available at ICRC, *ICRC Clarification Process on the Notion of Direct Participation in Hostilities under International Humanitarian Law (Proceedings)* www.icrc.org/eng/resources/documents/article/other/direct-participation-article-020709.htm (last accessed 26 February 2011).

[109] Note that the Interpretive Guidance limits its interpretation of the notion of direct participation in hostilities for the purposes of the conduct of hostilities only. It does not address the consequences of direct participation in hostilities for detention and trial.

[110] ICRC, *Interpretive Guidance*, 46.

3. the act must be specifically designed to directly cause the required threshold of harm in support of a party to the conflict and to the detriment of another (*belligerent nexus*).

The author considers that these criteria, as stated, provide a useful and workable standard. However, the second element, that there must be a direct causal link between the act and the harm caused, has been somewhat muddied by the description of the criterion given in the guidance. While the requirement that there be a direct causal link is uncontroversial, in attempting to delimit the boundary between direct and indirect participation the guidance specifies that the harm caused must be brought about 'in one causal step'.[111] This strict standard appears to be an innovation on the part of the ICRC authors of the study, as neither prior law nor the expert process contains any reference to a one-step test.[112] This is of particular significance in the case of computer network attacks, where the secondary effects of a particular act may in fact be the purpose of the attack. While this problem would be ameliorated somewhat by the second half of the criterion, namely the requirement that the act constitutes an integral part of a wider operation, that only applies in circumstances where the computer network attack forms part of a coordinated military operation. That said, the broader 'but for' standard suggested by Schmitt in his deconstruction of the requirement goes too far by 'extend[ing] participation as far up and downstream as there is a causal link', and arguably further than his earlier work discussed above positing a criticality test.[113] In the end, the lack of agreement on a more specific test leaves a general consensus that there must be a direct causal link, and a series of illustrative examples that have divided expert opinion. Thus the second criterion of direct causation will remain to be decided on a case-by-case basis.

[111] *Ibid.*, 53.

[112] Michael Schmitt suggests that, as the concept is not developed further by the guidance, it may in fact merely reflect poor drafting on the part of the author rather than an actual requirement: Schmitt, 'Interpretive Guidance', 30. However, the drafter of the report Nils Melzer has defended the criterion, indicating that despite the difficulties with this element it was at least intended: Nils Melzer, 'Keeping the Balance between Military Necessity and Humanity: A Response to Four Critiques of the ICRC's Interpretive Guidance on the Notion of Direct Participation in Hostilities' (2010) 42 *NYU J. Int'l L. & Pol.* 831.

[113] *Cf.* Michael N. Schmitt, 'Deconstructing Direct Participation in Hostilities: The Constitutive Elements' (2010) 42 *NYU J. Int'l L. & Pol.* 697, 739; Schmitt, 'Humanitarian Law and Direct Participation', 533.

2.2. *Offensive computer network attack*

Despite the need to examine the context of particular actions, it is clear that any civilian engaged in a proactive, offensive computer network attack against an adversary's networks or personnel would be taking a direct part in hostilities in much the same way that a civilian taking up conventional arms would be. This is the case regardless of whether the attack was designed to cause damage in and of itself, or whether it was designed to support a conventional attack such as, for example, disabling an enemy's air defence network prior to the launch of conventional air strikes. The fact that damage is caused by reason of computer manipulation rather than conventional arms is irrelevant. The ICRC Interpretive Guidance specifically envisages '[e]lectronic interference with military computer networks ... whether through computer network attacks (CNA) or computer network exploitation (CNE)' as an example of direct participation, and noted that during the expert meetings computer network attacks causing military harm to the adversary in a situation of armed conflict were clearly regarded as part of the hostilities.[114] The threshold of harm envisaged by the ICRC guidance is not limited to death, injury or destruction. Direct participation in hostilities may also be found where the act is likely to adversely affect the military operations or military capability of a party to the armed conflict. Thus the establishment and exercise of control by electronic means over military personnel and objects or over computer networks used by the adversary was considered direct participation in hostilities by a majority of experts involved in the process.[115] Likewise the assistance of civilian contractors in the design and execution of computer network attacks, whether as coders or systems specialists, would also constitute direct participation where it is linked to a particular attack. Such acts would form an integral part of the attack, regardless of it being carried out by military personnel. Likewise, participation in the inaptly named 'active defence' aspects of critical networks and systems would amount to direct participation, in so far as they essentially amount to counter-attacks.

2.3. *Computer network attack system support*

The issue becomes more complex when applied to civilians engaged in activities less immediately linked to hostilities. Computer and other

[114] ICRC, *Interpretive Guidance*, 48.

[115] ICRC, *Third Expert Meeting on the Notion of Direct Participation in Hostilities – Summary Report* (ICRC, Geneva, 2005), 13.

information technology support is a concept that seems ripe for civilian outsourcing; however there is a risk that support of systems and networks which are used to launch computer network attacks may be viewed as maintenance of a weapons system, placing the technicians who maintain these networks at risk of direct participation in hostilities. Clearly not every act which is performed by IT support staff will amount to direct participation in hostilities, and most would not meet the test of direct causation laid down by the ICRC Interpretive Guidance.

Knut Ipsen has argued that direct participation includes 'use of a weapons system in an indispensable function', although he gives no guidance as to which functions should be considered indispensable.[116] Two questions come out of this statement: first, whether a system used for launching computer network attacks is a weapons system; and second, whether maintenance and support of that network constitutes an indispensable function such that it would amount to direct participation in hostilities. The US Department of Defense dictionary of military terms defines a weapons system as '[a] combination of one or more weapons with all related equipment, materials, services, personnel, and means of delivery and deployment (if applicable) required for self-sufficiency'.[117] Depending on the type of computer network attack envisaged, the weapons involved may be malicious coding (in the case of Trojans, viruses and other kinds of backdoor attacks) or the network itself in the case of intrusion and sabotage. In either case it is quite clear from this definition that networks used to launch computer network attacks would fall within this definition, either as the weapon itself, or as the means of delivery and deployment.[118] Under Ipsen's analysis, maintenance of a weapons system would seem to be an act which has a direct causal relationship with the harm done to the enemy,[119] and yet the

[116] Ipsen, 'Combatants and Non-Combatants', 67.

[117] US Department of Defense, *Dictionary of Military and Associated Terms (as Amended through 20 March 2006)* (Washington DC, 2001)

[118] For a contrasting view see Gregory F. Intoccia and Joe Wesley Moore, 'Communications Technology, Warfare, and the Law: Is the Network a Weapons System?' (2006) 28 *Hous. J. Int'l L.* 467, 479. While arguing that the network should not be considered a weapons system, Intoccia and Moore base their opinion on an overly broad definition of the network which encompasses practically the entire range of communications equipment available both at national and international level.

[119] Note that this is not a universal view. Although disagreeing with the merits of it, Parks does not consider that the maintenance of the Swiss Air Force by civilian

maintenance of more and more military systems is being outsourced as part of the drive for cheaper, smaller, more streamlined armed forces. In 1988 the US Air Force reported to Congress that it now requires that maintenance and repair for all new critical weapons systems be under contractor support for at least four years and for life for non-critical systems.[120] This so-called 'factory to foxhole' support includes weapons systems such as the Patriot missile system, JSTARS,[121] data processing systems and the Fox nuclear chemical biological reconnaissance system, combat aircraft and the Abrams M1A1 tank.[122] Further, civilian contractors staff the entire information operations cell supporting the US Southern Command, which is responsible for defence operations in thirty-two countries in Central and South America and the Caribbean.[123] The trend to outsourcing is a global phenomenon, not restricted to the world's most powerful military.[124]

Despite this evidence of state practice, maintenance of these systems, particularly support and maintenance that takes place once a system has come under attack and are aimed against a direct intrusion, may well amount to direct participation in hostilities. How, then, to reconcile the increasing trend to outsource with the conclusion that some acts of support and maintenance are an indispensable function in the delivery of violence? Schmitt argues that immediate maintenance and support, that is, support not of a routine nature, may be seen as direct participation – parallels may be drawn with civilian aircraft engineers in charge of maintaining, loading and launching aircraft hundreds of

engineers would constitute direct participation under the terms of Additional Protocol I: Parks, 'Air War and the Law of War', fn. 397; but *cf.* Schmitt, 'Direct Participation', 508.

120 'Outsourcing and Privatization' 1988 Air Force Congressional Issue Papers Extract cited in Steven J. Zamparelli, 'What Have We Signed up For?: Competitive Sourcing and Privatization – Contractors on the Battlefield' (1999) 23(3) *AFJ Log.* 9.

121 JSTARS (Joint Surveillance Target Attack Radar System) offers an airborne, standoff range, surveillance, and target acquisition radar and a command and control centre to those managing a conflict. It possesses secure data links with air operations centre, army mobile ground stations and other military command, control and intelligence assets: Schmitt, 'Direct Participation', 512.

122 Zamparelli, 'What Have We Signed up For?', 13; Schmitt, 'Humanitarian Law and Direct Participation', 518.

123 Dan Verton, 'Navy Opens Some IT Ops to Vendors' (2000) *Federal Computer Week* www.fcw.com/fcw/articles/2000/0821/pol-navy-08-21-00.asp (last accessed 21 August 2000).

124 Singer, 'Outsourcing War'; P. W. Singer, *Corporate Warriors: The Rise of the Privatized Military Industry* (Cornell University Press, Ithaca, 2003).

miles away from a conflict zone.[125] Conversely, other routine maintenance falls into a different category and should not be so considered.

2.4. *Generic IT support*

Another question arises in the case of civilian technicians employed to maintain military networks that are not directly involved in offensive information operations, but which may subsequently become targets of computer network attacks. As noted above, the concept of direct participation in hostilities would not generally include routine computer maintenance any more than it would include laundry or catering services. Thus routine systems maintenance, security updates and other generic IT functions which are not related to hostilities (computer network attack or otherwise) would not be considered direct participation. Merely setting up security protocols on a network would be akin to civilians helping with the war effort to lay down barbed wire on the beaches in advance of a suspected landing.

However, as seen above, guarding a military objective against enemy action and defence of a military installation does comprise direct participation in hostilities. In highly technologically advanced militaries, particularly those that rely heavily on their networking capabilities for an advantage, disruption of the networks which link the various components of the military together will create a significant advantage for an opposing force. For example, any US military network, including those which utilise civilian assets either wholly or in part, becomes a valuable and legitimate target, subject always to the principle of proportionality. Nevertheless, in 2000 the US Navy opened up all defensive information operations, including information assurance and other defensive security operations, as 'non-inherently governmental' job functions and thus open to outsourcing.[126]

At what point does the technician cease to become a protected civilian merely supporting and maintaining a network (including network security measures), and become an active participant defending a military objective? The ICRC Interpretive Guidance on direct participation includes a wide criterion of threshold of harm; an act must be likely to adversely affect the military operations or military capacity of a party to an armed conflict.[127] During the course of the expert

[125] Schmitt, 'Direct Participation', 512.
[126] Verton, 'Navy Opens Some IT Ops to Vendors'.
[127] ICRC, *Interpretive Guidance*, 47.

meetings there was wide agreement that the causation of military harm as part of the hostilities did not necessarily pre-suppose the use of armed force or the causation of death, injury or destruction, but essentially included 'all acts that adversely affect or aim to adversely affect the enemy's pursuance of its military objective or goal'.[128] Defence of a network or system against an incoming computer network attack would definitely adversely affect the attacker's pursuance of its goal to compromise the military network. Thus, a conservative view would dictate that at the very moment the system comes under computer network attack, civilians must step away from their posts. In the context of computer network attacks, the author considers that this is both unrealistic and beyond what the law requires.

In contrast to defence of 'real world' military objectives, defence of a system or network may not involve the infliction of harm or additional threat to the attackers. For instance, when private contractors came under fire defending the Coalition Headquarters in Najaf, Iraq in 2004, a three-and-a-half hour fire-fight ensued that used thousands of rounds of ammunition and hundreds of grenades, and resulted in the wounding of three coalition personnel and an unspecified number of Iraqi casualties.[129] In contrast, defence of a computer network seldom results in casualties and solely defensive measures may not even comprise a use of force.[130] Although the threshold of harm is much broader than the application of force, it may be that this disjuncture from damage removes network defence sufficiently from the chain of causation to disqualify it from direct participation. In this respect it may be distinguished from the examples of civilian minesweepers, both naval and land based, cited by experts involved in the ICRC project during the discussion of the appropriate threshold of harm.[131] As with each of these scenarios, the specific circumstances of the civilian's actions will need to be assessed. However, it is also useful to note Michael Schmitt's comment that using civilians to perform defensive functions frees up soldiers for other combat missions, thereby further contributing to hostile action.[132]

[128] *Ibid.*, fn. 97.

[129] Priest, 'Private Guards Repel Attack on US Headquarters'. Coalition wounded comprised two contractors and one marine.

[130] See Chapter 2 above.

[131] See ICRC, *DPH Report 2005*, 31.

[132] Schmitt, 'Humanitarian Law and Direct Participation', 538.

Computer network operations pose an additional issue in relation to the difficulty in attributing attacks to a specific actor.[133] Civilians are entitled to defend property from criminals and looters without such actions constituting direct participation in hostilities; thus civilian technicians are entitled to defend military networks from regular hackers. Not only are military networks a prime target for enemy forces, they come under increasing attack from civilians during time of war as well.[134] Given the anonymous nature of the internet and the current lag in computer forensics in tracing the source of attacks, civilian technicians are unlikely to be able to determine whether or not they are directly participating in hostilities. They will be unable to ascertain immediately who is perpetrating the attack and in many cases, even where it is determined that a civilian is behind the attack, it will not be possible to determine if there is a sufficient nexus between the attack and any ongoing hostilities.[135]

2.5. Mercenaries

The increased numbers of civilian employees and contractors working for the military also raises issues regarding their possible classification as mercenaries. As set out in the section above, civilians accompanying the forces without being part of it are granted prisoner of war status under Article 4(4) Geneva Convention III; however, those directly participating in hostilities, for example where contractors are engaged to conduct proactive offensive computer network attacks, run the risk of being categorised as mercenaries. It is clear that both individual hackers and private military companies (PMCs) specialising in computer network operations have worked with states to provide computer network

[133] See Chapter 3 above.

[134] The 2008 computer network attacks launched against Georgian government and other websites are a prime example of this. Computer network attacks have also accompanied the Arab–Israeli conflict and Zapatista uprisings in Mexico, among other incidents. Interestingly, despite predictions of this behaviour in Iraq, the number of intrusions remained stable, although a large number of website defacements did occur: Peter Rojas, 'The Paranoia that Paid Off', *Guardian* (London), 24 April 2003, 27.

[135] Obviously in cases where attacks are accompanied by website defacement, such as have occurred in the above cases, the link with ongoing hostilities will be easier to ascertain. See, for example, Reuters, 'Cyber-War Rages over Iraq', *ZDNet News*, 31 March 2003; Izhar Lev, 'E-Intifada: Political Disputes Cast Shadows in Cyberspace' (2000) 12(12) *Jane's Intelligence Review* 16; Steve Mertl, 'Cyberspace Experts Await Full-Scale Attack', *Globe & Mail* (Canada), 27 December 2002, A11.

attack possibilities.[136] Although the use of mercenaries in warfare is an ancient practice, in post-colonial times mercenaries have largely fallen from favour and become *personae non grata* in international relations.

Under the Geneva Conventions, mercenaries (along with 'other militias') qualify as lawful combatants as long as they meet the conditions set out in Article 13(2).[137] However, by the implementation of the Additional Protocols in 1977 that position had changed. Adopted in the wake of nearly two decades of post-colonial struggles for self-determination in Africa, Additional Protocol I provides that mercenaries do not have the right to be combatants and deprives them of their right to be treated as prisoners of war if captured.[138] Significantly, given the historical context of the outlawing of mercenarism, Additional Protocol II relating to non-international armed conflicts does not contain any provisions relating to mercenaries.[139]

The definition of a mercenary is set out in Article 47(2) of Additional Protocol I:[140]

A mercenary is any person who:

(a) is specially recruited locally or abroad in order to fight in an armed conflict;
(b) does, in fact, take a direct part in the hostilities;
(c) is motivated to take part in the hostilities essentially by the desire for private gain and, in fact, is promised, by or on behalf of a Party to the conflict, material compensation substantially in excess of that promised or paid to combatants of similar ranks and functions in the armed forces of that Party;

136 See, for example, PBS Frontline 'Interview with Hacker' *Cyberwar!* (PBS Airdate 24 April 2003), www.pbs.org/wgbh/pages/frontline/shows/cyberwar/interviews/hacker.html (last accessed 16 January 2012); Ruth Alvey, 'Russian Hackers for Hire: The Rise of the E-Mercenary' (2001) 13(7) *Jane's Intelligence Review* 52.

137 The conditions are set out in section 1 above, i.e. responsible command, having a fixed distinctive sign, carrying arms openly and conduct in accordance with the laws and customs of war. See Dino Kritsiotis, 'Mercenaries and the Privatisation of Warfare' (1998) 22 *Fletcher Forum of World Affairs* 11, 16.

138 Art. 47(1), Additional Protocol I. Mercenaries are also banned under the 1977 OAU Convention for the Elimination of Mercenaries in Africa and the 1989 International Convention against the Recruitment, Use Financing and Training of Mercenaries.

139 For an in-depth analysis of the history and evolution of the prohibition against mercenaries see Leslie C. Green, 'The Status of Mercenaries in International Law' in L. C. Green (ed.), *Essays on the Modern Law of War* (Transnational Publishers, Dobbs Ferry, NY, 2000), 529; Todd S. Milliard, 'Overcoming Post-Colonial Myopia: A Call to Recognise and Regulate Private Military Companies' (2003) 176 *Mil. L. Rev.* 1.

140 The same definition can be found in the International Convention against the Recruitment, Use, Financing and Training of Mercenaries, UN GAOR 44th Sess., Supp. No. 43, at, UN Doc. A/RES/44/34 (1989).

(d) is neither a national of a Party to the conflict nor a resident of territory controlled by a Party to the conflict;
(e) is not a member of the armed forces of a Party to the conflict; and
(f) has not been sent by a State which is not a Party to the conflict on official duty as a member of its armed forces.

The conditions of the article are cumulative and it is widely considered that taken together the conditions are almost unworkable against contemporary 'soldiers for hire'.[141] It is particularly difficult to apply this six-part definition to modern-day private military companies. First, the corporate structure of the private military company offers protection to both the entity itself and the contractors employed by it. The definition in Article 47(2) applies only to natural persons, so the PMC itself cannot be held accountable. In addition, individuals are hired and paid as contractors by the company, not by a party to the conflict, enabling them to hide behind its corporate veil. Secondly, both the contractors and the PMCs are often hired for multiple purposes, rather than in respect of a specific conflict. Finally, nationality may be extended to contractors, or individual contractors may be incorporated into the armed forces of the state or given special status such as detectives, for the express purpose of avoiding the provisions.[142] All of these difficulties can be addressed as conditions of the contract of hire, to the extent that one commentator has been caused to remark that 'any mercenary who cannot exclude himself from this definition deserves to be shot – and his lawyer with him!'[143] However, in modern conflicts the increasing use of private contractors and the disjuncture between states' policy on the use of contractors in particular roles and their actual use in practice, means that despite the ability to manage the risk of mercenary status through contractual terms, some contractors may find themselves exposed.[144] As noted above, in an age of high-tech militaries, where the cost of training soldiers to operate increasingly complex and specialised systems is prohibitive, contractors are being used

[141] See, for example, Peter W. Singer, 'War, Profits, and the Vacuum of Law: Privatized Military Firms and International Law' (2004) 42 *Col. J. Trans. L.* 521, 524; Milliard, 'Post-Colonial Myopia', 42.

[142] For example a contract concluded between Sandline International (a PMC) and the government of Papua New Guinea, for provision of military assistance to deal with a secessionist rebel movement on Bougainville Island, provides that Sandline personnel are to be enrolled as 'special constables'. http://coombs.anu.edu.au/SpecialProj/PNG/htmls/Sandline.html (last accessed 5 July 2011).

[143] Geoffrey Best, *Humanity in Warfare: The Modern History of the International Law of Armed Conflicts* (Weidenfeld and Nicolson, London, 1980), 328.

[144] See PBS Frontline 'Interview with Peter Singer'.

to deliver, support and in some cases even operate systems. This brings them within the criteria for direct participation in hostilities and the recruitment specifically for the purposes of armed conflict. Thus, individual contractors who are foreign civilians, and are recruited for their specific computer network attack capabilities in respect of a particular conflict, are at risk of falling into this category. One final point of note, private contractors are routinely paid salaries in excess of that paid to their military counterparts, a fact that has caused resentment amongst serving military personnel, and despite concerns over determining the motivations of individuals engaged in contract work, many contractors operating in Iraq are open about the fact that they are there for financial gain.[145]

It should be noted that Additional Protocol I merely removes combatant privilege and rights to prisoner of war status from a mercenary (thus rendering them an unlawful combatant); it does not criminalise the status of mercenaries, nor does it make criminal the recruiting, training or financing of mercenaries.[146] Such acts are made criminal by both the OAU Mercenary Convention and the UN Mercenary Convention, and any contractors found to be in breach of those provisions and their hiring states may be punished.[147] However, as Dinstein points out, the UN Convention has not been widely ratified.[148] Mercenaries operating in countries which have not ratified the above Conventions may be punished under domestic or international law for their actions in the same manner as any other unlawful combatant.

3. Child soldiers

Literature on computer network attacks is full of examples of teenage hackers compromising critical infrastructure for personal reasons,[149]

[145] In 2005 in Iraq, guards working for private military firms could typically make US$400–600 per day, approximately twice the salary of a US soldier. Guards employed by Blackwater (an American PMC, charged with guarding US Ambassador Paul Bremer, the former head of the Coalition Provisional Authority) were paid up to US$1,000 a day: Frontline *Private Warriors*, 2005. (PBS: USA, 21 June 2005).

[146] Milliard, 'Post-Colonial Myopia', 41.

[147] *Ibid*., 19; Arts 1(2) and 4 OAU Convention for the Elimination of Mercenaries in Africa, OAU Doc. CM/433/Rev.L Annex 1; Arts 2–4 International Convention against the Recruitment, Use, Financing and Training of Mercenaries.

[148] Dinstein, *Conduct of Hostilities* (1st edn), 52.

[149] See, for example, the intrustions into the Arizona State Dam, Port Authority in Houston, and Operation Solar Sunrise in Appendix 1.

or engaging in acts of 'hacktivism'. While reports of states recruiting teenage hacker armies have featured in the popular press of many countries and are likely to be more the stuff of tabloid imagination than fact, in 2009 the involvement of young actors in an international computer network attack incident took a step forward. Konstantin Goloskokov, a 'commissar' of the Kremlin-supported Nashi youth movement in Russia, claimed responsibility, along with several associates, for the 2007 computer network attacks against Estonia.[150] There is no suggestion that the attacks against Estonia constituted an armed conflict; however, it emphasises a possibility and responsibility that states must consider when engaged in armed conflicts. If the same thing had happened during the Russia/Georgia conflict,[151] an issue would arise not only about the extent of Kremlin control involved in the attacks (both under the rules of state responsibility discussed above in order to assess whether the attacks may be attributed to the state, and also in terms of combatant status whether the attackers 'belong to a party to a conflict'), but also the implications for possible use of child soldiers: Nashi recruits members between seventeen and twenty-five years of age.

The Optional Protocol to the Convention on the Rights of the Child requires states to take all feasible measures to ensure that persons under the age of 18 are neither compulsorily recruited into the state's armed forces, nor take a direct part in hostilities if they are nevertheless members of their armed forces.[152] Armed groups separate from the armed forces may not under any circumstances recruit or use in hostilities any person under the age of eighteen; states are required to take all feasible measures to prevent such recruitment and use.[153] Both Additional Protocols to the Geneva Conventions contain a lower age limit. As with

[150] Clover, 'Kremlin-Backed Group Behind Estonia Cyber Blitz'.

[151] The cyber attacks that accompanied the military operations over South Ossetia have not been reliably attributed, there are no reports of specific Nashi involvement; however, many reports note the involvement of large numbers of patriotic hacker civilians and the Russian business network, encouraged by the state. See generally: David Hollis, 'Cyberwar Case Study: Georgia 2008' (2011) 7(1) *Small Wars Journal*, http://smallwarsjournal.com/blog/journal/docs-temp/639-hollis.pdf (last accessed 6 January 2011); US Cyber Consequences Unit, *Overview*.

[152] Arts 1 and 2, Optional Protocol to the Convention on the Rights of the Child on the involvement of children in armed conflict UN Doc. A/RES/54/263 (25 May 2000) entered into force on 12 February 2002. Where members of the armed forces are under the age of eighteen they are entitled to special protections: Art. 3.

[153] Art. 4, *ibid*. A compromise was reached in the drafting of the Optional Protocol to allow more flexibility for those states allowing the recruitment of persons under

the Optional Protocol, Additional Protocol I requires states to 'take all feasible measures' to ensure that children under the age of fifteen do not take a direct part in hostilities.[154] Additional Protocol II prohibits their participation in all hostilities, direct or indirect.[155]

The Appeals Chamber of the Special Court for Sierra Leone held that the prohibition against child recruitment had crystallised as customary international law entailing individual criminal responsibility by 1996.[156] The Commentary to Additional Protocol I states that the parties resolved to use the word 'feasible' as it was used elsewhere in the Protocols and thus it should be understood as meaning 'capable of being done, accomplished or carried out, possible or practicable'.[157] While it is certain that this will prohibit armed forces from actively recruiting minors, there is still scope for young volunteers to be utilised. Given the ease with which young hackers (many of whom are aged between 12 and 16) can now launch attacks, either through their own skill or by utilising another's coding, consideration must be given to what measures states may be required to put in place to ensure that minors do not participate by launching their own attacks on enemy forces. As Happold has pointed out, whether something is practicable is a question referring to whether, in the particular circumstances of the moment, the efforts required to do it are not disproportionate to the results obtained on having done so.[158] It seems likely that any effort

the age of eighteen into the armed forces. See UN Economic and Social Council, *Report of the Working Group on a Draft Optional Protocol to the Convention on the Rights of the Child on Involvement of Children in Armed Conflicts in Its Sixth Session*, UN Doc. E/CN.4/2000/74 (2000) paras 57–9.

154 Art. 77(2), Additional Protocol I.

155 Art. 4(3) Additional Protocol II. On this point the commentary to Additional Protocol II specifically lists the following examples, i.e., to participate in military operations such as gathering information, transmitting orders, transporting ammunition and foodstuffs, or acts of sabotage.

156 *Prosecutor* v. *Norman (Decision on Preliminary Motion Based on Lack of Jurisdiction (Child Recruitment))* (2004) Case No. SCSL-04-14-AR72(E)-131, Special Court for Sierra Leone, para. 53 (with Judge Robertson dissenting).

157 Pilloud, *et al.*, *Commentary*, 692,900. Note that the commentary on the Optional Protocol argues that a comparison of the French texts implies that the term feasible in the Optional Protocol should be interpreted more widely than in the Additional Protocols; however, it admits that its exact interpretation is uncertain and in any event will be controversial given the context of any particular case. Tiny Vandewiele, *Commentary on the United Nations Convention on the Rights of the Child, 46 Optional Protocol: The Involvement of Children in Armed Conflicts* (Martinus Nijhoff, Leiden; Boston, 2005), 27.

158 Matthew Happold, 'Child Soldiers in International Law: The Legal Regulation of Children's Participation in Hostilities' (2000) 47 *NIL Rev.* 27, 34.

to track down and prevent underage hackers would be vastly disproportionate to the result, particularly in circumstances where they are operating outside the battlespace. Certainly statements such as the general admonition issued by the US in respect of patriotic hackers 'joining-in' the conflict in Iraq would be an easy practical measure for states to implement.[159]

4. Conclusion

The combination of increased civilianisation and high-tech methods of warfare has raised some interesting challenges for the laws of armed conflict and in particular the determination of combatant status. The decreased relevance of time and proximity to the battlespace and the nature of the online environment have created problems for the relevance and interpretation of the law requiring distinction between civilian and combatant. While technical solutions are available, they are not without problems of their own. Other requirements of lawful combatancy merely require reinterpretation for the digital age.

One of the main classes of unlawful combatants, civilians engaged directly in hostilities, has also increased in numbers with the changes in the armed forces. While it seems clear that those engaged in proactive offensive computer network attacks will be considered to be taking a direct part in hostilities, the situation is more confused for those taking a less immediate role. Regardless of the difficulties and the need for case-by-case analysis, it is clear that a view must be reached in broad terms as to the line between legitimate support and direct participation, so that civilians are not unknowingly conceding their right to protection.

[159] Gallagher, 'Hackers; Government Tells Vigilantes Their "Help" Isn't Necessary'.

6 Targeting and precautions in attack

Computer network operations allow for both the precise targeting of particular systems vital to an adversary's war effort and the ability to cause widespread disruption to everyday life. Like any other military operation, when computer network attacks are employed in an armed conflict, the targets selected for attack must be legitimate military objectives, and the attack must conform to the principles of distinction, proportionality and necessity. However computer network attacks also raise a number of specific issues with regard to targeting and precautions in attack. Technological advances are adding to the pressures currently being exerted on the principle of distinction by modern conflict; in particular the ability to limit the effects of computer network operations has caused some commentators to question whether the same criteria should be applied. Most, although not all, targeting restrictions are based on the concept of an attack, and not all computer network attacks will rise to this level. Thus, the question is raised as to when a computer network attack becomes an attack for the purposes of international humanitarian law and what restrictions will operate in relation to those computer network attacks which do not reach that threshold. Computer network attacks also increase the opportunities for attacks by allowing the targeting of objectives which would otherwise be prohibited by the principle of proportionality. By minimising collateral damage and incidental injury, targets which would have been off-limits for neutralisation by traditional kinetic means may be made permissible by the simple expedient of turning them off. Further, the very nature of network design means that knock-on effects of attacks can be far reaching. Thus the level of network reconnaissance and understanding of the system's connectivity may be far above what might be required for a conventional attack in order to exercise the appropriate precautions in attack.

1. The principle of distinction

Article 48 of Additional Protocol I codifies the basic rule that parties must distinguish between civilian objects and military objectives:

> In order to ensure respect for and protection of the civilian population and civilian objects, the Parties to the conflict shall at all times distinguish between the civilian population and combatants and between civilian objects and military objectives and accordingly shall direct their operations only against military objectives.

This is a modern restatement of the principle of distinction which has been held by the International Court of Justice to constitute a 'cardinal principle' of the laws of armed conflict and one of the 'intransgressible principles of international customary law'.[1] The principle is further enshrined in the Statute of the International Criminal Court, which dictates that 'intentionally directing attacks against civilian objects, that is, objects which are not military objectives' constitutes a war crime in international armed conflicts.[2] Although neither Common Article 3 nor Additional Protocol II contain any specific requirements for parties to take precautions in attack,[3] the Appeals Tribunal in the *Tadic* case extended the application of the principle to conflicts not of an international nature. Citing with approval General Assembly resolutions on the protection of human rights in armed conflict, the Court recognised the principle of distinction as a principle of customary international law applicable 'in conflicts of any kind'.[4] The Hague Regulations of 1899 and 1907 also lay down certain protections for civilian property and undefended places;[5] however, the Regulations

[1] *Nuclear Weapons Case*, paras 78–9.

[2] Art. 8(2)(b)(ii), Statute of the International Criminal Court.

[3] Art. 13(1), Additional Protocol II contains a general principle that 'the civilian population and individual civilians shall enjoy general protection against the dangers arising from military oprations'. The commentary notes that this codifies a principle of customary international law, and notes that the implementation of such a protection requires that precautions are taken in both attack and defence: Pilloud, *et al.*, *Commentary*, 1448, para. 4772.

[4] *Tadic (Interlocutory Appeal)*, paras 112, 127.

[5] Art. 23(g) forbids the destruction and seizure of enemy property unless that action is imperatively demanded by the necessities of war. Art. 25 provides that the attack or bombardment, by whatever means, of towns, villages, dwellings, or buildings which

clearly imply that there is no conventional legal prohibition on the bombardment of civilians in defended places.[6] As methods of warfare developed, and particularly with the advancements in aerial warfare, the emphasis on differentiating between defended and undefended targets in the Hague Regulations was made obsolete. By the drafting of the 1923 Hague Rules of Air Warfare, although never ratified by states, the model had been replaced by the development of a definition of 'military objective', consideration of the concept of indiscriminate attacks and the introduction of a proportionality requirement.[7]

The additional targeting opportunities offered by computer network attack capabilities come at a time when there is already an increased tension in the laws of armed conflict over the continued pre-eminence of the principle of distinction in modern warfare. The tension is partially caused by changes in the types of armed conflicts and also by the emergence of effects-based operations as the dominant theory of warfare in advanced militaries. Two recent schools of thought have advocated the expansion of permissible targets to include previously prohibited civilian objects, but for fundamentally different reasons. The first approach is partially expounded in the United States' expanded definition of 'military objective' which includes war-sustaining objects discussed in section 2.2 below and finds its most extreme expression in the work of Charles Dunlap.[8] Dunlap argues:[9]

> We need a new paradigm when using force against societies with malevolent propensities. We must hold at risk the very way of life that sustains their depredations, and we must threaten to destroy their world as they know it if they persist. This means the air weapon should be unleashed against entire new categories of property that current conceptions of LOAC put off-limits.

This book will not attempt a full discussion of the arguments of Dunlap's paper, and this has been addressed extensively in other

are undefended is prohibited. Specific protections are also provided for cultural property in Art. 27, which are discussed in Chapter 7 below.

6 Richard R. Baxter, 'The Duties of Combatants and the Conduct of Hostilities (the Law of the Hague)' in UNESCO (ed.), *International Dimensions of Humanitarian Law* (Martinus Nijhoff, Dordrecht, 1988) 93–134, 115.

7 Judith G. Gardam, 'Proportionality and Force in International Law' (1993) 87(3) *AJIL* 391, 400.

8 Charles J. Dunlap, 'The End of Innocence: Rethinking Non-Combatancy in the Post-Kosovo Era' (2000) Summer *Strategic Review* 9. See also, Jeanne M. Meyer, 'Tearing Down the Facade: A Critical Look at the Current Law on Targeting the Will of the Enemy and Air Force Doctrine' (2001) 51 *AFL Rev.* 143.

9 Dunlap, 'End of Innocence', 14.

places;[10] however, it serves to illustrate the revival of a particular challenge to the principle of distinction. There are also those who attack the principle from the opposite side, arguing that the principle relies on an outdated world view in which large-scale interstate wars were the norm.[11] One such author, Gabriel Swiney, argues that strict adherence to the principle in an age of non-international armed conflicts violates the equal protection of laws, precluding the rule of law and endangering the lives of those whom it was designed to protect.[12] He argues that asymmetry handicaps insurgent parties in the conduct of non-international armed conflicts, thus forcing them to reject the laws of armed conflict because to do otherwise would mean defeat. Major powers bend the principle to suit their needs by utilising the civilian sector as contractors and using civilian settlers to effect their occupations (Swiney cites Israel and Sri Lanka as examples).[13] Both the approaches advocated by Dunlap and Swiney seek to undermine the principle of distinction, albeit for different reasons.

The principle of distinction has been under threat before. As Stephen Oeter points out, the principle of distinction has always belonged to the basic set of rules which in practice were put into doubt by belligerents who were not willing to restrict their use of violence as soon as such restrictions were perceived as being harmful to their strategies.[14] However, the present threat also stems from the multiple layers of movement currently happening in relation to armed conflict generally, many of which are both a product of, and instrumental in, the advances in military technology. One effect which is not, however, is at the politico-societal level, where a trend from conflicts occurring over resources and territory towards conflicts for the purpose of shaping decision-making processes is taking place.[15] This shift in the

[10] See, for example, contributions by Schmitt, Oeter, Parks in Wolff Heintschel von Heinegg and Volker Epping, *International Humanitarian Law Facing New Challenges: Symposium in Honour of Knut Ipsen* (Springer, Berlin; New York, 2007); Michael N. Schmitt, 'The Principle of Discrimination in 21st Century Warfare' (1999) 2 *Yale Hum. Rts & Dev. L.J.* 143.

[11] Gabriel Swiney, 'Saving Lives: The Principle of Distinction and the Realities of Modern War' (2005) 39 *International Lawyer* 733.

[12] *Ibid.*, 733. [13] *Ibid.*, 750.

[14] Stefan Oeter, 'Comment: Is the Principle of Distinction Outdated?' in W. H. von Heinegg and V. Epping (eds.), *International Humanitarian Law Facing New Challenges* (Springer, Berlin; New York, 2007) 53–65, 55.

[15] Smith, *Utility of Force*, 270; Gray, *Another Bloody Century*; Chapter 1 above. Although this may not represent a permanent trend, as predictions are that conflicts based on resources, particularly water, are expected to increase with environmental changes brought about by global warming. See generally, Klare, *Resource Wars.*

underlying purposes of conflict is also reflected in the changing goals of warfare in the modern era; a movement toward influencing political changes rather than outright military victory. The NATO intervention over Kosovo gives a prime example of such a 'compellence' or 'coercive' campaign, where force was applied in order to coerce Milosevic to abandon a policy of ethnic cleansing in the area.[16] Thus at the strategic level there is a movement in the dominant theory of warfare in the West from attrition warfare towards effects-based operations which are designed for coercive campaigns. Attrition warfare seeks to significantly weaken the military forces of the opposing side in order to destroy their physical capacity to wage war. Effects-based operations, however, attempt to coerce the other side using a combination of force and other non-forceful methods such as diplomacy to effect change at the political level.[17] Such a strategy may tempt military commanders to attack any target which will achieve the aims of the operation, and hence end the war, in the most effective manner possible; many times these will be civilian objects.[18] Effects-based operations also represent the operationalisation of network-centric warfare,[19] the move to which from the more traditional platform-centric warfare has increased the speed and effectiveness of the modern military, giving it a significant advantage. At the technological level, the advancement of weapons technology allows technologically sophisticated militaries to strike almost any target they choose in the battlespace. On top of this uneasily shifting structure sits computer network attack. Computer network attacks not only increase the number of targets that it is possible to attack by reducing the collateral damage and hence the proportionality equation in target selection, they also allow the possibility of operations which cause no physical damage but nonetheless destroy or

[16] Schmitt, 'Asymmetrical Warfare', 36–8. Attacks on industrial facilities in Bor and Smederevo were designed to put pressure on cronies of President Milosevic to influence him to withdraw his troops. William M. Arkin and Robert Windrem, 'The Other Kosovo War', *MSNBC News*, 29 August 2001; Marc J. Romanych and Kenneth Krumm, 'Tactical Information Operations in Kosovo' (2004) September–October *Military Review* 56.

[17] Effects-based operations are not new. Their roots can be traced back for centuries and are what good generals and statesmen have always attempted to do. When combined with network-centric thinking and technologies, however, such an operational approach offers a way of applying the power of the network to the human dimension of war and to military operations in peace and crisis, as well as combat. Smith, *Effects Based Operations*, xxiii.

[18] Schmitt, *High and Low-Tech Warfare*, 8.

[19] See generally, Smith, *Effects Based Operations*, 1.

merely neutralise the object or system in question. This is an attractive option for states, particularly in conflicts which will necessitate the reconstruction of the battlespace at the conclusion of hostilities; it is inefficient to bomb a power generating station if you can simply turn it off.

2. Legitimate military objectives

The principle of the military objective has become part of the customary international law for armed conflict, at sea as well as on land or in the air.[20] In 1977 the principle was codified and a definition of the term 'military objectives' was included in Article 52(2) of Additional Protocol I:[21]

> Attacks shall be limited strictly to military objectives. In so far as objects are concerned, military objectives are limited to those objects which by their nature, location, purpose or use make an effective contribution to military action and whose total or partial destruction, capture or neutralization, in the circumstances ruling at the time, offers a definite military advantage.

The definition contains several elements which raise interesting questions in relation to computer network attacks. The first is easily dealt with; the definition relates to objects (or in the French text 'biens'), which the ICRC commentary notes refers to something visible and tangible.[22] It could be argued that only material tangible things can be targets,[23] and thus the intangible nature of many computer network attack targets exclude them from that definition. However, it is clear from the text of the commentary that this definitional point is being made to distinguish the term object as a 'thing' from its use in the

[20] Horace B. Robertson, 'The Principle of the Military Objective' (1997–1998) 8 *USAF Acad. J. Legal Stud.* 35, 46; See also Yoram Dinstein, 'Legitimate Military Objectives under the Current Jus in Bello' in A. E. Wall (ed.), *Legal and Ethical Lessons of NATO's Kosovo Campaign* (Naval War College, Newport, RI, 2002) 139–73, 140.

[21] The definition of military objectives also appears in several subsequent instruments: Additional Protocols II and III, Annexed to the 1980 Conventional Weapons Convention, and the second protocol to the Cultural Property Convention. See Dinstein 'Legitimate Military Objectives', 141.

[22] Pilloud, *et al.*, *Commentary*, paras 2008–10.

[23] Marco Sassòli, 'Targeting: The Scope and Utility of the Concept of "Military Objectives" for the Protection of Civilians in Contemporary Armed Conflicts' in D. Wippman and M. Evangelista (eds.), *New Wars, New Laws? Applying the Laws of War in 21st Century Conflicts* (Transnational Publishers, Ardsley, NY, 2005) 181–210, 185.

sense of 'aim or purpose of an operation', rather than to exclude an intangible object from the definition. Thus any computer program, database, system or virtual network would still be a legitimate target if it meets the above definition, regardless of whether it has a tangible component or exists purely as lines of code.

2.1. *Nature, location, purpose or use*

Article 52(2) requires that any military objective must by its nature, location, purpose or use make an effective contribution to military action.[24] Those objects which contribute to military action by their nature include all objects directly used by the armed forces: weapons, equipment, transport, fortifications, depots, buildings occupied by the armed forces, staff headquarters, communications centres etc.[25] To meet this yardstick they must have something in their intrinsic character, an inherent attribute which contributes to military action.[26] In relation to prospective targets for computer network attacks this would include all weapons systems, sensor arrays, battlefield devices, military networks and databases, digital communications systems and any other digital device or system built to military specifications.[27]

There may also be objects which by their nature have no military function, but which by virtue of their location make an effective contribution to military action;[28] either to use it oneself or to prevent the other party from using it. Examples relating to computer network attack are not easily come by given the inherent nature of networks. The efficiency and robustness of a network comes from the existence of multiple pathways to the same destination, and in the case of a network such as the internet, the physical location of individual nodes is not of primary importance. However, there may be circumstances where it is important to deny a network or other object to the

[24] The formulation was influenced by that proposed in Art. 2 of the Edinburgh resolution of the Institute of International Law, which defined military objectives as facilities which by their 'very nature or purpose or use, make an effective contribution to military action, or exhibit a generally recognised military significance, such that their total or partial destruction in the actual circumstances gives a substantial, specific and immediate military advantage to those who are in a position to destroy them'. Bothe, *et al.*, *New Rules*, 321.

[25] Pilloud, *et al.*, *Commentary*, para. 2020.

[26] Dinstein, *Conduct of Hostilities* (1st edn), 88.

[27] Note that this would not include medical devices, databases or networks which are subject to measures of special protection. See Chapter 7 below.

[28] Pilloud, *et al.*, *Commentary*, para. 2021.

enemy where its location does play a role in computer network attacks. For example, a civilian wireless network may exist in a particular area that would enable the adversary's forces located in the area to piggy-back their military communications off the signal in the event of the military network being disabled or to usurp the use of the network entirely.[29] The network location of a particular node may also make that object a military objective. The primary nodes of states' internal telecommunications networks, connecting them to the internet backbone, make attractive targets depending on the level of connectivity of the particular state.[30] States have already demonstrated their willingness to shut down access to the internet during periods of internal unrest by ordering ISPs to shut down their routers. For example, in 2011 when protests erupted in Egypt the Mubarak government initiated a five-day shutdown of the internet and cell phone networks in an attempt to curb the protests.[31] Similarly, both Nepal in 2005 and Myanmar in 2007 have shut down their country's limited internet connection in response to national unrest.[32] In the United States, where shutting down access to the internet would require the cooperation (or coercion) of many hundreds of ISPs, a controversial bill has

[29] For example, some areas have entire cities connected with wireless networks, e.g. Toronto. These are not restricted to developed areas; wireless communications technology is fast becoming one of the steps in developing communications infrastructure in parts of Africa. See, for example Mark Cieslak, 'Bridging an African Digital Divide' *BBC Click Online*, 7 September 2007, http://news.bbc.co.uk/1/hi/programmes/click_online/6983397.stm (last accessed 16 January 2012). Art. 53 of the Hague Regulations allows an army of occupation to take possession of 'Railway plant, land telegraphs, telephones, steamers and other ships, apart from cases governed by maritime law, as well as depots of arms and, generally, all kinds of munitions of war, even though belonging to companies or to private persons, are likewise material which may serve for military operations, but they must be restored at the conclusion of peace, and indemnities paid for them.' This allows an occupying force to take control of all lines of communication including all networks.

[30] For example, New Zealand connects to the internet backbone through primary nodes connecting three undersea fibre-optic cables (the Pacrim, Tasman 2 and Southern Cross cables). Disabling these nodes would effectively cut New Zealand off from all but satellite communications. Many countries have similar limited connections to the backbone. For example, the Eastern African Submarine Cable System provides fibre-optic cable links for South Africa, Sudan, Mozambique, Madagascar, Tanzania, Kenya and Djibouti. Each of the nodes connecting to the national networks would present an attractive target. IRIN News Report, 'Africa: Getting Connected at Last', 24 January 2006, www.irinnews.org/report.aspx?reportid=57903 (last accessed 16 January 2012).

[31] 'Internet Blackouts: Reaching for the Kill Switch', *The Economist* 10 February 2011.

[32] OpenNet Initiative, *Pulling the Plug*.

been introduced to the Senate which mandates a remote 'kill switch' to remotely shut down connection to the internet.[33] During an armed conflict, denial of these methods of communication would have a distinct advantage, subject always to the principle of proportionality.

The criteria of 'purpose' and 'use' of a military objective in Article 52(2) are linked – purpose is concerned with the intended future use of an object, while its use is concerned with the object's present function.[34] Dinstein notes that the purpose of an objective is deduced from an established intention of a belligerent as regards future use.[35] However, he warns it must be predicated on intentions known to guide the adversary, not on those figured out hypothetically in contingency plans based on a 'worst-case scenario'.[36] The actual use of an object in a manner that makes an effective contribution to military action will likewise render the object a military objective and thus liable to attack. In the event that there is any doubt over whether an object which is normally a civilian object is being used to make an effective contribution to military action, Article 52(3) provides a presumption that it is not. Should a computer network attack be launched from a civilian computer system or network, that network would become a legitimate military objective and may be attacked (assuming that its destruction, capture or neutralisation also provides a definite military advantage to the attacking state).[37] This is significant for countries such as the United States, where a large percentage of all military communications are transmitted over civilian networks,[38] making them a potential target for computer network attacks.

33 'Internet Blackouts: Reaching for the Kill Switch'.

34 Pilloud, *et al.*, *Commentary*, para. 2022.

35 Dinstein, *Conduct of Hostilities* (2nd edn), 89.

36 Dinstein, *Conduct of Hostilities* (1st edn), 90. Dinstein cites the example of the attack on the Abbey of Monte Cassino in World War II as a warning against reliance on supposition backed by flimsy intelligence.

37 In much the same way, civilian taxis commandeered by the military governor of Paris to transport reservists to the front in 1914 became military objectives.

38 In 1995 more than 95 per cent of all US military communications were sent across civilian networks. A more recent figure has not been made publicly available. Aldrich, 'International Legal Implications', 105, citing Science Applications International Corporation, 'Information Warfare: Legal, Regulatory, Policy and Organizational Considerations for Assurance', research report for the chief, Information Warfare Division (J6K), Command, Control, Communications and Computer Systems Directorate, Joint Staff, The Pentagon, Washington DC, 4 July 1995.

2.2. *Effective contribution to military action*

The second part of the definition of military objective requires that the object must make an effective contribution to military action. While all parties accept this provision as a statement of customary international law, there is some variation in the interpretation of the scope of the contribution to military action required. Bothe *et al.* point out in their commentary to the Protocol that the requirement of effective contribution relates to military action in general, and there need be no 'direct connection' with specific combat operations such as that required of civilians who lose their immunity for directly participating in hostilities.[39] Despite being a broader requirement than 'direct participation', there remains an enduring disagreement over the level of connection to military action required.

In stating its position on military objectives, the United States substitutes the words 'war-fighting or war-sustaining capability' for the term 'military action' as used in the Protocol.[40] By extending the interpretation to include 'war-sustaining', the US argues that '[e]conomic targets of the enemy that indirectly but effectively support and sustain the enemy's war-fighting capability may also be attacked'.[41] The report on US practice provided for the ICRC work on customary international law explains that while the US accepts the customary nature of Article 52(2), the alternative formulation reflects its position that this definition is a wide one which includes areas of land, objects screening other military objectives and war-supporting economic facilities.[42] Other authors disagree with this position; for example, Dinstein argues that the reference to 'war-sustaining capability' goes too far, introducing a slippery slope to a situation in which just about every civilian

[39] Bothe, *et al.*, *New Rules*, 324.

[40] The latest military manual to be released in the United States is the *US Commander's Handbook on the Law of Naval Warfare* NWP 1-14M; it is useful as the most current expression of US policy in this area. The reference to war-sustaining capability is set out in para. 8.2. 'War-fighting' implies combat operations or activities closely associated thereto. 'War-supporting' activities are those which indirectly support combat operations, such as the production of armaments. 'War-sustaining' refers to activities that undergird a war effort, such as the export of items that generate income, allowing the eventual funding of the war effort. Schmitt, 'Deconstructing DPH', fn. 57.

[41] Michael N. Schmitt, 'Fault Lines in the Law of Attack' in S. Breau and A. Jachec-Neale (eds.), *Testing the Boundaries of International Humanitarian Law* (British Institute of International and Comparative Law, London, 2006) 277–307.

[42] Jean-Marie Henckaerts and Louise Doswald-Beck, *Customary International Humanitarian Law* (Cambridge University Press, 2005) Vol. 1, 31.

activity could be construed as indirectly sustaining the war effort.[43] The present author agrees; taken to its logical conclusion the US position means that any goods or services which support the economy of the country (and thus the ability of the government to wage war) would become legitimate targets. As Stefan Oeter has pointed out, the US targeting of war-sustaining capabilities moves the target of military operations away from the military effort of the enemy and onto the political command and control system and its resource base; this approach gives up the requirement of a close nexus between the target and ongoing military operations.[44]

Although some commentators have dismissed this disparity in interpretation as of primarily academic significance,[45] it will have particular relevance for computer network attacks against highly developed information societies where the propensity for damage is higher. Consider, for example, the damage which could be inflicted on a state which is dependent on the export of a natural resource such as oil or minerals. Destruction of the immediate processing plants of the state can be justified as making an effective contribution to the military effort; however, destruction of the resource itself would affect the long-term economic welfare of the state and be too far removed from military action to justify the attacks.[46] Contrast that with the economic meltdown which would be achieved in attacks aimed at the commercial heart of an information-based economy such as Taiwan. Taiwan is one of the most information technology dependent economies in the world; one of its greatest fears is that China will unleash a wave of computer network attacks which will completely shut down political and economic institutions in a matter of days.[47] Under the US interpretation such attacks

[43] Dinstein, *Conduct of Hostilities* (1st edn), 87. For an extreme application of the US position and an example of Dinstein's concerns regarding a slippery slope see Dunlap, 'End of Innocence'. Dunlap suggests that any object, civilian or military, not indispensable to the civilian population should be fair game.

[44] Oeter, 'Comment: Is the Principle of Distinction Outdated?', 56.

[45] W. Hays Parks, 'Asymmetries and the Identification of Legitimate Military Objectives' in W. H. von Heinegg and V. Epping (eds.), *Interntional Humanitarian Law Facing New Challenges* (Springer, Berlin, 2007) 65–116, 97.

[46] Schmitt, 'Fault Lines', 281.

[47] 'Taiwan Plays Cyber War Games', *BBC News*, 7 August 2000. See also David Lague, 'Chinese See Military Dependence on Computers as Weakness', *International Herald Tribune* (Paris), 29 August 2007, Asia Pacific. Although some analysts argue that Taiwan's advanced computing and information technology industry would allow the island's military to resist cyber attack more readily than China's mounting conventional firepower.

would be permissible as the economy, and therefore war-sustaining capability, of the target state would be affected. There is also a question about what would constitute an attack which is dealt with in section 4.1 below. However, a balance must be struck between legitimate economic measures in wartime, for example a sanctions regime or measures which may destabilise or devalue a state's currency or credit,[48] and military operations targeting legitimate military objectives. This is particularly so in the case of coercive campaigns where measures are designed to effect a change in the decision-making behaviour of the adversary. A similar effect, albeit on a smaller scale, was seen in the cyber attacks against Estonia in April and May 2007.[49] Estonia is one of the most wired societies in Europe, using the internet for everything from voting, filing taxes and paying for parking.[50] The prolonged distributed denial of service attacks, which lasted over a month, came close to shutting down the country's digital infrastructure, clogging the websites of several government agencies and several newspapers and forcing the country's main bank to cease operations. Under US targeting analysis such measures would be legally justified if they were to take place during the course of an armed conflict.

2.3. *Definite military advantage*

The definition of military objective in Article 52 also requires that the destruction, capture or neutralisation of the object in question must provide a 'definite military advantage'. Although there was much discussion in the working group that drafted the provision regarding the appropriate adjective to be applied to the term 'military advantage', on reporting back to the conference the Rapporteur commented that he was unable to draw any significance from the particular choice of 'definite'; Bothe, *et al.* conclude that the adjective is a word of limitation, denoting in this context a concrete and perceptible military advantage rather than a hypothetical and speculative one.[51]

[48] For example, the United States' refusal to back sterling during the Suez Crisis in 1956 led to a monetary crisis for the British economy, forcing it to call off its campaign.

[49] See Chapter 1 above and Appendix 1 for further details.

[50] Traynor, 'Russia Accused of Unleashing Cyberwar to Disable Estonia'; Mark Landler and John Markoff, 'Digital Fears Emerge after Data Seige in Estonia', *New York Times*, 29 May 2007, www.nytimes.com/2007/05/29/technology/29estonia.html (last accessed 15 July 2011).

[51] Bothe, *et al.*, *New Rules*, 326.

The advantage gained must also be military in nature and not, for example, purely political.[52] Thus, as Dinstein notes, forcing a change in the negotiating attitudes of the adverse party cannot be deemed a proper military advantage. That is not to say that once a potential target has met the criteria as a military objective, the choice between two competing objectives cannot be motivated by which one would produce a favourable political result. The targeting choices made by NATO forces in the Kosovo campaign as the conflict progressed provide a good illustration of a politically directed target selection process aimed to force Milosevic to capitulate to NATO demands.[53] Dinstein also notes that the scope of the advantage should be wider than purely tactical.[54] Australia, Canada and New Zealand have stated that the term 'military advantage' includes the security of the attacking forces.[55] In an age of coalition warfare, it should also be noted that the military advantage gained may also constitute a benefit for an allied force or the alliance as a whole.[56]

Military advantage refers to the advantage gained from an attack as a whole and not from its isolated or particular parts. Some confusion on this point appears to be introduced by the relatively specific concept of attack introduced by Article 49(1) of Additional Protocol I. Oeter notes that Additional Protocol I relies on a concept of 'attack' as an 'isolated ground operation by a specific unit', and argues that such an approach appears to ignore the problems resulting from modern strategies of warfare which are based on an integrated series of separate actions forming one ultimate compound operation.[57] In an age of network-centric warfare, the individual targeting of small parts

[52] Dinstein, *Conduct of Hostilities* (1st edn), 86.

[53] Although all targets selected were military objectives (under the US definition), the later part of the bombing campaign focused on industrial targets belonging to Milosevic's cronies. See generally Judith Miller's Comments in Andru E. Wall, *Legal and Ethical Lessons of NATO's Kosovo Campaign* (Naval War College, Newport, RI, 2002), 110. Stephen T. Hosmer, *The Conflict over Kosovo: Why Milosevic Decided to Settle When He Did* (RAND, Santa Monica, 2001); Benjamin S. Lambeth, US Air Force and Project Air Force, *NATO's Air War for Kosovo: A Strategic and Operational Assessment* (RAND, Santa Monica, 2001), 13.

[54] Dinstein, *Conduct of Hostilities* (1st edn), 86.

[55] Henckaerts and Doswald-Beck, *Customary International Humanitarian Law*, Vol. 1, 50.

[56] H. Meyrowitz, 'Le Bombardement Stratégique d'après le Protocole Additionnel I aux Conventions de Genève' (1981) 41 *ZaöRV* 1, 41, cited in Dinstein, *Conduct of Hostilities* (1st edn), 86.

[57] Stefan Oeter, 'Methods and Means of Combat' in D. Fleck (ed.), *The Handbook of Humanitarian Law in Armed Conflicts* (2nd edn, Oxford University Press, 2008) 120–235, 185, §444.

of an integrated system will accumulate to contribute to a military advantage that would not necessarily eventuate from neutralising a single part of the system. For example, the incapacitation of a particular communications network may give a significant military advantage to an attacking force; however, it may be necessary to neutralise all nodes in the network to achieve the desired outcome. Each node must be attacked separately, but without the neutralisation of all parts the advantage will not be gained. The attack on the Radio-Television Serbia (RTS) television centre in Belgrade as part of an attack on the Federal Republic of Yugoslavia (FRY) communications network during the Kosovo conflict is an example of this approach.[58] According to NATO reports the FRY command and control network was a complex web that could not be disabled in one strike; in actual fact, RTS was broadcasting again in just over three hours.[59] The ICTY investigating committee found that:[60]

> [w]ith regard to these goals, the strategic target of these attacks was the Yugoslav command and control network. The attack on the RTS building must therefore be seen as forming part of an integrated attack against numerous objects, including transmission towers and control buildings of the Yugoslav radio relay network which were 'essential to Milosevic's ability to direct and control the repressive activities of his army and special police forces in Kosovo' (NATO press release, 1 May 1999) and which comprised 'a key element in the Yugoslav air-defence network' (*ibid.*, 1 May 1999).

Rogers also addresses this point, noting that although a particular offensive may combine infantry, tanks, artillery, helicopters and other close support aircraft in coordinated actions, each would amount to an

[58] The bombing of the TV studio was part of a planned attack aimed at disrupting and degrading the C3 network. In coordinated attacks, on the same night, radio relay buildings and towers were hit along with electrical power transformer stations … The FRY command and control network was alleged by NATO to comprise a complex web that could thus not be disabled in one strike. As noted by General Wesley Clark, NATO 'knew when we struck that there would be alternate means of getting the Serb Television. There's no single switch to turn off everything but we thought it was a good move to strike it and the political leadership agreed with us', ICTY, *Final Report to the Prosecutor of the Committee Established to Review the NATO Bombing Campaign against the Federal Republic of Yugoslavia*, ICTY (2000) paras 72, 78, www.icty.org/x/file/About/OTP/otp_report_nato_bombing_en.pdf (last accessed 15 July 2011).

[59] Raising the issue of the importance of the military advantage gained by the attack *vis-à-vis* the civilian casualties. *Ibid.*, para. 78.

[60] *Ibid.*

attack, as would the whole.[61] Several states have commented on ratification of the Protocol stating that the military advantage anticipated from an attack is intended to refer to the attack as a whole and not from isolated or particular parts of the attack.[62] Although these statements have been made in respect of proportionality under Articles 51 and 57 of Additional Protocol I, for reasons of internal consistency it is likely that the concept expressed would apply equally to Article 52(2). Indeed, the ICRC commentary to the Article suggests that such a statement would be redundant: 'it goes without saying that an attack carried out in a concerted manner in numerous places can only be judged in its entirety'; the commentary goes on to say 'this does not mean that during such an attack actions may be undertaken which would lead to severe losses among the civilian population or to extensive destruction of civilian objects'.[63]

One further point deserves comment: military advantage refers to a commander's intent in determining whether or not an object constitutes a military objective rather than actual effect as determined subsequently.[64] As the ICRC *Customary International Humanitarian Law* study indicates, numerous states have pointed out that those responsible for planning, deciding upon or executing attacks necessarily have to base their decisions on the assessment of the information from all sources which is available to them at the time.[65] How much information is necessary for a computer network attack and how sophisticated the network intelligence should be, is addressed below in section 6.

3. Dual-use technology

The term dual-use target is not a term of international humanitarian law. It is a term which has become popular in various quarters to refer to an object that has concurrent civilian and military uses. While there may be policy reasons to refrain from attacking such objects, from the standpoint of international humanitarian law, once an object meets the definition of a military objective it becomes liable to attack. The discussion of any civilian aspect or purpose of that object or piece of

[61] Rogers, *Law on the Battlefield*, 29.
[62] See statements made by Australia, Belgium, Canada, France, Germany, Italy, the Netherlands, New Zealand, Nigeria, Spain and the United Kingdom.
[63] Pilloud, *et al.*, *Commentary*, para. 2218, cited in Rogers, *Law on the Battlefield*, 29.
[64] Parks, 'Asymmetries', 90.
[65] Henckaerts and Doswald-Beck, *Customary International Humanitarian Law*, Vol. 1, 50.

technology should therefore be considered as part of the proportionality equation rather than as an issue of distinction.

In discussions about dual-use targets one of the most often cited examples is the coalition bombing of the Iraqi electrical grid in the 1990–91 Gulf War. Yoram Dinstein effectively reviews the outcome of that campaign as follows:[66]

> Since the electrical grid in Iraq was totally integrated, attacks against it – and its installations – resulted not only in a tremendous military advantage (shutting down radar stations, military computers, etc.), but also extensive damage to civilians: hospitals stopped operating, water pumping facilities came to a standstill, etc. From a legal point of view, a 'dual-use' of Iraq's electrical grid did not alter its singular and unequivocal status as a military objective. There was, as usual with military objectives, the question of proportionality where collateral damage to civilians is concerned. But the extensive damage to civilians was not excessive in relation to the military advantage anticipated.

The issue is of particular relevance in relation to computer network attacks. Most computer technology, hardware and software, has become dual-use. Some systems initially designed for military use have become so integrated into civilian society that any interference or disruption caused by computer network attacks would have serious effects on civilians. For example, the Global Positioning System (GPS) is a US military system which has become integrated into many civilian applications from aircraft traffic control to cell phones and laptops and even the internet itself.[67] Disruption of the service through jamming or blocking, or spoofing the signal via computer network attack would cause massive disruption and potentially endanger civilian lives.[68] Other

[66] Comment by Yoram Dinstein in 'Discussion' in A. E. Wall (ed.), *Legal and Ethical Lessons of NATO's Kosovo Campaign* (Naval War College, Newport, RI, 2002) 211–22, 219.

[67] GPS uses two levels of signal: the military signal (Y-code) is more accurate and encrypted; the less secure civilian code (or P-code) is not and thus makes it more susceptible. The precision timing provided by the GPS system is needed for the accurate routing of information packets through computer networks.

[68] Jamming and blocking are both methods of electronic attack which will not be covered by this book. Spoofing the GPS signal involves feeding a GPS receiver a fake signal so that it computes the wrong time or location of the receiver. Spoofing can occur either through electronic means (such as broadcasting a fake GPS signal with a higher signal strength than the true signal via a GPS satellite simulator) or through the network to GPS receivers. Note that the military has specific anti-spoofing measures in place which encrypt the general civilian P-code signal into a more secure Y-code which only military receivers can use. All military GPS acquisitions post-2006 are required to have the Selective Availability Anti-Spoofing Module (SAASM) attached. See generally, Symmetricon, *Why Convert to a SAASM*

countries operate (or are in the process of developing) similar global navigation space systems which exhibit similar vulnerabilities.[69]

In the modern era of effects-based operations, dual-use targets become particularly attractive precisely because of their ties to both military and political objectives.[70] The attacker not only benefits from the destruction or neutralisation of the target's military value, but also from cumulative effects on the civilian population.[71]

4. Protection of civilian objects

In addition to Article 48 of Additional Protocol I outlined above, which provides the basic rule, Article 52(1) provides that 'civilian objects shall not be the object of attack or of reprisals'. The International Court of Justice stated in its *Nuclear Weapons* Advisory Opinion that 'States must never make civilians the object of attack'.[72] The Rome Statute makes it a war crime to intentionally direct attacks against the civilian population as such, individual civilians or civilian objects.[73] As noted in Chapter 5 above, civilians are defined negatively by Additional Protocol I as any person who does not belong to one of the categories of person referred to in Article 4(A)(1), (2), (3) and (6) of Geneva Convention III and in Article 42 of Additional Protocol I.[74] The civilian population is defined in Article 50(2) of Additional Protocol I as comprised of 'all persons who are civilians'; the presence of persons who are not civilians in the population does not deprive the population of its civilian character.[75] Similarly, civilian objects are negatively defined as all objects which are not military objectives.[76] Where there is doubt over

Based Global Positioning System (GPS)? (2006), www.symmetricom.com/resources/downloads/white-papers/GPS-Solutions/Why-Convert-to-a-SAASM-based-Global-Positioning-System-GPS/ (last accessed 7 July 2011); Scott Pace, *et al.*, *The Global Positioning System: Assessing National Policies* (RAND, Santa Monica, 1995). A 2001 report by the Royal Academy of Engineering in the UK has warned about over-reliance on these systems for life-critical systems with little or no non-GNSS based backups: Royal Academy of Engineering, *GNSS Report*.

69 For example, the Russian GLOSNASS system, Chinese Compass/Beidou system or the European Galileo system.

70 Schmitt, 'Targeting', 65. 71 *Ibid.*

72 *Nuclear Weapons Case*, para. 78.

73 Art. 8(2)(b)(i)–(ii) Additional Protocol I.

74 Art. 50(1) Additional Protocol I. In case of doubt whether a person is a civilian, that person shall be considered to be a civilian.

75 Arts 50(2) and (3) Additional Protocol I.

76 Art. 52(1) Additional Protocol I.

the civilian character of a person or an object, the doubt is resolved in favour of finding civilian status.[77]

As a general proposition, it is prohibited to direct computer network attacks against civilian objects in the same manner that a conventional attack would be prohibited. Thus attacks against oil pipelines, civilian air traffic control or rail networks, emergency response networks, financial institutions and other civilian objects are prohibited. During the NATO action over Kosovo, there were legal concerns expressed inside the US administration regarding proposals to conduct information operations such as inserting viruses into Serbian computer systems or hacking bank accounts thought to contain funds plundered from Serbian businesses by Milosevic's cronies.[78] However, the nature of computer network attacks raises some interesting dilemmas for modern armed forces. In addition to the issues raised by the 'war-sustaining' phraseology of the United States' interpretation, which are discussed above, computer network attacks raise two particular issues for the law of targeting. First, computer network attacks do not necessarily result in physical damage, death or injury, thus opening a range of possible targets for attack which might otherwise be unreachable due to excessive collateral damage. A second and related point is that computer network attacks can be designed to result in a range of outcomes, allowing the attacker to merely disable or neutralise a particular target without causing permanent damage or destruction. Such computer network operations may not even rise to the level of an 'attack' at all.

4.1. *Attacks and operations*

The ability of a computer network attack to neutralise or destroy target systems without causing physical damage raises an interesting question with regard to the legitimate targets of such attacks. Although the basic rule of distinction laid down in Article 48 of Additional Protocol I is general in nature,[79] the majority of the provisions prohibiting the

[77] Arts 50(1) and 52(3) Additional Protocol I.

[78] Arkin and Windrem, 'The Other Kosovo War'.

[79] Art. 48 provides: 'In order to ensure respect for and protection of the civilian population and civilian objects, the Parties to the conflict shall at all times distinguish between the civilian population and combatants and between civilian objects and military objectives and accordingly shall direct their operations only against military objectives.'

targeting of civilians and civilian objects are phrased in terms of the prohibition of 'attacks'. Attacks are defined in Article 49 of Additional Protocol I as 'acts of violence against the adversary, whether in offence or defence'.[80] Bothe *et al.*'s commentary to Article 49 states that the term 'acts of violence' denotes physical force and thus the concept of attacks does not include dissemination of propaganda, embargoes or other non-physical means of psychological, political or economic warfare.[81] However, it should be noted that any act of violence fills this requirement: not only massive air attacks or artillery barrages but also small-scale attacks such as a sniper firing a single round.[82] This has led to a disagreement between commentators, particularly in the area of computer network attacks, regarding the legality of directing non-violent computer network attacks against civilian objects. The problem is particularly relevant in respect of modern armed conflicts, not only because of the technology now available to modern militaries but also because of the transformation in the character of warfare towards effects-based operations and the increased importance in the use of force to influence decision-making patterns rather than for territorial gain or control of resources.[83]

It is common ground that computer network attacks which result in physical damage to civilian property, injury or death to civilians constitute attacks under international humanitarian law and are thus prohibited. However, the status of computer network attacks directed at civilian objects that do *not* result in such deleterious effects remains the subject of debate. Knut Dörmann argues that physical damage is not a requirement of an attack.[84] He points out that the definition of a military objective refers to neutralisation of an object as the possible outcome of an attack, and thus concludes that the mere disabling of an object, such as shutting down an electricity grid without destroying it, should also qualify as an attack.[85] However, Dörmann's argument

[80] Note that this only applies to objects on land: Art. 49(3).

[81] Bothe, *et al.*, *New Rules*, 289.

[82] Dinstein, *Conduct of Hostilities* (1st edn), 141.

[83] See Chapter 1 above, also see generally, Gray, *Another Bloody Century*; Smith, *Utility of Force*. This is by no means to suggest that the use of force against civilians or civilian objects to influence their governments is a creation of the modern military. World War II bombing campaigns are a prime example of operations specifically targeting civilian morale.

[84] Dörmann, 'Additional Protocols', 142–3.

[85] *Ibid.* Dörmann uses Bothe, *et al.*'s commentary which provides (in full): 'The term "neutralization", insofar as it deals with bombardment, refers to an attack for

relies on the location of the definition of 'military objective' in the section dealing with attacks against civilian objects. This is problematic for two reasons. First, neutralisation is referred to solely in relation to military objectives, not with regard to the prohibition of attacks against civilians. Secondly, the argument fails to acknowledge that the use of the term 'military objective' in Additional Protocol I is not restricted to articles and paragraphs detailing the permissible objects of attacks. The term is used in several places to discuss military operations more generally.[86]

Although the concept of neutralisation was added to the definition without much explanation by the drafting committee, it is clear that an object may be neutralised by a conventional attack, that is, one which causes death, injury or destruction, therefore the inclusion of the term cannot shed any light on whether or not an operation which does not cause such effects meets that definition.

Michael Schmitt takes an alternative view. As noted above, the definition of attacks refers to acts of violence, leading Bothe *et al.* to conclude that the concept of attacks does not include dissemination of propaganda, embargoes or other non-physical means of psychological, political or economic warfare; such a conclusion is also supported by state practice.[87] Schmitt acknowledges the disparity in the terminology between Article 48, which relates to 'military operations', and Article 52(2) and surrounding, which are all framed in terms of 'attacks'. However, he argues that the commentary on Article 48 specifies that the 'word "operation" should be understood in the context of the whole of the Section; it refers to military operations during which violence is used.' Schmitt concludes that, in light of this interpretation, computer network operations which are not designed to, nor would foreseeably

the purpose of denying the use of an object to the enemy without necessarily destroying it. For example, a specific area of land ... might be neutralised by laying landmines on it, thus denying its use to the enemy. Enemy artillery or surface-to-air missiles may be neutralized for a sufficient time to prevent their interference with a planned operation by firing antipersonnel munition at such targets in an attempt to force gun crews to take shelter. Such an attack would not be likely to destroy their intended target, but it would neutralize the target for the limited time required by the attacker.' Bothe, *et al.*, *New Rules*, 325.

[86] Art. 48 refers to military operations directed against military objectives; Art. 51(7) also refers to military objectives in the context of military operations.

[87] Bothe, *et al.*, *New Rules*, 289. See, for example, the US intervention in Suez as an example of economic warfare; examples of propaganda from World War II onwards; sanctions etc. as acts of political warfare.

cause, injury, death, damage, or destruction, may be directed against non-military objectives.[88] Should this be the case, it opens up a range of targets for the military which may not be attacked but may be targeted in other ways; computer network attack technology could open up large swathes of permissible targets which may better serve the coercive element of effects-based operations. However, the assertion that the distinction allows states to deliberately target, but not 'attack', civilian objects in ways not designed to cause injury, death etc. must be examined further.

While the present author agrees that there is a distinction between military operations and attacks, it does not follow that non-violent computer network attacks may be therefore conducted against civilian objects. Article 48, setting out the basic rule, prohibits 'operations' rather than attacks, and while it is agreed that the term operations should refer to *military* operations (as opposed to any other activity supporting the war effort), the term is in no way synonymous with attacks. The ICRC Commentary to Article 48 sets out that the word operations should be understood in the context of the whole of the section. Although, as Schmitt notes, the Commentary refers to 'military operations during which violence is used', it does so in order to distinguish operations of a military nature rather than 'ideological, political or religious campaigns'. The broader nature of the definition is confirmed by the following sentences of the commentary, which then sets out the dictionary definition of military operation which 'refers to all movements and acts related to hostilities that are undertaken by the armed forces'.[89] This interpretation is echoed by the commentary provided by Bothe *et al.* who note:[90]

> As used in Protocol I, this term deals generally with those aspects of military operations that are likely to cause civilian casualties or damage to civilian objects. Generally, the provisions of this section regulating attacks and other violent phases of military operations do not necessarily affect movement or manoeuvres by which a military unit secures or exercises dominion and control over key terrain features, lines of communication and avenues of approach … The discussion in the preceding paragraph of the operation to take a non-defended place which is open to occupation without resistance is equally relevant to the term 'military operation'. Nevertheless, such operations when

[88] Schmitt, 'Wired Warfare', 194.

[89] Pilloud, *et al.*, *Commentary*, para. 1875. This is also reiterated in the commentary to Art. 51 (Protection of the Civilian Population).

[90] Bothe, *et al.*, *New Rules*, 286.

carried out in areas containing a dense concentration of civilians present the adverse Party with a military target and invoke the obligation of Art. 58 to take feasible and appropriate precautions.

The reference to the discussion in the preceding paragraph refers to a discussion on the permissibility of entering and occupying a non-defended locality and exercising 'dominion and control' over such an area without contravening the prohibition on 'attacks by any means whatsoever'. Thus the commentary states the need to take precautions in attack when conducting military operations in dense concentrations of civilians. Schmitt argues that the general acceptance by the international community of psychological operations as an element of warfare, operations which he describes categorically as 'military operations', suggests that the term is shorthand for attacks in which physical violence is the consequence. The commentaries would appear to suggest otherwise. The fact that acts associated with the application of violence, such as movement and manoeuvre, do not necessarily result in violent consequences of their own does not exempt them from the requirement to take precautions in attack when they take place in dense concentrations of civilians, a requirement of a military operation, not just an attack.

Contrary to Schmitt's argument, Article 51 dealing with protection of the civilian population and Article 57 relating to precautions in attack both support this reading. Both articles set out the general rule relating to military operations in their first paragraphs and continue in subsequent paragraphs to relate specifically to attacks, the most likely method by which civilians would come to harm. However, the commentary to both articles emphasises the wider meaning of 'military operations'. In Article 51 it means 'all the movements and activities carried out by armed forces related to hostilities'.[91] Likewise in Article 57 'the term "military operations" should be understood to mean any movements, manoeuvres and other activities whatsoever carried out by the armed forces with a view to combat'.[92] Therefore, wherever an activity takes place in conjunction with hostilities it must be restricted to military objectives. Such an approach is also consistent with the position adopted by several states parties to the Protocol which made reservations or declarations interpreting the military advantage to be gained from an attack as a whole not merely

[91] Pilloud, *et al.*, *Commentary*, 1936.
[92] *Ibid.*, para. 2191.

a part.[93] Thus to fall within the definition of a military operation, the computer network attack must be associated with the use of physical force, but it does not have to result in violent consequences itself. The proximity calculation is similar to that used to determine direct participation in hostilities for civilians but is not as strict.[94] Where it does so, the computer network attack may not be directed against a civilian object.

This raises particular issues with attacks in which civilian networks or systems are targeted and used as gateways to get inside networks which are legitimate military objectives. A civilian network may be infected with malware, however it may not execute its payload until it reaches a network with the targeted network configuration.[95] The civilian system is essentially unharmed (albeit infected with a virus or other malware which is a nuisance to remove), but it passes on the malware to the targeted system where the payload executes, causing damage, injury or death. The legal difficulty here is that, unlike tanks passing through a village on their way to shell a military objective, the civilian network has been deliberately targeted by the attacker as part of its military operation. A query remains whether this is sufficient to hold that the state is directing its military operations against the civilian target, contrary to Article 48; or whether the operation is directed against the final, legitimate military objective and the minimal damage anticipated to the civilian target is merely taken into account within the usual proportionality equation. As with all computer network operations, the legality of the particular attack will depend on what it is designed to do. Further, this also raises a question whether, if a civilian object is targeted by computer network attack in order to achieve a physical force strike, was it making a sufficient contribution to military action to become a military objective in its own right?

93 Several states have indicated that in their target selection they will consider the military advantage to be anticipated from an attack as a whole and not from parts thereof. See, for example, the statements of Australia, Canada, France, Germany, Italy, New Zealand, Spain and the United States. Certain military manuals also consider that the anticipated military advantage can include increased security for the attacking forces or friendly forces. Henckaerts and Doswald-Beck, *Customary International Humanitarian Law*, Vol. 1, 31.

94 Making the calculation between operations and hostilities generally rather than in respect of computer network attacks: Bothe, *et al.*, *New Rules*, 324.

95 Stuxnet is an example of this style of attack. See, for example, Kim Zetter, 'Report: Stuxnet Hit 5 Gateway Targets on Its Way to Iranian Plant', *Wired*, 11 February 2011, www.wired.com/threatlevel/2011/02/stuxnet-five-main-target/# (last accessed 21 March 2011).

This difference between attacks and operations leaves an anomaly in the law and may place the military in the unenviable situation of having a course of action available to it before an armed conflict is embarked upon, only to have that same course of action denied during the course of conflict. Although Schmitt's analysis, that the meaning of the general provision is to be taken from the specific provisions relating to attack which follow it, is appealing in order to avoid this problem, at present it is not correct in law and is not without difficulty of its own. Once the ability to target civilian objects is permitted, it crosses the fundamental philosophical line enshrined in the 1868 Declaration of St Petersburg, which states 'the only legitimate object which States should endeavour to accomplish during war is to weaken the *military forces* of the enemy'. While the argument has been made that the character of war has changed in the twenty-first century such that this bright-line distinction between civilian and military is no longer appropriate,[96] until a legitimate forum decides that it is appropriate, then it is the present author's opinion that targeting civilian objects as part of a military operation remains prohibited by Article 57. This is also in conformity with the approach of the Appeals Chamber of the ICTY in *Kupreškić* which held that the Martens Clause should be used to interpret Articles 57 and 58 to protect civilians and limit attacks.[97] It is possible that the issue will be circumvented by the use of civilian contractors (or government operatives) engaging in those parts of the overall strategy which require that undertaking anyway.

4.2. *Indiscriminate attacks*

While, in general, computer network attacks allow for great precision in their application, some forms of malicious code are designed to spread from computer to computer without discrimination. The prohibition against indiscriminate attacks is a rule of customary international law and is expressed in Article 51(4) of Additional Protocol I:

Indiscriminate attacks are prohibited. Indiscriminate attacks are:

(a) those which are not directed at a specific military objective;
(b) those which employ a method or means of combat which cannot be directed at a specific military objective; or

[96] See section 1 of this chapter above.
[97] *Prosecutor* v. *Kupreškić* (2000) Case No. IT-95-16-T, International Criminal Tribunal for the Former Yugoslavia, para. 525.

(c) those which employ a method or means of combat the effects of which cannot be limited as required by this Protocol; and consequently, in each such case, are of a nature to strike military objectives and civilians or civilian objects without distinction.

As Dinstein points out, indiscriminate attacks differ from direct attacks against civilians in that 'the attacker is not actually trying to harm the civilian population', the injury to the civilians is merely a matter of 'no concern to the attacker'.[98]

Viruses and worms are two methods of computer network attack which are particularly likely to fall into this category as their effects are often not limited by their creators.[99] Although a virus or worm may be inserted into a closed system, or be programmed to check for certain conditions existing on the network or system prior to infection, many virus programmers design their creations to propagate indiscriminately. Both forms of malicious code can be designed to carry a payload which may cause a variety of effects from mere annoyance to compromising the system by leaving a backdoor for an attacker (in order to access or control the computer), or deleting or rewriting code on the infected system to varying effect.[100] Where the payload is designed to cause effects of such a magnitude as to constitute an attack, both types of malware may fall foul of Article 51(4) above, as they are methods of distribution which do not discriminate between civilian and military computers.[101] Interestingly, when the Stuxnet virus was discovered and reverse-engineered in 2010, researchers discovered that although the virus had been designed to propagate and spread fairly

[98] Dinstein, *Conduct of Hostilities* (2nd edn), 127.

[99] Viruses are programs or bits of malicious code which are attached to a program or file and spread from computer to computer as they are passed between users. Generally they cannot infect computers without being opened or run by the user. Worms, although similar in design, are self-replicating and take advantage of email or other information transport systems on the system to travel unaided. For example, the worm may replicate and send itself to everyone on the user's email contacts list; this is how several of the more high-profile email worms of the past few years (Slammer/Sapphire, Mydoom and Nimda) have propagated.

[100] Obviously rewriting the code which controls value pressure on an oil pipeline system or adjusting centrifuge speeds will have far more serious effects than the 'nuisance' viruses which generally circulate.

[101] From a technical standpoint it is possible to create a virus with a specific dialling protocol which will only dial specific IP address ranges. The IP address ranges of US and other military and intelligence agencies are freely available on the internet. However most virus writers are trying to get the maximum coverage possible, therefore the Slammer worm utilised a random dialling algorithm.

indiscriminately within a network, it had been coded to only execute its payload where specific conditions were fulfilled that would indicate it was on the targeted system.[102] Thus, although the method of warfare appears indiscriminate, the consequences which would constitute an attack are limited to particular systems.

Article 57(5) then sets out two examples of indiscriminate attacks: target area bombing, and disproportionate attacks causing excessive collateral damage.[103] It is difficult to envisage a computer network attack equivalent of target area bombing, as each attack must be conducted against a specific system or node in that system. No matter how 'high-level' the target system or node is in the victim network, and whatever the subsequent effects of its destruction or neutralisation, each node must be assessed on its own merits to qualify, or not, as a military objective. For example, the DNS root servers which run the internet have come under attack twice in recent years.[104] The denial of service attacks used in those incidents merely shut down the servers (or slowed them); however, in both incidents the individual servers targeted are either military objectives (for example, the case of root server G, maintained by the US Department of Defense) or civilian objects. Where the particular node (in this case, a server) is the reference for both military and civilian sources, it is a so-called dual-use target; the question of the effects of the attack will consequently become more important.[105]

[102] Where the conditions were not fulfilled the virus just finds a way to spread to other computers in search of its target. Falliere, OMurchu and Chien, *Stuxnet Dossier*; Zetter, 'Gateway Targets'.

[103] Art. 57(5) provides: 'Among others, the following types of attacks are to be considered as indiscriminate: (a) an attack by bombardment by any methods or means which treats as a single military objective a number of clearly separated and distinct military objectives located in a city, town, village or other area containing a similar concentration of civilians or civilian objects; and (b) an attack which may be expected to cause incidental loss of civilian life, injury to civilians, damage to civilian objects, or a combination thereof, which would be excessive in relation to the concrete and direct military advantage anticipated.'

[104] In both 2002 and 2007 attacks were launched against the DNS root servers. In 2002 all 13 root servers were attacked; however, in 2007 the attacks were limited to three of the servers including the server operated by the US Department of Defense. Ryan Naraine, *Massive DDoS Attack Hit DNS Root Servers*, www.internetnews.com/bus-news/article.php/1486981 (last accessed 6 September 2007); Roger A. Grimes, 'Security Adviser: DNS Attack Puts Web Security in Perspective', *InfoWorld* 8, 19 February 2007, www.infoworld.com/article/07/02/16/08OPsecadvise_1.html (last accessed 7 July 2011).

[105] See section 3 above.

One of the problems brought on by the interconnectedness of the internet is that the knock-on effects of computer network attacks may have even further reaching consequences than they do with conventional kinetic attacks. For example, it has been reported that US officials may have rejected launching a planned cyber attack against Iraqi financial computers because Iraq's banking network is connected to a financial communications network also located in Europe.[106] Similarly, the Iraqi oil pipeline communications network is reportedly cross-linked with the fibre-optic Tiger Song air defence network.[107] Such close linkages reportedly frustrated attempts by US Forces to design a computer network attack that could be limited to military objectives solely in Iraq.[108] Although exacerbated by computer network technology, this problem is certainly not unique to computer network attacks. As noted above, both the 1990–91 Gulf War and the NATO campaign over Kosovo faced similar problems when the attacks on the power supply networks resulted in water pumping stations being closed.[109]

5. Proportionality

The prohibition on indiscriminate attacks discussed above, as well as other provisions of the Protocol, are underpinned by the principle of proportionality. While the principle is considered part of the customary law of armed conflict, its codification and exact scope have not been without debate.[110] The principle is codified in Article 51(5)(b) which prohibits as indiscriminate:

> an attack which may be expected to cause incidental loss of civilian life, injury to civilians, damage to civilian objects, or a combination thereof, which would be excessive in relation to the concrete and direct military advantage anticipated.

[106] Clay Wilson, *Information Operations, Electronic Warfare, and Cyberwar: Capabilities and Related Policy Issues*, Congressional Research Service, RL31787 (2007) 5.

[107] Charles R. Smith, 'U.S. Information Warriors Wrestle with New Weapons', *NewsMax.com*, 13 March 2003, www.newsmax.com/archives/articles/2003/3/12/134712.shtml (last accessed 7 July 2011).

[108] *Ibid.*

[109] In the case of the attacks on Iraqi power networks, knock-on effects affected hospitals, refrigeration as well as water supplies. 'NATO Denies Targeting Water Supplies', *BBC News*, 24 May 1999.

[110] *Nuclear Weapons Case*, per Higgins J in her dissenting opinion (dissenting on other grounds); Henckaerts and Doswald-Beck, *Customary International Humanitarian Law*, Vol. 1, 46.

Indiscriminate acts in violation of the rule of proportionality constitute 'grave breaches' under Article 85 of Additional Protocol I. The wording is repeated in Article 57(2)(iii), requiring those planning an attack to refrain from deciding to launch attacks which would cause 'excessive' damage to civilians in relation to the concrete and direct military advantage anticipated.

Ruth Wedgwood has addressed the proportionality equation for computer network attacks and suggested that greater damage to civilian objects may be tolerated in order to eliminate a security threat, so long as that damage is reversible.[111] This view seems to be based on the idea that an attack may not cause destruction but mere incapacity. Although this view corresponds with Knut Dörmann's approach outlined in section 4.1 above,[112] it does not accord with the requirement of the Protocol's definition of attacks as 'acts of violence'. Taken to its logical conclusion, Wedgwood's argument would seem to infer that any damage is permissible as long as the damage can be reconstructed at the conclusion of the conflict.

Michael Schmitt has also queried the extent to which specialised computer expertise must be available during the targeting process to assess possible collateral damage and incidental injury.[113] As he points out, in traditional kinetic attacks, properly trained mainstream military officers can usually conduct reliable collateral damage estimates based on their knowledge of the weapons system involved and its effects; however, in computer network attacks highly specialised expertise would be required.[114] While this is undoubtedly true, the present author considers that where states wish to engage in the kind of sophisticated networked warfare that computer network attacks are designed for, the analysis of targets, connectivity and likely collateral damage would likewise have to increase in sophistication. Commanders are obliged to make their decisions on the basis of all the information available to them at the relevant

111 Ruth G. Wedgwood, 'Proportionality, Cyberwar, and the Law of War' in M. N. Schmitt and B. T. O'Donnell (eds.), *Computer Network Attack and International Law* (Naval War College, Newport, RI, 2002) 219–32, 228.

112 Dörmann, 'Additional Protocols', 142–3.

113 Michael N. Schmitt, 'CNA and the Jus in Bello: An Introduction' (Paper presented at the International Expert Conference on Computer Network Attacks and the Applicability of International Humanitarian Law, Stockholm, 17–19 November 2004) 101–25, 117.

114 *Ibid.*

time; however, they are also under an obligation to obtain the best possible intelligence,[115] and to act with due diligence and in good faith.[116] Where commanders do not possess the requisite knowledge themselves they must obtain it from a specialist. If, in establishing their intelligence-gathering networks regarding the likely collateral damage of an attack, the practical and logical solution is to have a network expert involved in the targeting process then so be it. Given the general nature of computer network attacks and the likelihood that such attacks will be planned and launched away from an immediate combat zone in circumstances where there is more time for considered decision making, it does not seem unduly onerous to require commanders to consult network specialists to ensure they have all the relevant information. Although the obligations are addressed to those who plan and decide on an attack, should individual cyber warriors become aware that the proportionality rule will be breached during the implementation of an attack, they are then obligated to suspend or cancel the attack.

Two major problems for modern proportionality judgments, especially with respect to computer network attacks, are the extent to which the knock-on effects of attacks must be incorporated into the calculation and the effects of increasingly dual-use technological systems on that calculation.

5.1. Knock-on effects

As Christopher Greenwood has noted, the Gulf War of 1990–91 has highlighted the fact that knock-on effects of attacks cause more harm to civilians than the direct effect of the attack itself.[117] Application of the proportionality test today, at least at the strategic level, requires that less immediate damage of this kind must also be taken into account, although the difficulty of doing so is apparent.[118] While this problem is not unique to computer network attacks (both the 1990–91

[115] Henckaerts and Doswald-Beck, *Customary International Humanitarian Law*, Vol. 1, 50, 54.

[116] Michael Bothe, 'Legal Restraints on Targeting: Protection of Civilian Population and the Changing Faces of Modern Conflict' (2001) 31 *Israel Y.B. Hum. Rts* 35 regarding the military character of an objective selected for attack.

[117] Christopher Greenwood, 'The Law of Weaponry at the Start of the New Millennium' in M. N. Schmitt and L. C. Green (eds.), *The Law of Armed Conflict: Into the Next Millennium* (Naval War College, Newport, RI, 1998), 185–231, 202.

[118] *Ibid.*

Gulf War and the NATO action in Yugoslavia illustrated the knock-on effects of targeting the electricity networks),[119] the problem is exacerbated by the nature of computer systems and linkages between military and civilian systems. Certainly the attacker will be required to have conducted some sort of mapping of the target network or system to ascertain what ancillary networks or systems are connected to the target. It is unclear how many levels of these cascading effects will need to be taken into account by the planners of the attacks and those executing them. Michael Schmitt argues that those effects that are reasonably foreseeable, no matter what 'tier' of effect they may be, must be factored into the proportionality calculation.[120] This fits with the language of the article which refers to the *expected* consequences. Ironically, the move towards militaries buying off-the-shelf technology and systems may aid attackers in correctly predicting the effects of certain attacks.

5.2. *Dual-use systems*

To what extent does the fact that a state has deliberately integrated civilian and military systems together impact the proportionality calculation? For example, the Iraqi Tiger Song air defence network was cross-wired with the Iraqi oil pipeline communications network;[121] additionally a large majority of US military communications travel across civilian networks. While such actions undoubtedly expose the networks to attack as military objectives, can the argument be made, as with the case of voluntary human shields,[122] that if the defenders have integrated their military and civilian systems such that the military system may not be attacked without impacting the civilian, that the civilian impact should be excluded from the proportionality calculation? The answer must be negative, as the protection only exists

[119] In the Gulf War of 1990–91, Allied forces disabled the Iraqi power distribution networks using a variety of tactics including carbon-fibre filament munitions. The unintended (and apparently unexpected) side effects of these attacks were to deny electricity to the sewerage and water treatment facilities supplying the civilian population: William M. Arkin, 'Cyber Warfare and the Environment' (2001) 25 *Vermont Law Review* 779, 781, citing Daniel T. Kuehl, 'Airpower Vs Electricity' (1995) 18 *Journal of Strategic Studies* 28. Similarly, when NATO forces attacked Yugoslavia's electrical supply network, water pumping stations were affected: 'NATO Denies Targeting Water Supplies'.

[120] Schmitt, 'Fault Lines', 296; Schmitt, *High and Low-Tech Warfare*, 10.

[121] Smith, 'U.S. Information Warriors Wrestle with New Weapons'.

[122] Schmitt, 'Fault Lines', 298.

for civilians not civilian objects. In addition, for the most part civilians would be unaware that the systems were so intertwined; thus the parallel would be drawn with involuntary human shields which must definitely be taken into account when conducting the targeting analysis. Further, as Schmitt points out, there are instances where protected objects lose their protected status due to the adversary's misconduct; a hospital housing combatants (who are not otherwise *hors de combat*) may be attacked once a warning to desist has been ignored.[123] Rogers argues that a tribunal that is considering the criminal liability of an attacker in respect of death or injury to civilians would be entitled to consider the extent to which the defenders had flouted their obligations to separate military objects from civilian objects.[124]

6. Precautions in attack

Article 57 of Additional Protocol I requires attackers to take precautionary measures in carrying out military operations and attacks. The ICTY has recognised the customary nature of these precautions in both the *Kupreškić* and *Tadic* cases.[125] In particular, the Appeals Tribunal in *Tadic* cited with approval the UN General Assembly Resolution 2675 that 'all necessary precautions should be taken to avoid injury loss or damage to civilian populations', stating that the resolution represented customary international law 'in armed conflicts of any kind'.[126] The judgment confirms that the rule extends to non-international armed conflicts despite a lack of specific provisions in Common Article 3 or Additional Protocol II. Oeter notes that the rule is deduced from the general principle of military necessity and, in the case of non-international armed conflicts, the humanitarian minimum standard laid down in Common Article 3 of the Geneva Conventions.[127]

Article 57(1) provides that in the conduct of military operations, constant care must be taken to spare the civilian population, civilians

[123] *Ibid.*, 300.

[124] Rogers, *Law on the Battlefield*, 129. Rogers considers this proportionality approach would redress the balance which might otherwise be tilted in favour of the unscrupulous.

[125] *Tadic (Interlocutory Appeal)*, para. 111–12; *Kupreškić*, para. 524.

[126] *Tadic (Interlocutory Appeal)*, paras 111–12, citing GA Res 2675 (XXV) Basic Principles for the Protection of Civilian Populations in Armed Conflicts, UN GAOR, 25th Sess., Supp. No. 28, 76, UN Doc. A/8028 (1971) of 1 December 1970.

[127] Oeter, 'Methods and Means of Combat', 208.

and civilian objects.[128] As discussed in section 4.1 above, military operations are a broader concept than attacks and the general rule thus applies more widely than the specific rules which follow relating to that part of the operation that constitute an attack. Boivin has usefully summarised the measures to be taken by those who plan or decide on attacks as follows:[129]

1. do everything feasible to verify that the objectives to be attacked are military objectives;[130]
2. take all feasible precautions in the choice of means and methods of warfare;[131]
3. do everything feasible to assess whether the attack may be expected to cause excessive collateral damage;[132]
4. do everything feasible to cancel or suspend an attack if it becomes apparent that the proportionality rule will be breached, or that the target is not a military objective or that it is subject to special protection;[133]
5. give effective advance warning prior to an attack that is likely to affect the civilian population, unless the circumstances do not permit;[134] and
6. where a choice between several military objectives is possible, choose the one that will cause the least danger to civilian lives and civilian objects.[135]

Computer network attacks raise several issues with regard to the requirement to take precautions in attack. However, two general matters should be dealt with before addressing these specific concerns. First, it should be noted that, as with the obligations regarding targeting set out in Article 49 and following, the majority of provisions relating to the precautions in attack refer to 'attacks'. That is, other than the general rule expressed in Article 57(1) requiring that constant care be taken to spare the civilian population, individual civilians and civilian objects in the course of military operations, the specific obligations

[128] Art. 57(1), Additional Protocol I.

[129] Alexandra Boivin, *The Legal Regime Applicable to Targeting Military Objectives in the Context of Contemporary Warfare*, University Centre for International Humanitarian Law, 2 (2006) 36, www.adh-geneva.ch/docs/publications/collection-research-projects/CTR_objectif_militaire.pdf (last accessed 7 July 2011).

[130] Art. 57(2)(a)(i), Additional Protocol I.

[131] Art. 57(2)(a)(ii), Additional Protocol I.

[132] Art. 57(2)(a)(iii), Additional Protocol I.

[133] Art. 57(2)(b), Additional Protocol I.

[134] Art. 57(2)(c), Additional Protocol I.

[135] Art. 57(3) Additional Protocol I.

listed above will only apply to those computer network attacks which result in physical damage, injury or death.[136]

Secondly, the first four precautionary measures all refer to 'feasibility', a measure which has been interpreted by many states, and defined in the Commentary, as 'those precautions that are practicable or practically possible, taking into account all the circumstances ruling at the time, including humanitarian and military considerations'.[137] This wording has also been adopted by Protocols II, III and amended Protocol II to the Conventional Weapons Treaty.

6.1. *Verification of military objectives*

The law requires that commanders do everything feasible to verify that the target is not protected from attack and that it is a military objective.[138] As most targeted computer network attacks (as opposed to indiscriminate attacks, such as untargeted worms and viruses) require fairly extensive system surveillance and scanning to determine an entry point to the system, this obligation should not prove difficult for pre-selected targets. However verification of targets of opportunity may prove more difficult.

Attacks refer to acts of violence, both in offence and in defence. Where a commander wishes to respond to an adversary's computer network attack, they must verify, first, the source of the attack, and secondly, that it is a military objective. As discussed in Chapter 3, the problem of accurate attribution of computer network attacks is made difficult by the tendency of attackers to spoof the origin of the attack, that is, to deliberately mislead the adversary as to the source of the attack. Note that the deception must not extend to a violation of Article 51(7) by the dissemination of false intelligence reports intended to induce the enemy to attack civilians and civilian objects, in the mistaken belief that they are military objects.[139]

The extent of knowledge to be expected from the commander in assessing possible targets may prove problematical, and echoes the

136 For a discussion of the difference between attacks and operations and what this may mean for computer network attacks, see section 4.1, above.

137 See, for example, UK Declaration of Understanding, corrected letter of 28 January 1998 sent to the Swiss government by Christopher Hulse, HM Ambassador of the United Kingdom.

138 Art. 57(2)(a)(i), Additional Protocol I.

139 Bothe, *et al.*, *New Rules*, 363, giving the example of British intelligence in World War II sending out false intelligence reports thus inducing the Luftwaffe to bomb civilian areas believing they were bombing strategic military objectives.

concerns held by many parties to Additional Protocol I regarding the level at which commanders' decisions were to be taken. As discussed in relation to proportionality above, Michael Schmitt has also raised this issue, querying to what extent computer expertise must be available during the targeting process to assess possible collateral damage and incidental injury.[140] However, military commanders are not expected to have personal knowledge of every target they attack and must rely on intelligence reports for much of their information regarding conventional attacks. Military commanders have to make their decisions on the basis of the information from all sources that are available to them at the time.[141] Many military manuals stress that the commander must obtain the best possible intelligence, including information on concentrations of civilian persons, important civilian objects, specifically protected objects, the natural environment and the civilian environment of military objectives.[142] In its Final Report to the Prosecutor, the Committee Established to Review the NATO Bombing Campaign Against the Federal Republic of Yugoslavia described the obligation thus:[143]

> A military commander must set up an effective intelligence gathering system to collect and evaluate information concerning potential targets. The commander must also direct his forces to use available technical means to properly identify targets during operations. Both the commander and the aircrew actually engaged in operations must have some range of discretion to determine which available resources shall be used and how they shall be used.

It follows that the same level of reliance on information must occur in respect of computer network attacks. In practical terms, it is also unlikely that any computer network attacks will be organised from a field position located in a combat zone, but would be the task of dedicated teams of computer technicians elsewhere in the battlespace where the requisite expertise is available.

6.2. *Choice of weapons*

Ironically, the requirement for an attacker to take all feasible precautions in the choice of means and methods of warfare may require states

[140] Schmitt, 'CNA and the Jus in Bello', 117.

[141] Henckaerts and Doswald-Beck, *Customary International Humanitarian Law*, Vol. 1, 50, 54, citing the military manuals of Algeria, Australia, Austria, Belgium, Canada, Equador, Egypt, Germany, Ireland, Italy, the Netherlands, New Zealand, Spain, the United Kingdom and the United States.

[142] *Ibid.*, 55.

[143] ICTY, *Final Report to the Prosecutor*, para. 29.

that have the ability to launch computer network attacks to use that ability in preference to more traditional means. A similar argument has been advanced in relation to the use of precision-guided missiles, particularly with respect to warfare in urban environments.[144] Article 57(2)(a)(ii) of Additional Protocol I requires those deciding on attacks to take all feasible precautions in the choice of means and methods of attack with a view to avoiding, and in any event to minimising, incidental loss of civilian life, injury to civilians and damage to civilian objects. The ICRC study has concluded that state practice establishes this as a norm of customary international law applicable in both international and non-international armed conflicts.[145] As Kalshoven points out, the primary obligation in the provision is to 'avoid' damage to the civilian population; the goal of 'minimizing' such damage will come into play only when total avoidance is not feasible.[146] These factors tend to promote the use of computer network attack methods which do not have an inherent risk to civilian objects and do not necessarily cause destruction. As Schmitt notes:[147]

> Whereas in the past physical destruction may have been necessary to neutralize a target's contribution to the enemy's efforts, now it may be possible to simply 'turn it off'. For instance, rather than bombing an airfield, air traffic control can be interrupted. The same is true of power production and distribution systems, communications, industrial plants, and so forth.

One of the difficulties of certain computer network attack techniques is the fact that once they have been used for the first time, they may be guarded against and may not work again.[148] Does this impact on a commander's obligation to field such weapons? The limited availability of weapons has been discussed with respect to the use of precision-guided munitions. Thus, Dinstein argues that the legal position is quite simple: the law of armed conflict instructs the planners to take whatever steps are necessary in order to avoid or minimise collateral damage to

144 See generally, Stuart Walters Belt, 'Missiles over Kosovo: Emergence, Lex Lata, of a Customary Norm Requiring the Use of Precision Munitions in Urban Areas' (2000) 47 *Naval Law Review* 115. For an alternative view see also, D. L. Infeld 'Precision guided Munitions' cited in Dinstein, *Conduct of Hostilities* (1st edn), 126.

145 Henckaerts and Doswald-Beck, *Customary International Humanitarian Law*, Vol. 1, 57.

146 Frits Kalshoven and Liesbeth Zegveld, *Constraints on the Waging of War* (ICRC, Geneva, 2001), 108.

147 Schmitt, 'Wired Warfare', 394.

148 In the same way that security flaws in software may be patched for domestic applications, ports may closed, patches installed and anti-virus software updated.

civilians: 'The availability of precision-guided munitions by no means forecloses alternative precautionary options.'[149]

Can such weapons be held in reserve in case they are needed further down the line? With regard to the Gulf War of 1990–91, Christopher Greenwood has noted that the United States did not invariably use precision-guided munitions whenever an attack involved a risk of collateral damage, on the grounds that supplies of these weapons were limited and they might have to be conserved for attacks on other objectives later in the campaign. Greenwood states that this approach involves a broad (though not untenable) interpretation of the duty to take 'feasible' precautions.[150] Certainly, the position is stronger where the weapons are held in reserve for an attack which is definitely planned down the line. The US Department of Defense Final Report to Congress on the conduct of the Gulf War of 1990–91 makes it clear that the United States did not regard itself as bound always to select the method or means of attack which would cause the least danger to civilians, but was entitled to take account of the risk to coalition aircrews and the likelihood of successfully destroying the target. The report states:[151]

> To the degree possible and consistent with allowable risk to aircraft and aircrews, aircraft and munitions were selected so that attacks on targets within populated areas would provide the greatest possible accuracy and the least risk to civilian objects and the civilian population. Where required, attacking aircraft were accompanied by support mission aircraft to minimize attacking aircraft aircrew distraction from their assigned mission.

Greenwood argues that this approach is consistent with the interpretation placed on the word 'feasibility' in Article 57 by several states on signature or ratification of the Protocol and that it is inconceivable that any state would fail to take such factors into account.[152] Certainly the United Kingdom, which interprets the term 'feasible' to mean 'all measures practicable under the circumstances ruling at the time', lists the risks to a commander's own troops under the various options open

[149] Dinstein, *Conduct of Hostilities* (1st edn), 126–7.

[150] Christopher Greenwood, 'Customary International Law and the First Geneva Protocol of 1977 in the Gulf Conflict' in P. J. Rowe (ed.), *The Gulf War 1990–1991 in International and English Law* (Routledge, London, 1993) 63–88, 85–6.

[151] US Department of Defense, *Conduct of the Persian Gulf War: Final Report to Congress*, US Department of Defense (1992) 697–8.

[152] Greenwood, 'Customary International Law in the Gulf', 85, 375, fn. 117.

to him as a factor to be considered when choosing what means and methods of warfare to employ.[153]

It must also be noted that the choice of weapons available to the warfighter will vary, depending on the level of the decision making. A military commander will have a greater ability to choose between various methods of attack than a unit commander located in the battlespace or an individual combatant. As Kalshoven has pointed out, a combatant simply cannot be equipped with a wide array of weapons for all kinds of situations, as the golf player is with his bag of clubs.[154] Interestingly, the advent of networked militaries and the concept of network-centric warfare actually allows a unit on the ground access to more technologies than have been previously available. Real-time communications allow the soldier on the ground to call in air strikes, 'paint' targets with laser sights for laser guided missiles or plant GPS locator beacons, all increasing the options available to them rather than being limited to the contents of their backpack. These increased options will require commanders in the field to assess the appropriate options for attack.

Even more than the military commanders, the authorities who decide on the armament of the armed forces have the option to select suitable weapons. Although Kalshoven has noted that considerations of military efficiency will tend to carry more weight in the deliberations of those authorities, he considers that 'at the same time they will fail in their duty if they lose sight of the humanitarian requirement of minimisation of human suffering'.[155] However, humanitarian law does not contain any obligation to acquire military capabilities that provide greater protection to civilians; instead, it limits itself to imposing a duty to use capabilities once in the inventory.[156]

[153] UK Ministry of Defence, *UK Manual*, 83–4. The factors are as follows: (1) the importance of the target and the urgency of the situation; (2) intelligence about the proposed target – what it is being or will be used for and when; (3) the characterisitcs of the target itself, for example, whether it houses dangerous forces; (4) what weapons are available, their range, accuracy and radius of effect; (5) conditions affecting the accuracy of the targeting, such as terrain, weather, and time of day; (6) factors affecting incidental loss or damage, such as the proximity of civilians or civilian objects in the vicinity of the target or other protected objects or zones and whether they are inhabited, or the possible release of hazardous substances as a result of the attack; (7) the risks to his own troops under the various options open to him.

[154] Frits Kalshoven, 'The Soldier and His Golf Clubs' in C. Swinarski (ed.), *Studies and Essays on International Humanitarian Law and Red Cross Principles in Honour of Jean Pictet* (Martinus Nijhoff, The Hague; Boston, 1984) 369–85, 385.

[155] *Ibid.* [156] Schmitt, *High and Low-Tech Warfare*, 11.

6.3. Choice of targets

Article 57(3) of Additional Protocol I provides that where there is a choice of several military objectives for obtaining a similar military objective, the objective chosen should be the one that causes the least danger to civilian lives and to civilian objects. Christopher Greenwood notes that, although this Article may have gone beyond the customary law as it stood at 1977, it certainly represented customary international law by the 1990–91 Gulf War.[157] The obligation is particularly relevant for computer network attacks, as the form of attack opens multiple options to achieve the same effect. For example, a system may be neutralised by disabling an essential component of the system so that it is unable to function, attacking the system as a whole, attacking the network on which that system resides, or by shutting off the electrical supply to the target system. All would achieve the same result, i.e. denying the target system to the adversary. This is, to a certain extent, a natural extension of the obligation to choose means and methods of attack which minimise harm to civilians; however, as computer network attacks increase the ability to break target networks into their component systems, the targeting analysis will likewise become more refined. Thus the obligation is to choose 'the lesser of two evils'; the example given in the ICRC Commentary is the bombing of railways lines rather than stations which are primarily located in urban areas.[158] Of course the obligation, as with the other precautions in attack, is to take feasible measures, therefore the question will be dependent on the ability to access the networks, the ability to determine the effects of neutralising a particular component, the desired effect of the attack and whether the systems can be cracked in time for the purposes of the operation. Although the specifics of this provision will only apply to those computer network attacks which amount to attacks as discussed above, the general obligation under Article 57(1) would oblige attackers to take these factors into consideration as part of any military operation.

At first glance, the failure to sufficiently refine the target, particularly in relation to dual-use targets, appears to raise possible parallels with target area bombardment which is prohibited by Article 51(5)(a):[159]

[157] Greenwood, 'Customary International Law in the Gulf', 83.

[158] Bothe, *et al.*, *New Rules*, paras 2227–8.

[159] Art. 51(5)(a), Additional Protocol I provides: 'The presence or movements of the civilian population or individual civilians shall not be used to render certain

An attack by bombardment by any method or means which treats as a single military objective a number of clearly separated and distinct military objectives located in a city, town, village or other area containing a similar concentration of civilians or civilian objects.

If a system or network is disrupted too far 'upstream' from the ultimate objective, it will affect the not only the objective, but all other nodes or systems (including any civilian systems) on the network. However, the analogy is a false one; any node or system attacked will qualify as a military objective or not on its own merits (with any knock-on effects included in the proportionality equation), rather than treating a number of objectives as a single objective.

7. Precautions against the effects of attacks

The main prohibition addressed to defenders in respect of the civilian population or individual civilians is expressed in Article 51(7) of Additional Protocol I which prevents the use of civilians as human shields for military objectives.[160] However, this does not raise many issues for computer network attacks, other than to comment that counter-attacks against computer network attacks may not necessarily be returned in kind, and the use of civilian contractors to defend against intrusions into military networks does not prevent them from being targeted or attacked by other means. Of more interest is Article 58, which provides that:

The Parties to the conflict shall, to the maximum extent feasible:

(a) without prejudice to Article 49 of the Fourth Convention, endeavour to remove the civilian population, individual civilians and civilian objects under their control from the vicinity of military objectives;
(b) avoid locating military objectives within or near densely populated areas;
(c) take the other necessary precautions to protect the civilian population, individual civilians and civilian objects under their control against the dangers resulting from military operations.

points or areas immune from military operations, in particular in attempts to shield military objectives from attacks or to shield, favour or impede military operations. The Parties to the conflict shall not direct the movement of the civilian population or individual civilians in order to attempt to shield military objectives from attacks or to shield military operations.'

160 Sassòli, 'Targeting', 206.

These rules represent customary international law;[161] however, as Christopher Greenwood points out, in most conflicts little more than lip service appears to have been paid to this rule.[162] The wording of the provision clearly indicates that these obligations are weaker than those of the attacker.[163] Unlike the obligations of the attacker, failure to comply with the provision does not constitute a grave breach of the Protocol. Similarly, defenders' obligations only have to be taken 'to the maximum extent feasible', and the defender is only required to 'endeavour to remove' the civilian population and to 'avoid' locating military objectives nearby.

What does this mean for computer network attacks? Paragraph (a) requires that parties endeavour to remove civilians and civilian objects under their control from the vicinity of military objectives, to the maximum extent feasible.[164] This would require parties (where practicable) to extricate military systems and networks from civilian ones and to avoid using civilian networks for military communications. However, as has been pointed out throughout this volume, the increasing civilianisation of the military and widespread networking of modern militaries has led to the opposite happening. Examples are the integration of civilian and military technology such as civilian use of the GPS system, and military use of civilian communications satellites and networks.[165] The lack of partitioning between Iraqi military and civilian systems caused difficulties for the US in achieving some of their computer network attacks aims in the 2003 Iraq conflict.[166] Given the lack of feasibility of

[161] Henckaerts and Doswald-Beck, *Customary International Humanitarian Law*, Vol. 1, 67–76. The Trial Chamber in *Kupreškić* considered that both Art. 57 API (pertaining to precautions in attack) and Art. 58 are now part of customary international law, not only because they specify and flesh out general pre-existing norms, but also because they do not appear to be contested by any state, including those which have not ratified the Protocol. *Kupreškić*, para. 524.

[162] Greenwood, 'Customary International Law in the Gulf', 374, fn. 122.

[163] Sassòli, 'Targeting', 207.

[164] As with other articles in the Protocol which incorporate the word 'feasible', a number of delegations have indicated that the word feasible means that which is practicable or practically possible, taking into account all the circumstances at the time, including those relevant to the success of military operations. Bothe, *et al.*, *New Rules*, 373.

[165] Latest figures estimate 98 per cent of US government communications travel over civilian-owned and operated networks and systems. Michael McConnell, Former Dir. of Nat'l Intelligence, Keynote Address at the Texas Law Review Symposium: Law at the Intersection of National Security, Privacy, and Technology (4 Feburary 2010) cited in Jensen, 'Precautions'.

[166] Smith, 'U.S. Information Warriors Wrestle with New Weapons'.

separating military and civilian networks, states must therefore adopt measures to protect the civilian population and civilian objects.

Article 58(c) of Additional Protocol I imposes a general obligation on parties to the conflict to take other necessary precautions to protect the civilian population, individual civilians and civilian objects under their control from dangers resulting from military operations. It is unclear how far this obligation will go in an age of computer network attack. For one thing, most critical infrastructure and networks which might be exposed to computer network attacks through military usage or proximity are in fact under the control of the private sector. Secondly, the steps taken by states in relation to traditional threats indicate construction of shelters, digging of trenches, direction of traffic, guarding of civilian property and the mobilisation of civil defence organisations.[167] Such measures in relation to digital property may include ensuring that all publicly administered digital property is properly backed up and all systems have built-in redundancy, so that any loss of systems or information can be restored. Similar digital disaster planning programmes were put in place on a wide scale in preparation for the year 2000 changeover following predictions of catastrophic electronic failures.

[167] See summary of practice reports submitted to the ICRC customary international humanitarian law study. Henckaerts and Doswald-Beck, *Customary International Humanitarian Law*, Vol. II, 419–29.

7 Measures of special protection

Special protection is granted to certain personnel and objects under the laws of armed conflict. Although even the most advanced information society is yet to deploy technology which would enable direct attacks against protected personnel using computer network attacks, other objects have become sufficiently incorporated into computer networks to make them vulnerable to a computer network attack. The environment, cultural property, installations containing dangerous forces (namely dams, dykes and nuclear power plants), hospitals and other medical units are all granted special protection from attack over and above the general protection granted to civilian infrastructure. As much of the developed world's critical infrastructure is now controlled using computer networks, this protection will extend to prohibit computer network attacks against such objects. Dams, power stations, chemical plants, water and sewage, gas and oil pipelines are all controlled by networked systems such as Supervisory Control and Data Acquisition (SCADA) systems, thus making them vulnerable to computer network attacks.[1] Vast digitisation projects and the birth of 'born-digital' cultural property not only help to preserve and share such works with a wider audience but expose them to attack and incidental damage via computer network attacks. In addition, information societies now rely on digital information and networks for fast and reliable access to up-to-date information; this trend is also seen in the medical sector where medical records and other information are stored and transmitted over computer networks, thus also leaving them susceptible to computer network attack. While there is no question that

[1] SCADA systems are organisational systems which control other networks and automated processes.

these installations, works and data remain protected by their status regardless of the means or method of warfare adopted against them, some new issues arise in light of new technology.

1. The environment

Harm caused to the environment during periods of armed conflict, both directly and as a by-product of war, has been an unfortunate inevitability of conflict throughout history; post-modern warfare is no exception. For example, the 2006 armed conflict between Hezbollah and Israel resulted in severe damage to the Lebanon coastline after the Israeli bombing of a power station caused a massive oil spill;[2] likewise the oil fires resulting from the 1990–91 Gulf War caused substantial damage to the ecology of Kuwait and Iraq.[3] In the latter incident, the deliberate damage caused to the environment by Iraqi troops shocked the world and prompted a flurry of legal commentary on the degree of protection provided by the law pertaining to environmental warfare.[4]

At the time of writing, no reports of incidents of wartime environmental damage using computer network attack exist in the public domain. However, in April 2000 a domestic case emerged in Queensland, Australia, solving a mystery that had perturbed authorities for months, and showed the potential of this new type of attack for environmental damage.[5] Vitek Boden was arrested after being caught using a stolen computer and radio transmitter to gain access to a sewerage treatment system. Over the previous two-month period Boden had accessed the system forty-six times, gaining complete control of the treatment of the region's sewerage and drinking water facilities and dumping 250 million tonnes of putrid sludge into the area's rivers and parks, killing wildlife and plants. Although Boden was acting for personal reasons

[2] Mark Kinver, '"Damage Is Done" to Lebanon Coast', *BBC News*, 8 August 2006, http://news.bbc.co.uk/1/hi/sci/tech/5255966.stm (last accessed 15 July 2011).

[3] Oil well fires were greater in number than all well fires in previous history put together. Oil slicks were more than two to three times the size of the word's previously largest oil spill, from the *Exxon Valdez*. Gushing oil wells, pipes and storage tanks left rivers and lakes of spilled oil, more than 90 million barrels covering over 50 per cent of Kuwait's land area. This huge amount of exposed oil released toxic substances, heavy metals and unequalled emissions of hydrocarbons. Arkin, 'Cyber Warfare and the Environment'.

[4] Eric Talbot Jensen, 'The International Law of Environmental Warfare: Active and Passive Damage During Armed Conflict' (2005) 38 *Vand. J. Transnat'l L.* 145.

[5] *R* v. *Boden*.

in his attacks,[6] the case illustrates the potential for intrusion and establishment of control over infrastructure utilising SCADA systems, a tactic which could easily be adopted for military purposes. Could a repeat of the 1990–91 Gulf War oil disaster occur as a result of computer network attacks? Based on current technology, the answer is undoubtedly yes. For example, approximately 30 per cent of the United Kingdom's oil output runs over one pipeline system in the North Sea, pumping a volume of 2.5 million gallons of oil a day. This, the Forties pipeline, is controlled by a SCADA system similar to that which controlled the Queensland water and sewerage treatment plant infiltrated by Boden. By resetting the valves on the North Sea Forties oil pipeline it may be possible to cause a hammering effect in the lines;[7] the resulting rupture of the pipeline would be likely to cause severe damage to the ecology of the area (including an area of special scientific interest) and cripple the UK oil supply for weeks.

Specific protection of the environment in armed conflict has been rising in prominence since the Vietnam War and has come to the forefront of legal attention since the 1990–91 Gulf War. However, as with all of the issues outlined in this book, the general principles of the laws of armed conflict continue to apply to computer network attacks despite the application of new technology to cause harm. Those general principles also protect the environment indirectly, even where specific environmental prohibitions do not apply.[8]

1.1. *Additional Protocol I*

Additional Protocol I contains two innovative articles containing measures of direct protection for the environment during international armed conflict: Articles 35(3) and 55. There is no equivalent provision in Additional Protocol II relating to non-international armed conflicts.[9]

[6] Evidence at his trial suggested that Boden was motivated either by a desire for vengeance or that he hoped to be re-employed by the company running the system in a consulting capacity to solve the problem he had caused: *ibid.*; Gellman, 'Cyber-Attacks by Al Qaeda Feared'.

[7] See also the account of the 'Farewell Dossier' incident in Appendix 1, although an explosion of that magnitude would be unlikely in a predominantly seabed system.

[8] For example, the general protection granted to civilian objects, protection of objects indispensable to the survival of the civilian population and simple application of the principles of proportionality and necessity will provide protection for many parts of the environment.

[9] A proposal was made at the diplomatic conference to introduce into Additional Protocol II a provision analogous to Art. 35(3) and Art. 55 of Additional Protocol I, but the idea was ultimately rejected. Antoine Bouvier, 'Protection of the Natural

Although the ICRC customary international humanitarian law study characterises these rules as customary, considerable debate remains over whether this is in fact the case.[10] Both articles contain broad prohibitions against any means or method of warfare which is intended or may be expected to cause damage to the environment;[11] thus computer network attacks which meet the requisite criteria for damage are also prohibited.

Both Article 35(3) and Article 55(1) are restricted in terms of the intended or foreseeable consequences of the attack on the environment to damage which is 'widespread, long-term and severe'.[12] However, unlike Article 35(3), Article 55(1) is phrased in terms of protection of the population, in that its prohibitions are directed at environmental damage that may be prejudicial to the health or survival of the population, rather than protection of the environment itself.[13] Article 55(1) also contains a general obligation of care to protect the natural environment against 'widespread, long-term and severe' damage. Article 55(2) of Additional Protocol I contains a blanket prohibition of any attacks against the natural environment by way of reprisals; this con-

Environment in Time of Armed Conflict' (1991) 285 *IRRC* 567, www.icrc.org/web/eng/siteeng0.nsf/html/57JMAU (last accessed 19 September 2006).

[10] Karen Hulme, 'Natural Environment' in E. Wilmshurst and S. Breau (eds.), *Perspectives on the ICRC Study on Customary International Humanitarian Law* (Cambridge University Press, 2007) 204–37. For a more in-depth discussion of the obligations outlined in this section see Karen Hulme, *War Torn Environment: Interpreting the Legal Threshold* (Martinus Nijhoff Publishers, Leiden, 2004); Michael N. Schmitt, 'Green War: An Assessment of the Environmental Law of International Armed Conflict' (1997) 22 *Yale J. Int'l L.* 1 (particularly with reference to the US position on these provisions).

[11] Art. 35(3) states: 'It is prohibited to employ methods or means of warfare which are intended, or may be expected, to cause widespread, long-term and severe damage to the natural environment.' Art. 55 states: 'Care shall be taken in warfare to protect the natural environment against widespread, long-term and severe damage. This protection includes a prohibition of the use of methods or means of warfare which are intended or may be expected to cause such damage to the natural environment and thereby to prejudice the health or survival of the population.'

[12] Note that under Additional Protocol I, a means or method of warfare must cause damage which cumulatively fulfils all three conditions for it to be rendered unlawful. For an examination of the meaning of the terms 'widespread, long-term and severe' under the Protocol, see Hulme, *War Torn Environment*, 91–100.

[13] The word 'health' was included to indicate that the provision was also concerned with acts which could seriously prejudice health, such as congenital defects, degenerations or deformities. Pilloud, *et al.*, *Commentary*, 663–4. The distinction in focus between the environment and the population reflects a split in the negotiating parties. See Schmitt, 'Green War'.

stitutes an absolute standard of zero harm to the environment in the case of reprisals.[14]

1.2. *ENMOD Convention*

Like Additional Protocol I, the 1977 Convention on the Prohibition of Military or Any Other Hostile Use of Environmental Modification Techniques (ENMOD) does not limit itself to specific means and methods of warfare.[15] Article 1 of the Convention prohibits states parties from military or any other hostile use of environmental modification techniques which result in 'widespread, long-lasting or severe' effects as a means of affecting any other state party to the Convention.[16]

Environmental modification techniques are defined broadly as:[17]

> Any technique for changing – through the deliberate manipulation of natural processes – the dynamics, composition or structure of the Earth, including its biota, lithosphere, hydrosphere and atmosphere, or of outer space.

Possible methods mooted for manipulating the environment are: triggering earthquakes, generating tsunamis, triggering landslides, fluidising thixotropic soils, activating volcanoes, breaching water containments, melting polar ice, disrupting permafrost soils, seeding clouds to create rain and flooding, creating holes in the ozone layer, and creating drought conditions.[18] Other than breaching water containment facilities, a topic which is dealt with further in the following

[14] Hulme, *War Torn Environment*, 73.

[15] Convention on the Prohibition of Military or Any Other Hostile Use of Environmental Modification Techniques, 18 May 1977.

[16] The terms 'widespread, long-lasting or severe' are used deliberately to mirror the terminology in Additional Protocol I, which was negotiated in the same time-frame. Note, however, the use of 'or', rather than 'and', which gives the ENMOD Convention a broader application. Although the terminology is almost identical, the terms are not used synonymously. While not defined in the Convention, the terms have been given definition by a set of 'Understandings' which were drafted at the same time by the Committee and are attached to the Convention (although not officially incorporated into it). 'Widespread': encompassing an area on the scale of several hundred square kilometres; 'long-lasting': lasting for a period of months, or approximately a season; severe: involving serious or significant disruption or harm to human life, natural and economic resources or other assets.

[17] Art. 2, ENMOD Convention.

[18] Hallan C. Noltimier, 'Techniques for Manipulating the Geosphere' in A. H. Westing (ed.), *Environmental Warfare: A Technical, Legal and Policy Appraisal* (Taylor & Francis, London, 1984) 25–31; Ernö Mészáros, 'Techniques for Manipulating the Atmosphere' in A. H. Westing (ed.), *Environmental Warfare: A Technical, Legal and Policy Appraisal* (Taylor & Francis, London, 1984), 13.

section on installations containing dangerous forces, it is difficult to see how a computer network attack could directly manipulate the environment. However, where a computer network attack *is* capable of performing such a function, the provisions of the ENMOD Convention are broad enough to prohibit it.

With the current state of technology, the most likely scenario involves hostile manipulation of existing peacetime environmental modification techniques put in place by states to combat environmental problems.[19] Such solutions are likely to be controlled by SCADA systems or other networked computer systems and may therefore be susceptible to appropriation and manipulation for hostile purposes. For example, the Thames Barrier, a 523-metre-long gated barrier across the River Thames in London, is an example of an environmental modification technique used for peaceful purposes. The Barrier was created to protect London and the Thames Estuary from flooding caused by rising tide levels and surge tides.[20] However, the massive hydraulic gates of the Barrier are ultimately controlled via a computer system and are therefore theoretically vulnerable to manipulation by computer network attack. For example, preventing the system from closing the gates during a tidal surge would allow the natural flow of flood waters to cause damage to a substantial part of London. It would also be possible to use the Barrier itself to amplify the deleterious effects of such tides on the city. By closing the gates during a surge tide and then opening them at the height of the tide, a wall of water would flood central London, causing immense loss of life and property damage. Such damage would certainly meet the 'severe' criteria of the Convention.

The ENMOD Convention is not limited to international armed conflicts; however, its application to non-international armed conflicts is limited by the requirement that damage must be caused to another state party. The Convention would nevertheless cover the situation where an environmental modification technique was used intentionally against a domestic opponent, but caused cross-border environmental damage to another state party.[21]

19 Peaceful use of environmental modification techniques are specifically excluded from the ambit of the ENMOD Convention under Art. 3(1).

20 See generally The Environment Agency, *The Thames Barrier: Flood Defence for London*, www.environment-agency.gov.uk/regions/thames/323150/335688/341764/ (last accessed 29 November 2006).

21 Dinstein, *Conduct of Hostilities* (1st edn), 189.

1.3. *Other protections*

The environment is also protected by general rules relating to the protection of civilian objects, proportionality and military necessity, all of which apply equally to computer network attacks. The relevance of the general principles of proportionality and necessity for the protection of the environment were underscored by the International Court of Justice in its advisory opinion on *Nuclear Weapons*:[22]

> States must take environmental considerations into account when assessing what is necessary and proportionate in the pursuit of legitimate military objectives. Respect for the environment is one of the elements that go to assessing whether an action is in conformity with the principles of necessity and proportionality.

In addition, the UN General Assembly has stated that the destruction of the environment not justified by military necessity and carried out wantonly, is clearly contrary to existing international law.[23] The International Court of Justice cited this passage in the *Nuclear Weapons* case, noting that although General Assembly resolutions are not binding as such, 'they provide evidence of the existence of a rule or the emergence of *opinio juris*'.[24] This has led some commentators to confirm that the protection of the environment is a norm of customary international law applicable in both international and non-international armed conflicts.[25]

While the incorporation of environmental protections during wartime was innovative at the time of drafting the Additional Protocols, the concept had gained sufficiently in importance that the inclusion of an environmental war crime in the Rome Statute of the International Criminal Court was not contentious.[26] The International Criminal Court has jurisdiction in respect of war crimes that consist of '[i]ntentionally launching an attack in the knowledge that such attack will

[22] *Nuclear Weapons Case*, para. 30.

[23] Protection of the Environment in Times of Armed Conflict, UN GAOR, 47th Sess., Agenda Item 136, UN Doc A/Res/47/37 (1992).

[24] The Court also noted that: 'Addressing the reality that certain instruments are not yet binding on all States, the General Assembly in this resolution "*[a]ppeals* to all States that have not yet done so to consider becoming parties to the relevant international conventions."' *Nuclear Weapons Case*, para. 32.

[25] See, for example, Henckaerts and Doswald-Beck, *Customary International Humanitarian Law*, Vol. 1, 143; Dinstein, *Conduct of Hostilities* (1st edn), 193. However *cf.* Schmitt, 'Green War' for an alternate view of the customary status in relation to the environment.

[26] Hulme, 'Natural Environment', 205.

cause incidental ... widespread, long-term and severe damage to the natural environment which would be clearly excessive in relation to the concrete and direct overall military advantage anticipated'.[27]

Protection of the environment is also inferred from the provisions relating to protection of civilian objects and protection of objects indispensable to the civilian population (discussed below). Civilian objects are defined as all objects which are not military objectives, and while in some cases the environment may constitute a military objective through its use (for example by providing cover for troops), as a general matter it is likely to be considered a civilian object; in case of doubt the presumption that it is not a military objective must be observed.[28] In some circumstances the environment may also be considered an object indispensable to the civilian population (for example, natural water reservoirs) and thus be provided protection under Article 54 and Article 14 of Additional Protocol I and Additional Protocol II respectively. Article 14 of Additional Protocol II is perhaps of more importance in protecting the environment as Additional Protocol II, unlike Additional Protocol I, does not protect civilian objects in general.[29] The ICRC has also issued a set of Guidelines for Military Manuals and Instructions on the Protection of the Environment in Times of Armed Conflict.[30] While not formally approved by the UN General Assembly, the Assembly did invite all states to give due consideration to the possibility of incorporating the guidelines into their military manuals and other instructions addressed to their military personnel.[31] As with any other attack, the environmental consequences of computer network attacks must be factored into the planning stages of any operation.

2. Cultural property

The destruction and looting of cultural property has taken place in almost every conflict since ancient times, whether as a deliberate act or

[27] Art. 8(2)(b)(iv) Rome Statute of the International Criminal Court 1998 (entered into force 1 July 2002). Note that knowledge in this instance means actual knowledge not reasonable forseeability (Art. 30(3) of the Rome Statute).

[28] Art. 52(3), Additional Protocol I.

[29] Pilloud, *et al.*, *Commentary*, para. 4794.

[30] ICRC, 'Guidelines for Military Manuals and Instructions on the Protection of the Environment in Times of Armed Conflict' (1996) 311 *IRRC* 230, www.icrc.org/eng/resources/documents/misc/57jn38.htm (last accessed 15 July 2011); annex to UN Doc. A/49/323 (1994).

[31] GA Res 49/50, 9 December 1994.

as an incidental result of the conduct of military operations. The idea of protecting cultural property during times of war has gradually developed over time, and through the devastation of many wars, to the point that cultural property is now protected by both specific cultural property conventions and the more general framework of the laws of armed conflict.[32] In all cases, the principle of distinction applies and cultural property remains protected by virtue of being civilian property. While a computer network attack is unlikely to bring down the Sphinx, destroy a Rembrandt or delete the Great Mosque at Mecca, where it *is* possible to cause physical damage to existing cultural property the use of technology to undertake the attack will not detract from the illegality of the action under the laws of armed conflict. However, in the modern era, more and more cultural monuments, libraries and scientific collections are being digitised and stored on information systems, in some cases becoming the only surviving record of a lost art, language or culture. Other works exist only in digital form and are commonly referred to as 'born-digital'.[33] Where such digitisation or creation takes place, the resulting records and collections become vulnerable to the effects of computer network attacks, either by destruction or damage, or through misappropriation of these cultural works. Computer network attacks raise two interesting issues in relation to cultural property: the first is what constitutes protected cultural property in the digital age where every copy made is identical and may be backed up; and second, what acts against digital cultural property will be prohibited?

2.1. The legal framework

The Hague Regulations contain the first formal treaty protection for cultural property during armed conflict, although the term is not so

[32] For a more complete historical background to the development of the laws relating to the protection of cultural property during armed conflict than is possible here, see generally Jirí Toman, *The Protection of Cultural Property in the Event of Armed Conflict* (Dartmouth: UNESCO, Aldershot; Brookfield, VT, 1996), 71; Patrick J. Boylan, *Review of the Convention for the Protection of Cultural Property in the Event of Armed Conflict* (UNESCO, Paris, 1993), 28; Kevin Chamberlain, *War and Cultural Heritage: An Analysis of the 1954 Convention for the Protection of Cultural Property in the Event of Armed Conflict and Its Two Protocols* (Institute of Art and Law, Leicester, 2004).

[33] Jean-Michel Rodes, Geneviève Piejut and Emmanuèle Plas, *Memory of the Information Society* (UNESCO, Paris, 2003), 39. Born digital works include art works completed and shown in digital format, documents and archives stored only in electronic format, digital recordings, etc. It incorporates those digital works associated with a physical medium on which the file is recorded and stored, and also those works whose constituent parts are stored on physical media but where the work in question only reconstitutes itself in the digital environment.

defined. The first of these provisions, protecting all 'buildings dedicated to religion, art, science, or charitable purposes, historic monuments, hospitals, and places where the sick and wounded are collected', is qualified by the dictates of military necessity, and is limited to buildings rather than their contents.[34] It should also be noted that buildings are defined in terms of their purpose rather than their cultural importance.[35] In terms of computer network attacks, this provision of the Hague Regulations does not raise any specific issues; in short, the servers or other devices which may contain digital cultural property are not buildings.[36] Article 56 of the Regulations also provides that the property of such institutions is to be treated as private property and that all seizure, destruction or wilful damage to the institution, historic monuments or works of art or science is forbidden.[37]

The Geneva Conventions do not specifically address the protection of cultural property; however, they do provide some protection by virtue of cultural property being civilian property. In addition to its general protection of civilian objects, Article 53 of Additional Protocol I also includes specific protections for cultural property.

Without prejudice to the provisions of the Hague Convention for the Protection of Cultural Property in the Event of Armed Conflict of 14 May 1954, and of other relevant international instruments, it is prohibited:

(a) to commit any acts of hostility directed against the historic monuments, works of art or places of worship which constitute the cultural or spiritual heritage of peoples;

[34] Art. 27 provides: 'In sieges and bombardments all necessary steps must be taken to spare, *as far as possible*, buildings dedicated to religion, art, science, or charitable purposes, historic monuments, hospitals, and places where the sick and wounded are collected, provided they are not being used at the time for military purposes. It is the duty of the besieged to indicate the presence of such buildings or places by distinctive and visible signs, which shall be notified to the enemy beforehand.' (emphasis added), Convention Respecting the Laws and Customs of War on Land and Its Annex: Regulations Concerning the Laws and Customs of War on Land, 18 October 1907.

[35] Chamberlain, *War and Cultural Heritage*, 28.

[36] Even by analogy, the difficulties are numerous. Where a place of worship or other protected institution is situated on one particular website, sites are generally hosted by an ISP on servers which host multiple sites, not merely protected cultural ones. Thus while individual sites may qualify for protection, the server itself would not.

[37] Art. 56, Hague Regulations. Note that the definition does not extend to archives and it has been concluded that an occupying power thus had the right to seize archives and military plans. Toman, *Protection of Cultural Property*, 47.

(b) to use such objects in support of the military effort;
(c) to make such objects the object of reprisals.

Article 16 of Additional Protocol II provides almost identical wording.[38] It should be noted that both the commentary to the Protocol and the ICRC customary international humanitarian law study consider that Article 53 was only meant to cover a limited amount of very important cultural property.[39]

2.1.1. Cultural Property Convention 1954 and its Protocols

The primary convention for the protection of cultural property in times of war is, and remains, the Hague Convention for the Protection of Cultural Property in the Event of Armed Conflict of 1954 (Cultural Property Convention).[40] Both the subsequent 1999 Protocol II to the Convention and the relevant sections of both Additional Protocols to the Geneva Conventions are drafted 'without prejudice' to its terms.[41] The Cultural Property Convention defines cultural property (irrespective of origin or ownership) as 'movable or immovable property of great importance to the cultural heritage of every people'.[42] The definition includes 'monuments of architecture, art or history, whether religious or secular; archaeological sites; groups of buildings which, as a whole, are of historical or artistic interest; works of art; manuscripts, books and other objects of artistic, historical or archaeological interest; as well as scientific collections and important collections of books or archives or of reproductions of the property defined above'. The definition also includes 'buildings whose main and effective purpose is to preserve or exhibit the movable cultural property' defined above,[43]

[38] Art. 16, Additional Protocol II provides: 'Without prejudice to the provisions of the Hague Convention for the Protection of Cultural Property in the Event of Armed Conflict of 14 May 1954, it is prohibited to commit any acts of hostility directed against historic monuments, works of art or places of worship which constitute the cultural or spiritual heritage of peoples, and to use them in support of the military effort.'

[39] Dinstein, *Conduct of Hostilities* (2nd edn), 179–80; Roger O'Keefe, *The Protection of Cultural Property in Armed Conflict* (Cambridge University Press, Cambridge; New York, 2006), 209–14.

[40] Convention for the Protection of Cultural Property in the Event of Armed Conflict, 14 May 1954, 249 UNTS 240.

[41] Art. 53, Additional Protocol I; Art. 16, Additional Protocol II; Roger O'Keefe, 'The Meaning of "Cultural Property" under the 1954 Hague Convention' (1999) 46 *NIL Rev.* 26, 31.

[42] Art. 1(a), Cultural Property Convention.

[43] Such as museums, large libraries and depositories of archives, and refuges intended to shelter cultural property in the event of armed conflict: Art. 1(b) *ibid.*

and 'centres containing a large amount of cultural property'.[44] This definition is used by both Protocols to the Convention and the two Additional Protocols to the Geneva Conventions.

There are two main types of digital cultural property falling within this definition that may be at risk from computer network attacks; works which are digital reproductions of pre-existing cultural property and those which are 'born-digital' and exist only in digital form.[45] A significant part of the current digital heritage consists of the products of digital reproduction of pre-existing works.[46] Where the original works represent cultural property covered by the conventions above, their digital counterparts should be considered reproductions and will be covered by the Cultural Property Convention where they are held in important collections. In discussions regarding the protected status of reproductions of cultural property at the conference which adopted the Convention, the French and Swiss delegates emphasised the importance of protecting reproductions, pointing out that they become of even more importance where the original is destroyed.[47] However, not all reproductions are intended to be protected by the Conventions; no one would suggest that the millions of souvenir copies of Michelangelo's David sold in Florence each year are protected under the Convention despite the undoubted protected status of the original. An additional problem is that not all digital reproductions reproduced for the purpose of conservation form part of collections, but are nonetheless important reproductions of protected and highly fragile cultural objects.[48]

A second type of digital cultural property that is particularly susceptible to computer network attacks characterises the cultural property of digital societies. 'Born-digital' works result from an all-digital process of initial production, the work being digitally encoded at the moment of its creation – for example, a collection of digital photographs

[44] Art. 1(c), *ibid.*

[45] Rodes, Piejut and Plas, *Memory of the Information Society*, 39. [46] *Ibid.*, 37.

[47] Intergovernmental Conference on the Protection of Cultural Property in the Event of Armed Conflict, *Records of the Conference Convened by the United Nations Educational, Scientific and Cultural Organisation and Held at the Hague from 21 April to 14 May 1954* (Government of the Netherlands, The Hague Staatsdrukkerij en Uitgeverijbedrijf, 1961) paras 214–15, cited in Toman, *Protection of Cultural Property*, 134.

[48] For example, an institution may create one digital reproduction for a specific project relating to one particular piece of cultural property, i.e. the Sphinx Mapping Project.

of planet earth.[49] Such works include art works completed and shown in digital format,[50] documents and archives stored only in electronic format, digital recordings, etc. It incorporates those digital works associated with a physical medium on which the file is recorded and stored, and also those works whose constituent parts are stored on physical media but where the work in question only reconstitutes itself in the digital environment.[51] Art museums now exhibit new digital art works by artists who specialise in the digitised virtual medium,[52] and many film makers now shoot only in digital mediums.

Under the Cultural Property Convention, defending states are obliged to take measures they consider appropriate during peacetime to safeguard the cultural property situated in their territory against the foreseeable effects of armed conflict.[53] With respect to digital cultural property, it would seem that ensuring sufficient backups and format-proofing of protected works would meet this obligation. Further, the Convention also grants special protection to a limited number of refuges intended to shelter movable cultural property and centres containing monuments or other immovable cultural property of very great importance.[54] The protection of refuges will apply equally to facilities established to store digital cultural property and has the advantage of being able to be set up and functioning before any armed conflict begins, and without disrupting the normal functioning of the institution to which the property belongs.

Both attacking and defending states must refrain from using any such property and surroundings in a manner that might expose the property to destruction or damage in the event of an armed conflict.[55] Likewise, states must refrain from any 'act of hostility' directed against such property. A number of other regimes, including the two protocols to the Cultural Property Convention (adopted in 1954 and 1999

[49] Rodes, Piejut and Plas, *Memory of the Information Society*, 39.

[50] See, for example, the works of artists submitted for the UNESCO digital art awards http://portal.unesco.org/culture/en/ev.php-URL_ID=29021&URL_DO=DO_TOPIC&URL_SECTION=201.html (last accessed 7 July 2011)

[51] Rodes, Piejut and Plas, *Memory of the Information Society*, 39.

[52] Guy Pessach, *Digital Art Museums – Legal Perspectives* (2006) http://islandia.law.yale.edu/isp/writing%20paper/digital_art.htm (last accessed 9 June 2006).

[53] Art. 3, Cultural Property Convention.

[54] Arts 8 and 9, *ibid*. Cultural property under special protection is designated as such by registering the property in the International Register of Cultural Property under Special Protection, from which time the property will enjoy immunity from attack and use for military purposes.

[55] Art. 4(1), *ibid*.

respectively) and the Roerich Pact adopted by the Pan American Union, also offer protections to cultural property during armed conflict.[56] The following sections discuss two of the most likely ways that computer network attacks will impact upon digital cultural property: attacks and damage against digital works; and theft or misappropriation.

2.2. Attacks on, and damage to, digital works

International humanitarian law prohibits attacks on, and damage to, cultural property except in cases of imperative military necessity.[57] This prohibition will apply despite translation to a digital environment. However, while digital works may still constitute cultural property, the digital environment operates in a different way from the physical one. The degree and necessary consequences of the attacks required to constitute an infringement may need re-analysis. The integrity of digital records is more fragile than that of their physical counterparts, thus making the risk, and consequences, higher. The digital environment provides a forum in which records can be restored to their original form from backups without any trace that the record was ever changed. While this is one of the great strengths of digital media, it also represents one of its greatest threats; it means that records may also be changed unnoticed without leaving a trace of their amendments, leaving open the possibility of an Orwellian rewriting of history.[58] The defacement of websites has occurred on an increasing scale during diplomatic incidents over the past decade. While such actions would not be sufficient to constitute an attack, could drawing a moustache on a digital copy of the Mona Lisa be sufficient to breach the cultural property conventions once an armed conflict is under way? Will altering the work constitute damage – is it still an offence when digital

[56] Protocol I deals mainly with the protection of property in occupied territory. The Roerich Pact, adopted by the Pan American Union in 1935 (Treaty on the Protection of Artistic and Scientific Institutions), is in force between ten American states. For these parties, the Hague Convention is supplementary to the Pact, which requires respect and protection of 'historic monuments, museums, scientific, artistic, educational and cultural institutions' and establishes a distinctive flag for identification.

[57] Art. 51, Hague Regulations; Art. 51, Additional Protocol I; Art. 4, Cultural Property Convention.

[58] George Orwell's dystopian novel *1984* describes a department of a government ministry (the Ministry of Truth) which is responsible for rewriting segments of history (for example newspaper reports) which no longer fit with official policy. The slogan of the regime being 'Who controls the past controls the future; who controls the present controls the past'.

property is not harmed permanently but can easily be restored to the original standard using backups? As seen in Chapter 6 above, it is no defence to argue that the work or site in question was not protected.

The wording contained in the relevant conventions is fairly broad. Article 56 of the Hague Regulations provides that property belonging to institutions dedicated to religion, charity and education, and the arts and sciences are protected as private property and further 'seizure of, destruction or wilful damage' to these institutions, historic monuments, works of art and science is forbidden.[59] The protection is part of customary international law, and presumably extends to the digital assets of those institutions and to digital monuments, works of art and science as well as to any tangible works.

Article 4(1) of the Cultural Property Convention and both Article 53 of Additional Protocol I and Article 16 of Additional Protocol II require states parties to the Conventions to refrain from 'any act of hostility' directed against cultural property. The obligation under the Cultural Property Convention may be waived only when military necessity imperatively requires such a waiver.[60] The broad language contained in the Additional Protocols was discussed in the ICRC commentary on the Additional Protocols and confirmed by the Trial Chamber in *Jokic*, which held that, according to the Additional Protocols, it is prohibited to direct attacks against 'historic monuments, works of art or places of worship which constitute the cultural or spiritual heritage of peoples' whether or not the attacks result in actual damage.[61] However, the Appeals Chamber in *Kordic and Cerzec* subsequently held that, while recognising that attacks in violation of Articles 51 and 52 of Additional Protocol I are clearly unlawful even without causing serious harm,[62]

[59] Art. 56, Hague Regulations.

[60] Art. 4(2), Cultural Property Convention. Note that the territoriality principle was specifically removed from the article to affirm that cultural property is to be respected wherever it is situated. The amendment was tabled by Belgium, France, the Netherlands and Switzerland: 'Our amendment has been designed to break with the territorial concept and to affirm the principle that cultural property, wherever situated, must be respected by all States. It is important to break away from the notion of frontiers as, in time of war, military vicissitudes may lead to a State's overflowing its frontiers', Records, 136, para. 247, cited in Toman, *Protection of Cultural Property*, 69.

[61] ICRC Commentary to Additional Protocol I, para. 2067, 2069–72; *Prosecutor* v. *Miodrag Jokic (Sentencing Judgment)* (2004) Case No. IT-01–42/1-S, International Criminal Tribunal for the Former Yugoslavia – Trial Chamber I, para. 50.

[62] *Prosecutor* v. *Dario Kordic and Mario Cerkez (Appeal)* (2004) Case No. IT-95–14/2-A, International Criminal Tribunal for the Former Yugoslavia, Appeals Chamber, para. 65.

the broad wording used in the above articles has been tempered by attaching individual criminal liability only to those acts which result in damage or destruction of the property:[63]

> ... deliberate attacks on civilian objects such as historic monuments, works of art and places of worship are considered to be grave breaches of the Additional Protocol only insofar as the attack results in extensive destruction.

It would appear that while attacks on digital works would constitute a breach of the Convention and/or Protocols, they would not rise to the level of grave breaches (and thus incur individual criminal liability) without resulting in substantial damage. However, one of the defining aspects of digital works is that a copy, for example a backup copy, is identical to the original, meaning that in many instances a digital work may be restored completely with no lasting damage. It remains to be seen whether damage to a digital work must be irreparable, but it would seem in line with the reasoning of the Appeals chamber in *Kordic and Cerzec* that it would not constitute a grave breach to damage a digital work which can be completely and identically restored from backup copies.[64] The maintenance of such backups would presumably fall within the measures taken by the victim state under its obligation to safeguard cultural works.[65] Whether or not those measures had been taken by the victim state would have to fall within the foreseeable harm analysis of the attacking state. Recent domestic criminal hacking cases have attempted to raise the lack of security on compromised systems in mitigation of the expected severity of the sentence.[66] However, such a defence would have the opposite effect in the case of cultural property. Article 85 of Additional Protocol I makes it a grave breach of the Protocol to make a non-defended locality the object of an attack.

2.3. *Theft, pillage or misappropriation of digital works*

The Cultural Property Convention 1954 and its later Protocol contain absolute prohibitions against 'any form of theft, pillage or misappropriation' of cultural property;[67] the prohibition may not be derogated from through military necessity. Further, the Hague Regulations make

63 *Ibid.*, para. 65.; Art. 85(4)(d), Additional Protocol I; Art. 3(d), Statute of the International Criminal Tribunal for the Former Yugoslavia.

64 *Ibid.* 65 Art. 3, Cultural Property Convention.

66 'UK Hacker "Should Be Extradited"', *BBC News*, 10 May 2006, Technology.

67 Art. 4(3), Cultural Property Convention; Art. 15, Second Cultural Property Protocol.

all property belonging to cultural institutions private property and forbids the seizure of any such institution, historic monument or work of art or science; pillage is prohibited absolutely.[68]

As discussed in Chapter 8 below regarding pillage generally, the intangible nature of digital works will not create an issue for international law; courts and tribunals have had no problem in finding pillage involving intangible property such as shares and property rights.[69] Thus where a state has digital reproductions of cultural property or born-digital works which constitute cultural property, those works would also be subject to the prohibition against theft, pillage and misappropriation.

2.3.1. Unauthorised copying of works

Once stored in digital format, not only do works constituting cultural property become exposed to the risk of damage from computer network attacks as seen above, they may also be copied for use for the attacking state's own purposes. As the 'Titan Rain' incidents show,[70] vast amounts of such copying can occur in minutes. One of the defining characteristics of digitally stored information is that any number of copies may be made, at negligible cost, without in any way degrading the original.[71] This feature makes unauthorised copying of digital cultural property perhaps the most likely form of theft or misappropriation to affect digital works. Not only is it now possible that the reproduction of Nefertari's tomb may be copied in its entirety to be reproduced and displayed in a museum somewhere else, a more mundane use may be made of such digital works. Take the following hypothetical example. Arcadia is at war with Ruritania and subsequently occupies a portion of the territory of Ruritania. During the occupation, an enterprising group of soldiers copy digital art works from Ruritania's

[68] Arts 47 and 56, Hague Regulations.

[69] See, for example, *Trial of Carl Krauch and Twenty-Two Others (I.G. Farben Trial)* (1948) X *Law Reports of Trials of War Criminals* 1, United States Military Tribunal, Nuremburg; *Trial of Alfied Felix Alwyn Krupp von Bohlen und Halbach and Eleven Others (Krupp Trial)* (1948) X *Law Reports of Trials of War Criminals* 69, United States Military Tribunal, Nuremburg.

[70] See Appendix 1.

[71] In fact, one of the greatest paradoxes of digital cultural property is that preservation relies on multiple copies being made. Computing reverses those very propositions which seemed the most certain: the survival of a document is not dependent on how long the medium carrying it will last, but on the capacity of that document to be transferred from one medium to another as often as possible: Rodes, Piejut and Plas, *Memory of the Information Society*, 35.

national gallery collection and place them on carpets, selling them in Arcadia. The art works copied are some of the most sacred and important cultural works of Ruritania's indigenous people and the resulting carpets cause great offence by having the 'enemy' literally walk on the spiritual traditions of the occupied territory.[72]

Leaving aside the issue of residual copyright in a digital reproduction as a work itself,[73] the question must be asked whether mere copying of a digital work, original or reproduction, leaving the digital 'master' unaltered and unharmed, will constitute theft, pillage or misappropriation. Understandably, the conference that drafted the Convention did not consider the meaning of the words theft, pillage and misappropriation – all presumably deemed self-explanatory.[74] In determining whether illegal copying may amount to theft, pillage or misappropriation, two questions must be answered. First, must the owner of the property be deprived of it entirely, or is it sufficient that the owner's property rights, namely the right to control the use made of the work, are infringed? And second, with regard to pillage, must the property be acquired through threats or use of violence? This second question is examined in Chapter 8 below in relation to pillage generally, as misappropriation is a wider term.

The International Criminal Tribunal for the Former Yugoslavia held in the *Celebici* case that 'the prohibition against the unjustified appropriation of public and private enemy property is general in scope, and extends both to acts of looting committed by individual soldiers for their private gain, and to the organized seizure of property undertaken within the framework of a systematic economic exploitation of occupied territory.'[75] The Nuremburg cases, applying the Hague Regulations, tended to group property crimes together under a general heading of

[72] This example is a variant on the facts of an Australian copyright case of *Milpurrurru* v. *Indofurn Pty Ltd* (1994) 54 FCR 240 in which Indofurn appropriated and simplified sacred aboriginal designs and reproduced them on carpets manufactured in Vietnam.

[73] A digital work is likely to have the protection of the relevant intellectual property act of the jurisdiction of origin. The unauthorised copying of the work would thus be considered a breach of copyright and be actionable under the domestic jurisdiction of the relevant state. However, this section will deal only with the digital work's relevance as cultural property for the purposes of the laws of armed conflict.

[74] The word misappropriation was added to replace the term 'removal of property' in regard to a party's obligation not to requisition property; however, its basic meaning was not discussed. Toman, *Protection of Cultural Property*, 71.

[75] *Prosecutor* v. *Zejnil Delalic et al. (Celebici)* (1998) Case No. IT-96–21-T, International Criminal Tribunal for the Former Yugoslavia, para. 590.

spoliation, and in some cases it can be very difficult to establish which crime forms the basis of the charges.[76] However, the phrase 'other misappropriation' contained in the Cultural Property Convention is a similarly broad-based charge which would encompass the crimes set out by the military tribunals which included the offence contained in Article 46 of the Hague Regulations, namely respect for private property.

The *Flick Trial* convicted Friedrich Flick of crimes against property in a case which perhaps maps most closely to unauthorised copying of a digital work. Flick's offence consisted of operation of a plant in occupied territory of which he was not the owner and without the consent of the owner.[77] It is interesting to note that the Tribunal regarded his acts as illegal despite the fact that (a) 'the original seizure may not have been unlawful'; (b) Flick had nothing to do with the expulsion of the owner; (c) the property was left 'in a better condition than when it was taken over'; (d) there was 'no exploitation either for Flick's personal advantage or to fulfil the aims of Goering', there being no proof that the output of the plant went to countries other than those who benefited before the war.[78] In a situation such as the hypothetical scenario above, where the property is exploited for the personal benefit of the perpetrators, it seems certain that a court would have no difficulty in finding that the work had been misappropriated. Indeed the commentary to the *Krupp Trial* states that the prosecution was probably correct in claiming that violation of Article 46 of the Hague Convention (respect for private property):[79]

> need not reach the status of confiscation. Interference with any of the normal incidents of enjoyment of quiet occupancy and use, we submit is forbidden. Such incidents include, *inter alia*, the right to personal possession, control of

[76] For example, it is not completely clear which offence against private property Flick is found guilty of. The acts in question related to the seizure and operation of the Rombach plant in occupied France; however, the Tribunal held that none of the defendants were shown to have been 'responsible for any act of pillage as that word is commonly understood … Flick's acts and conduct contributed to a violation of Hague Regulation 46, that is that private property must be respected. Of this there can be no doubt. But his acts were not within his knowledge intended to contribute to a programme of "systematic plunder" conceived of by the Hitler regime.' The analysis concludes that Flick must therefore have been found guilty either of an offence other than spoliation or a particular type of pillage: 'it may be that Flick's offence is to be regarded as an offence against property in occupied territories other than plunder or spoliation.' *Trial of Friedrich Flick and Five Others (Flick Trial)* (1947) IX *Law Reports of Trials of War Criminals* 1, United States Military Tribunal, Nuremburg, 40.

[77] *Ibid.* [78] *Ibid.* (footnotes removed). [79] *Krupp Trial*, 16, fn. 5.

the purpose for which the property is to be used, disposition of such property, and the right to the enjoyment of the income derived from the property.

Finally, in 2003 UNESCO adopted the Charter on the Preservation of the Digital Heritage, recognising that the digital heritage consists of resources which have 'lasting value and significance, and therefore constitute a heritage that should be protected and preserved for the current and future generations'.[80] The Charter does not refer to the protection of the digital heritage in armed conflict; however, it sets out measures for the protection and preservation of digital heritage and emphasises the threat of loss and the need for action in protecting it.[81] It should be noted that cultural heritage is a broader concept than cultural property,[82] and any digital property for which protection is claimed would also need to meet the definition of cultural property set out above.

3. Installations containing dangerous forces

Until 2002, the idea of a country being attacked through its computer networks as a coordinated act of war was considered remote and largely dismissed as panic-mongering.[83] Although United States intelligence agencies were monitoring China, Russia and other states on the threat to US information systems, the threat from non-state actors was largely underestimated.[84] Then, in 2002, troops clearing the cave system in the Tora Bora region of Afghanistan uncovered an Al Qaeda laptop which indicated a strong interest in computer network attacks. Computer forensics showed that the laptop had made multiple visits to sites offering sabotage handbooks, software and programming instructions on

[80] Art. 1, Charter on the Preservation of the Digital Heritage, Adopted at the 32nd session of the General Conference Paris, France, 17 October 2003.

[81] Arts 3–9, Charter on the Preservation of the Digital Heritage.

[82] See generally, Manlio Frigo, 'Cultural Property v. Cultural Heritage: A "Battle of Concepts" In International Law?' (2004) 86(854) *IRRC* 367. The Cultural Property Convention notes 'damage to cultural property belonging to any people whatsoever means damage to the cultural heritage of all mankind, since each people makes its contribution to the culture of the world'. Thus cultural property is a subset of cultural heritage.

[83] See, for example, PBS Frontline, *Interview with John Hamre: Cyberwar!* 24 April 2003, www.pbs.org/wgbh/pages/frontline/shows/cyberwar/interviews/hamre.html (last accessed 16 January 2012); PBS Frontline, *Interview with James Lewis: Cyberwar!* 24 April 2003, www.pbs.org/wgbh/pages/frontline/shows/cyberwar/interviews/lewis.html (last accessed 16 January 2012).

[84] Gellman, 'Cyber-Attacks by Al Qaeda Feared'.

SCADA systems, and other 'cracking' tools. In combination with the Mountain View surveillance programme, which had been uncovered the year before,[85] officials became increasingly concerned about Al Qaeda's computer network attack capabilities. In January 2002, another computer was seized at an Al Qaeda office in Kabul, Afghanistan. The computer contained models of a dam, made with structural architecture and engineering software, and also included geological soil identification software which would enable the planners of an attack to simulate the dam's catastrophic failure and plot the consequences of a breach.[86] Although the authorities declined to say whether the schematics related to a particular targeted dam, the use of cyberspace to infiltrate a dam is not unprecedented. In 1998, a 12-year-old hacker, exploring for fun, broke into the computer system that controls the Roosevelt Dam in Arizona, USA.[87] Although he was unaware of the fact, federal authorities claim the boy had complete control of the SCADA system which controls the dam's massive floodgates and the 489 billion gallons of water which it contains. Unleashed, the water would have coursed down the Salt River and over a downstream flood plain (home to an estimated population of one million people) before reaching the state capital, Phoenix.

Dams are not the only installations containing dangerous forces to have been compromised via the internet. In January 2003, the Davis-Besse nuclear power plant in Ohio, USA, was hit by the Slammer worm, disabling a safety system for nearly five hours and a process computer for nearly six hours.[88] Fortunately, the power plant was offline at the time; however, the incident provided a stark reminder of the vulnerability of such installations, prompting a review of safety protocols. In the UK, the Bradwell nuclear power plant was likewise compromised in June 1999 by a security guard who attempted to alter sensitive information, and succeeded in deleting records from one of the systems.[89]

[85] In 2001 Mountain View, California, police began investigating a suspicious pattern of surveillance against silicon valley computers. The visitors were studying emergency telephone systems, electrical generation and transmission, water storage and distribution, nuclear power plants and gas facilities. While some probes indicated planning for a conventional attack, others homed in on the digital devices which run critical infrastructure. *Ibid.*

[86] *Ibid.*

[87] *Ibid.* See also Appendix 1 for queries that have been raised regarding the veracity of some of the facts of this account.

[88] Kevin Poulsen, 'Slammer Worm Crashed Ohio Nuke Plant Network', *Security Focus*, 19 August 2003, www.securityfocus.com/news/6767 (last accessed 7 July 2011).

[89] Kevin Maguire, 'Guard Tried Sabotage at Nuclear Reactor: Security Checks Tightened after High-Level Alert', *Guardian* (London), 9 January 2001, 2.

As with the Davis-Besse incident, the Bradwell incident prompted a review of security and a subsequent change of procedures.

Additional Protocol I to the Geneva Conventions grants special protection to installations containing dangerous forces, namely dams, dykes and nuclear power stations. Article 56(1) provides as follows:

> Works or installations containing dangerous forces, namely dams, dykes and nuclear electrical generating stations, shall not be made the object of attack, even where these objects are military objectives, if such an attack may cause the release of dangerous forces and consequent severe losses among the civilian population. Other military objectives located at or in the vicinity of these works or installations shall not be made the object of attack if such attack may cause the release of dangerous forces from the works or installations and consequent severe losses among the civilian population.

The first sentence of this Article is repeated verbatim in Article 15 of Additional Protocol II relating to non-international armed conflicts. The provisions represent an innovation in the laws of armed conflict, and are a reflection of the attempt to limit the extent of permissible collateral damage.[90] In light of the 2010 Stuxnet worm, which allegedly targeted the Iranian nuclear enrichment programme, it should also be noted that the list of installations is *not* merely illustrative. The commentary to the Article notes that it was only after the special protection was limited to dams, dykes and nuclear power generating stations that it was possible to draw up a text that was generally acceptable.[91] Thus the Natanz nuclear enrichment facility, one of the apparent targets of the Stuxnet virus, would not be subject to special protection during an armed conflict under this provision. Although reports indicate that the worm also superficially infected the Bushehr nuclear power plant,[92] a target which would undoubtedly fall within the scope of this provision, the worm had not impacted the plant's control systems and thus was not at risk of causing the release of dangerous forces.

An interesting question arises in the respect of computer network attacks as to whether the concept of military objectives 'located at or in the vicinity of' such works and installations will extend to network vicinity as well as physical proximity. As has been illustrated in previous

90 Oeter, 'Methods and Means of Combat', 194.

91 Pilloud, *et al.*, *Commentary*, para. 2147.

92 Gregg Keizer, 'Iran Admits Stuxnet Worm Infected PCs at Nuclear Reactor', *Computerworld*, 27 September 2010, www.computerworld.com/s/article/9188147/Iran_admits_Stuxnet_worm_infected_PCs_at_nuclear_reactor (last accessed 10 April 2011).

chapters, in information-age warfare, physical distance is no longer a useful yardstick for the amount of damage that can be inflicted. While a strict reading of the text of Article 56(1) would tend to indicate that the physical location of the objective is the only prerequisite, if the Article is to maintain its utility in the internet age it would seem that it should extend to network proximity as well, the operative part of the prohibition being 'if such an attack may cause the release of dangerous forces … and consequent severe losses among the civilian population'. As Yoram Dinstein points out, the guiding consideration is the protection of the civilian population from catastrophic collateral damage.[93] Where a computer network attack is designed to disable an adjacent system or network, such that it would have a knock-on effect onto a dam, dyke or nuclear generator, causing that installation to release its forces, it should not matter that the system or network is not physically located in the vicinity of that installation. Where it is reasonably foreseeable – using normal network reconnaissance techniques – that the target system is connected to the installation, such that making the former the object of an attack would affect the latter, the prohibition should stand.[94] The requirement to take suitable precautions in attacks is dealt with in more detail in Chapter 6 above. Note that this issue will only arise in relation to international armed conflicts, as the relevant sentence is omitted from Article 15 of Additional Protocol II.

While the advent of computer network attacks may remove some military objectives from the permissible target list by virtue of their being in close network proximity to a work or installation containing dangerous forces, others may become open to attack. The commentary to Article 56 cites the case of a hydroelectric power station incorporated in a dam or located in the immediate vicinity as an example of a military objective which cannot be attacked because of its proximity to the dam.[95] Computer network attacks may allow the attacking force to disengage the power station from the dam to deny the opposing force electricity, without running the risk of destroying the dam. Such

[93] Dinstein, *Conduct of Hostilities* (2nd edn), 194. See also Oeter, 'Methods and Means of Combat', 195.

[94] Note, for example, the loss of civilian water distribution, purification and sewerage facilities which followed the US destruction of the Iraqi electricity grid in 1991. Given US intelligence and reconnaissance capabilities at the time, such a consequence should have been reasonably foreseeable; however, it appears that the result was unexpected: Arkin, 'Cyber Warfare and the Environment', 781, citing Kuehl, 'Airpower Vs Electricity'.

[95] Pilloud, *et al.*, *Commentary*, para. 2156.

an action would have severe consequences for those countries where the main source of electricity is hydroelectric power.[96]

It should be noted that, as with dams, dykes and nuclear power generators, military objectives located at or in the vicinity of the works or installations lose their special protection only if they are used in regular, significant and direct support of military operations (a higher threshold than that of effective contribution to military effort) and if such an attack is the only feasible way to terminate such support.[97] Parties also have an obligation to endeavour to avoid locating military objectives in the vicinity of works or installations.[98]

4. Objects indispensable to the survival of the civilian population

Computer network attacks against systems and networks which are indispensable to the survival of the civilian population are prohibited under Additional Protocols I and II in the same manner that conventional attacks would be prohibited against such objects. Article 54(2) of Additional Protocol I provides:

> It is prohibited to attack, destroy, remove or render useless objects indispensable to the survival of the civilian population, such as foodstuffs, agricultural areas for the production of foodstuffs, crops, livestock, drinking water installations and supplies and irrigation works, for the specific purpose of denying them for their sustenance value to the civilian population or to the adverse party, whatever the motive, whether to starve out civilians, to cause them to move away, or for any other motive.

Article 14 of Additional Protocol II provides equivalent wording. The list of objects is merely illustrative and the Commentary to the provision notes that 'it cannot be excluded that as a result of climate or other circumstances, objects such as shelter or clothing must be considered as indispensable'.[99] As has been demonstrated by the Australian domestic case of Vitek Boden,[100] drinking water installations are particularly

[96] For example, Norway produces virtually all of its electricity from hydroelectric sources, while Iceland (83 per cent), Austria and Canada (both over 70 per cent) would be hugely affected by denial of hydroelectric sources. Notably, China is the world's largest producer of hydroelectric power.

[97] Art. 56(2), Additional Protocol I.

[98] Art. 56(5), Additional Protocol I.

[99] Pilloud, *et al.*, *Commentary*, para. 2103.

[100] *R* v. *Boden*. See section 1 above and Appendix 1.

susceptible to computer network attacks; likewise irrigation works have been tampered with in a domestic case,[101] and any other works primarily controlled by SCADA systems are also vulnerable.

Unlike ordinary civilian objects, Michael Schmitt's argument regarding the permissibility of targeting objects with computer network attacks not severe enough to amount to attacks, cannot be applied to objects indispensable to the civilian population.[102] After lengthy discussions during the drafting conference, the words 'remove' and 'render useless' were specifically added to the terms 'attack' and 'destroy' in Articles 54(2) and 14 in order to cover all possibilities.[103] However, all such actions against objects indispensable to the civilian population are only prohibited where they are conducted for the 'specific purpose of denying them for their sustenance value to the civilian population'.[104] Likewise the Rome Statute only considers depriving civilians of objects indispensable for survival a war crime where it constitutes intentional starvation as a method of warfare.[105]

5. Hospitals and other medical units

Hospitals, medical units and medical transports (including hospital ships and aircraft) all receive special protection under international humanitarian law. Indeed, they form the basis of the origins of the modern laws of armed conflict, and are protected by custom as well as specific *lex scripta*. For the most part this protection does not raise any additional issues in the event of computer network attacks; the protection remains regardless of the method of attack. However, two issues, the location of and access to medical databases and the encryption of communications to and from hospital ships, require some thought in the information age.

5.1. Location of and access to medical databases

In line with the trend towards networked services in both civilian and military life, the supply of operational support services to the military,

[101] Dan Goodin, 'Electrical Supe Charged with Damaging California Canal System', *The Register*, 30 November 2007, www.theregister.co.uk/2007/11/30/canal_system_hack/ (last accessed 7 July 2011).

[102] See Chapter 6, section 4.1 above.

[103] Pilloud, *et al.*, *Commentary*, paras 2100–1.

[104] *Ibid.*

[105] Art. 8(2)(b)(xxv), Rome Statute.

such as medical treatment facilities, has also benefited from increased network connectivity. For example, the United States military has instituted an information system to provide electronic access to medical information and provide the ability to electronically access and update medical records of serving personnel.[106] The system allows for integrated patient care which can keep pace with the patient's progression from the medic located in the field, through combat support hospitals, to medical centres situated far away from the fighting.[107] The handheld devices, laptops and database which comprise the system undoubtedly form part of the material and supplies of the medical units and thus are protected by the Geneva Conventions,[108] Additional Protocols and customary international law.[109] However, protection only remains in place while the system is used exclusively for the treatment of the wounded or sick and for the prevention of disease.[110] Care must be taken, therefore, that the medical database and the associated information systems are not used for any other purpose. For example, in addition to the standard access to medical records for treatment, commanders also use the system for medical situational awareness (for instance, to access information on incidence of illness in order to assess the need for vaccinations), a use which is covered by the disease prevention arm of the protections. However, the same broad spectrum

106 For a description of the system and its component applications see Sandra Basu, 'Military Electronic Medical Records Support Quality Treatment Abroad', *US Medicine* (Washington DC), February 2006.

107 Field medics are given handheld devices to capture medical data about a casualty in theatre. The device is then connected to a laptop where it uploads the information to a centralised database. The database can be accessed by treatment facilities anywhere in the world, allowing doctors to see exactly what treatment has been provided and what still needs to be done. *Ibid.*

108 Art. 19, Geneva Convention I requires respect and protection for fixed establishments and mobile medical units. Art. 33 provides specific protections for the material and stores of the units, which must remain available to the medical personnel to enable them to perform their functions; they may not be intentionally destroyed.

109 Art. 12, Additional Protocol I; Art. 11, Additional Protocol II.

110 Art. 21, Geneva Convention I states: 'The protection to which fixed establishments and mobile medical units of the Medical Service are entitled shall not cease unless they are used to commit, outside their humanitarian duties, acts harmful to the enemy. Protection may, however, cease only after a due warning has been given, naming, in all appropriate cases, a reasonable time limit and after such warning has remained unheeded.' Art. 34, Geneva Convention II relating to hospital ships has similar wording, as does Art. 19, Geneva Convention IV relating to civilian hospitals.

use of the database for other purposes would not be covered by the Conventions. For example, using the database to research the effects of new weapons systems, a standard part of weapons development research,[111] would risk discontinuance of protection of the systems and expose the database to targeting by computer network attack.

In addition, paragraph 2 of Article 19 of Geneva Convention I contains an obligation on parties to ensure that medical units are, as far as possible, situated away from military objectives. Bearing in mind that military networks have become targets for computer network attack, medical databases and associated information systems will need to be isolated from systems which are now considered legitimate targets. Marking such systems as medical systems and informing the opposing side of their existence would also be required.[112] The problem of adapting identification techniques to modern methods of warfare is not new; the issue was previously addressed at the time of drafting the Additional Protocols in respect of medical aircraft. Methods of electronic marking of aircraft were discussed and a secondary radar system of transponders (to automatically transmit an allocated identification code) was adopted.[113] This system appears easily adaptable to the computer environment. The communications standard over the internet revolves around the TCP/IP protocol. Under this system every packet communicated over the internet contains data about the originating network and destination network, as well as multiple layers of information about the data itself.[114] Incorporating information regarding the medical nature of the data would not be difficult. Another alternative discussed in Chapter 5 above is the designation of certain IP addresses as military networks, although, as discussed, this proposal is not without difficulty as it may mark a network for targeting.[115] However, the idea may translate more easily for networks that

[111] Data on the effects of new weapons are used as part of the so-called Solferino cycle, a development cycle which includes providing the observation and documentation of the effects of weapons both to weapons designers and to international humanitarian lawyers. See generally Robin M. Coupland, 'The Effects of Weapons and the Solferino Cycle: Where Disciplines Meet to Prevent or Limit the Damage Caused by Weapons' (1999) 319(7214) *BMJ* 864.

[112] Art. 39 Geneva Convention I requires the emblem to be displayed on all equipment employed in the medical service.

[113] Pilloud, *et al.*, *Commentary*, paras 4203–5.

[114] For further explanation see the entry for 'TCP/IP' and 'Communication Protocol' in J. Ashley Roach, 'The Law of Naval Warfare at the Turn of Two Centuries' (2000) 94 *AJIL* 64.

[115] See Chapter 5, section 1.1 above.

are specifically protected, such as medical networks. The success of such a solution would depend on the network having a fixed IP address; a situation which may not prove technologically practical in modern-day field operations.

5.2. *Hospital ships*

Article 34(2) of Geneva Convention II states that hospital ships may not possess or use secret codes for their wireless or other means of communication. Dinstein points out that this injunction against secret codes (forcing hospital ships to send and especially receive all messages in the clear) has severe practical problems.[116] As J. Ashley Roach notes, technology has changed since 1949; all messages to and from warships, including unclassified messages, are now automatically encrypted when sent and decrypted when received by communications equipment that includes the crypto function.[117] This is also true for all network communications with naval craft and thus leaves hospital ships with the unattractive alternatives of being precluded from reports of movements of the fleet (and particularly advance notice of military operations likely to require their services),[118] or being in breach of their obligations under the second Geneva Convention. The San Remo Manual now moves in the direction of allowing hospital ships to use cryptographic equipment while prohibiting the transmission of intelligence data.[119] Without such encryption the hospital ships could provide a dangerous backdoor into military networks, allowing an attacker to gain access to other networks to which the ship is connected.

6. Non-defended localities and demilitarised zones

Article 59 of Additional Protocol I prohibits parties to a conflict attacking any non-defended locality by any means whatsoever. Article 60 prohibits the extension of military operations to zones which have been agreed between the parties as demilitarised zones. While these provisions are undoubtedly broad enough to apply to computer network attacks against physical areas, the characterisation of

[116] Dinstein, *Conduct of Hostilities* (2nd edn), 192.

[117] Roach, 'The Law of Naval Warfare at the Turn of Two Centuries', 75.

[118] Louise Doswald-Beck, 'Vessels, Aircraft and Persons Entitled to Protection During Armed Conflicts at Sea' (1994) 65 *BYBIL* 211, 251; cited in Dinstein, *Conduct of Hostilities* (2nd edn), 192.

[119] Dinstein, *Conduct of Hostilities* (2nd edn), 192.

non-defended localities or demilitarised zones raises an interesting issue in the computer age. Can a computer system or network ever be considered in itself a non-defended locality or be designated a demilitarised zone?

6.1. *Non-defended localities*

Article 59 codifies and confirms customary international law and reiterates almost entirely Article 25 of the Hague Regulations.[120] Article 59 defines the concept of a non-defended location as an 'inhabited place' near or in a zone where armed forces are in contact and which is open for occupation. Likewise the Hague Regulations refer to 'towns, villages, dwellings, or buildings', all places of human occupation, which cannot easily be equated with computer systems. Traditionally, customary international law has referred to open towns and undefended areas,[121] and may provide more scope for translation to a computer network. However, given the ambiguous nature of the characterisation, the parties would be wise to agree between themselves that a particular computer system or network constituted such a locality and what existing network protections may remain in place without compromising its designation as non-defended. The customary law prohibition of attacks against non-defended localities is also applicable in non-international armed conflicts.[122]

6.2. *Demilitarised zones*

Of greater interest in computer-age conflict is the parties' ability to agree to grant particular zones the status of demilitarised zones.[123] The subject of the agreement can be any zone expressly agreed by the parties, although the Commentary to the Additional Protocols states that the essential character of the zones is humanitarian and not political – they are specially intended to protect the population living there against attacks.[124] While the concept was designed for physical locations, there is no reason why the parties could not agree to designate a particular

[120] Pilloud, *et al.*, *Commentary*, para. 2263; Art. 25, Hague Regulations states: 'The attack or bombardment, by whatever means, of towns, villages, dwellings, or buildings which are undefended is prohibited.'

[121] Henckaerts and Doswald-Beck, *Customary International Humanitarian Law*, Vol. I, 124.

[122] For example, Art. 3, Statute of the International Criminal Tribunal for the former Yugoslavia provides for prosecution of the violations of the laws and customs of war including 'attack, or bombardment, by whatever means, of undefended towns, villages, dwellings, or buildings'.

[123] Art. 60, Additional Protocol I.

[124] Pilloud, *et al.*, *Commentary*, para. 2303.

network or system as a demilitarised zone.[125] Once designated as a demilitarised zone, Article 60(1) of Additional Protocol I prohibits parties to the conflict from 'extend[ing] their military operations to zones on which they have conferred by agreement the status of demilitarised zone, if such extension is contrary to the terms of this agreement'. This wording is broader than the protection provided under Article 59 relating to non-defended localities, discussed above, which only prohibits attacks against the locality. The Commentary to the Additional Protocols states that the expression 'military operations' should be understood as all movements and activities related to hostilities, carried out by armed forces.[126] Thus all use of a particular system for *any* activity relating to hostilities would be prohibited.

[125] Art. 60(3) sets out the general outline for the terms of such an agreement; however, as indicated by the inclusion of the term 'normally', it can be adapted for specific situations. Art. 60(3) provides: 'The subject of such an agreement shall normally be any zone which fulfils the following conditions: (a) all combatants, as well as mobile weapons and mobile military equipment, must have been evacuated; (b) no hostile use shall be made of fixed military installations or establishments; (c) no acts of hostility shall be committed by the authorities or by the population; and (d) any activity linked to the military effort must have ceased.'

[126] Pilloud, *et al.*, *Commentary*, para. 2304.

8 Means and methods of warfare

While predominantly viewed as a method of attack, computer network attack scenarios may represent both means and method of attack, depending on the attack being executed. According to the ICRC Commentary to the Additional Protocols, the term 'means of combat' or 'means of warfare' generally refers to the weapons being used, while the expression 'methods of combat' generally refers to the way in which such weapons are used.[1] For example, a worm in and of itself is usually a method of attack: it is designed as a means to distribute malicious code, or alternatively to cause generalised damage by overwhelming the networks with packets of information causing massive denial of service. To take one instance, the Slammer/Sapphire worm contains an algorithm designed to self-replicate and randomly generate addresses in order to spread itself as rapidly as possible.[2] Other forms of attack contain specific malicious code which is designed to cause damage to the computer, system or network directly. Trojans designed specifically to destroy hard drives or information are examples. A more indirect example is the code inserted into the Canadian pipeline software which resulted in the 'Farewell Dossier' incident, a huge explosion involving the Soviet Trans-Siberian gas pipeline.[3] Such malware fits the

[1] Pilloud, *et al.*, *Commentary*, para. 1957. Note, however, that the term method and means of warfare includes weapons in the widest sense, as well as the way in which they are used; para. 1402.

[2] The Slammer/Sapphire worm hit the internet in January 2003 and to date is the fastest spreading worm found in the wild. It infected more than 90 per cent of vulnerable hosts (at least 75,000 hosts) in ten minutes. David Moore, *et al.*, *The Spread of the Sapphire/Slammer Worm*, (2003) www.caida.org/publications/papers/2003/sapphire/sapphire.html (last accessed 7 July 2011). Other viruses may have different propogation techniques.

[3] See Appendix I.

definition of a weapon far more closely than merely a method of attack. The more recent attacks have combined malicious code with a means of propagation such as a worm to form blended attacks. For example, the Storm worm, which was discovered 'in the wild' in January 2007, combines multiple components: a component to steal email addresses and redistribute itself; a backdoor Trojan to allow subsequent access to the compromised machine; a bot recruiter to incorporate the machine into the Storm botnet; coding to allow remote control of the compromised machine across peer-to-peer networks (which morphs every few minutes to avoid detection); and a denial of service attack tool in a rootkit payload.[4] Likewise, the Stuxnet worm discovered in 2010, the most sophisticated piece of malware to come to light to date, utilises four different vulnerabilities in systems to spread and a vast array of component parts to maximise its chances of success.[5]

1. Law of weaponry

Article 22 of the Hague Regulations sets out the basic principle that 'the right of belligerents to adopt means of injuring the enemy is not unlimited'.[6] This concept of limited warfare, repeated almost verbatim in Article 35(1) of Additional Protocol I and the preamble to the Conventional Weapons Convention, forms the basis for the legal regulation of means and methods of warfare.[7] As with the rest of the laws of armed conflict, the law of weaponry represents a balancing act between the principle of military necessity, that is, what is required to efficiently conduct military operations, and what is required by humanitarian considerations. In the words of the preamble of the St Petersburg Declaration of 1868, to fix 'the technical limits at which the necessities of war ought to yield to the requirements of humanity'.

[4] For further details of the Storm worm see Joe Stewart, *Storm Worm DDoS Attack*, Secure Works (1997), www.secureworks.com/research/threats/storm-worm (last accessed 7 July 2011). Thorsten Holz, *et al.*, 'Measurements and Mitigation of Peer-to-Peer-Based Botnets: A Case Study on Stormworm' (Paper presented at the First USENIX Workshop on Large-Scale Exploits and Emergent Threats (LEET '08), San Francisco, http://honeyblog.org/junkyard/paper/storm-leet08.pdf (last accessed 7 July 2011).

[5] This includes zero-day exploits, a Windows rootkit, the first ever PLC rootkit, antivirus evasion techniques, complex process injection and hooking code, network infection routines, peer-to-peer updates, and a command and control interface. Falliere, OMurchu and Chien, *Stuxnet Dossier*, 2.

[6] Art. 22, Hague Regulations. [7] Oeter, 'Methods and Means of Combat', 126.

The law of weaponry consists of both general principles, such as those prohibiting indiscriminate weapons or unnecessary suffering, and a number of specific rules prohibiting or limiting the use of certain weapons or methods of warfare.[8] As Christopher Greenwood has remarked, the general principles tend to refer to the effects produced by the use of weapons or methods of warfare, whereas the specific provisions usually concentrate on the means employed.[9] Although only a few of the specific provisions are relevant to computer network attacks, the same general principles apply regardless of the style of weapon employed. As demonstrated by the discussion of the legality of nuclear weapons by the International Court of Justice, these principles established in the last century are capable of being applied well into the next, even to methods of warfare undreamed of when those principles were formulated.[10]

1.1. *General principles*

Flowing from the principle of 'limited warfare' several sub-principles have developed historically, giving the rule of military necessity its specific contours.[11] Not all of these will raise specific issues for computer network attacks; however, a number require additional consideration as to their interpretation in the modern battlespace, in particular the principle regarding unnecessary suffering and the prohibition against indiscriminate weapons. Other general principles which form part of the law of weaponry are examined elsewhere in the relevant sections of this volume, for example those dealing with environmental protection in Chapter 7 above, and perfidy in section 2 below.

1.1.1. Superfluous injury and unnecessary suffering

The International Court of Justice has confirmed that the prohibition of means and methods of warfare which are of a nature to cause superfluous injury or unnecessary suffering is one of the cardinal principles of international humanitarian law.[12] Article 35(2) of Additional Protocol I represents one of the most recent statements of the prohibition:[13]

[8] Greenwood, 'Law of Weaponry', 192. [9] *Ibid.* [10] *Ibid.*, 186.
[11] Oeter, 'Methods and Means of Combat', 127.
[12] *Nuclear Weapons Case*, para. 238.
[13] Art. 35(2) Additional Protocol I; the language is repeated in the preamble to the Conventional Weapons Treaty.

It is prohibited to employ weapons, projectiles and material and methods of warfare of a nature to cause superfluous injury and unnecessary suffering.

It follows and clarifies similar statements in the St Petersburg Declaration of 1868 and Article 23(e) of the Hague Regulations.[14] The prohibition of *methods* of warfare which are of a nature to cause superfluous injury or unnecessary suffering was first introduced in Additional Protocol I; however, the commentaries note that this change of language does not alter the principle which represents customary international law.[15] The Appeals Chamber of the ICTY has held that the principle is also applicable in conflicts not of an international character, stating 'what is inhumane, and consequently proscribed, in international wars, cannot but be inhumane and inadmissible in civil strife'.[16]

In its advisory opinion in the *Nuclear Weapons* case, the International Court of Justice defines unnecessary suffering as 'harm greater than that unavoidable to achieve legitimate military objectives'.[17] Inescapably, the test is valid only for weapons designed exclusively for anti-personnel purposes, inasmuch as anti-materiel weapons may be expected to cause injury to personnel in the vicinity of the target that would be more severe than necessary to render the combatants *hors de combat*.[18] Likewise, Greenwood notes that, while there is general agreement that the character and effect of anti-materiel weapons differ from those commonly used against personnel, such weapons do not violate the 'unnecessary suffering' principle, because the advantages they offer (for example, to destroy materiel) means that the additional suffering they may cause cannot be classed as unnecessary.[19] In this respect, when viewed as a weapon, computer network attacks are inherently

[14] The wording 'of a nature to cause' clarifies the English translations in previous versions of this provision, which used the words 'calculated to' and had occasionally been taken to mean the unnecessary suffering would have to be proved intentional. Hans Blix, 'Means and Methods of Combat' in UNESCO (ed.), *International Dimensions of Humanitarian Law* (Martinus Nijhoff, Dordrecht, 1998) 135–51, 138.

[15] See, for example, Bothe, *et al.*, *New Rules*, 194. Pilloud, *et al.*, *Commentary*, para. 1417.

[16] *Tadic (Interlocutory Appeal)*, para. 119. See also Henckaerts and Doswald-Beck, *Customary International Humanitarian Law*, Vol. 1, 237.

[17] *Nuclear Weapons Case*, para. 238.

[18] Bothe, *et al.*, *New Rules*, 196; Dinstein, *Conduct of Hostilities* (1st edn), 60. Although the deliberate employment of an anti-materiel weapon against people, might also fall foul of this principle.

[19] Greenwood, 'Law of Weaponry', 196. In this respect, Greenwood notes the use of inflammable bullets which came to be accepted as lawful against aircraft, despite the effect that they may have on an aircrew, but remain unlawful in a simple anti-personnel role: 225, fn. 61.

anti-materiel weapons, as malicious code cannot (at the current state of technology) act directly on an individual, but rather on the physical and technological environment in which that person is situated. As noted above, computer network attacks may be either a means or a method of warfare depending on their use; however, the prohibition against unnecessary suffering refers to both, therefore the use of computer network attacks as a method of warfare must be assessed.[20]

However, even in relation to weapons designed for other than anti-personnel purposes, the application of the unnecessary suffering principle requires a balancing of sorts. As Boothby notes, 'unnecessary' and 'superfluous' are comparative not absolute concepts; however, the appropriate comparison to be made remains the subject of disagreement.[21] Some states and commentators assert that the correct comparison is between the military effectiveness or military advantage which may result from the use of a particular weapon (in its normal, intended use), and the degree of injury and suffering which it is likely to cause.[22] Greenwood adds to this approach, arguing that a determination of whether or not a weapon's use would violate the unnecessary suffering principle requires a balancing of the likely military advantages and the likely human suffering which its use in the future will entail, and then a comparison between that balance and what would result from the use of alternative weapons.[23]

The former approach has been criticised by other commentators, such as Dinstein, who argues that the question of military advantage is inextricably linked to questions of proportionality relating to collateral damage to civilians and civilian objects and has nothing to do with injury or suffering sustained by combatants.[24] Rather, he argues that the injunction against 'superfluous injury or unnecessary suffering' hangs on a determination of whether injury/suffering

[20] See, for example, a 'real-world' equivalent in the discussion relating to the use of earthmovers and ploughs mounted on Abrams battle tanks to destroy battle trenches and bunkers, and burying alive the Iraqi soldiers inside them: Françoise J. Hampson, 'Means and Methods of Warfare in the Conflict in the Gulf' in P. J. Rowe (ed.), *The Gulf War of 1990–1991 in International and English Law* (Routledge, London, 1993) 89–110, 92–3.

[21] William H. Boothby, *Weapons and the Law of Armed Conflict* (Oxford University Press, Oxford; New York, 2009), 62.

[22] See Greenwood, 'Law of Weaponry', 195; Boothby, *Weapons*, 62–3; Bothe, *et al.*, *New Rules*, 200.

[23] Greenwood, 'Law of Weaponry', 199.

[24] Dinstein, *Conduct of Hostilities* (2nd edn), 65.

is avoidable or unavoidable.[25] In this, he agrees with the second limb of Greenwood's comparison test, namely a comparison between the weapon in question and other options with sufficient military effectiveness. On balance, the present author considers that the correct view is that offered by Greenwood's approach. The question of whether the suffering caused to combatants is necessary must be weighed against the military effectiveness and advantage expected from the weapon, and then compared with that caused by appropriate alternatives. As each computer network attack may be varied as to its use and execution, each attack will need to be assessed separately to ensure that the balance is maintained.

As it is unlawful to use a weapon which causes more suffering or injury than another that offers the same or similar military advantage, the classification necessarily involves a comparison between different weapons systems.[26] This is perhaps one of the most interesting aspects of law of weaponry in relation to computer network attacks, namely the effect that access to these techniques will have on the legality of the parties' choice of other weapons. One of the advantages of computer network attack is that it allows neutralisation and destruction of targets, with fewer casualties and less physical destruction, and in many cases more accuracy, than conventional weapons. For example, there is no need to bomb an electrical grid if you can simply turn it off for the desired period of time. This ability may have the effect of making other, more conventional methods of warfare illegal, as the damage caused by those methods becomes subsequently 'unnecessary'. However, as Greenwood points out, it is not enough simply to compare the immediate effects of the two weapons (or methods of warfare);[27] the availability (including the expense) of both types of weapon, the logistics of supplying the weapon and its ammunition at the place where it is to be used, and the security of the troops which employ it, are all additional factors to be taken into account. While these would tend to resolve in favour of using computer network attack methods, there are also factors which may advocate against using such techniques in a given situation. As discussed in relation to precautions in attack in Chapter 6 above, given that some computer network attack methods can only be used once before effective countermeasures are put in place, the likely future need for such an attack will also be a factor to

[25] *Ibid.* [26] Blix, 'Means and Methods of Combat', 138–9.
[27] Greenwood, 'Law of Weaponry', 198.

be considered by commanders, as would the difficulty in ascertaining knock-on effects of a particular attack.

1.1.2. Indiscriminate weapons

The International Court of Justice, in its advisory opinion on the *Threat or Use of Nuclear Weapons*, held the principle of discrimination as a cardinal principle of international humanitarian law and held that states 'must consequently never use weapons that are incapable of distinguishing between civilian and military targets'.[28] The key distinction here is between weapons that are inherently unable to discriminate between military targets and civilian objects, rather than the indiscriminate use of a particular weapon, which is discussed in Chapter 6 above. Article 51(4) of Additional Protocol I prohibits indiscriminate attacks. Subparagraph (b) of that Article prohibits means and methods of attack which cannot be directed at a specific military objective. The classic example of this is the long-range missile with a guidance system which cannot accurately target a specific objective.[29] Although some computer network attacks are targeted very specifically against the particular system or network to be attacked, most computer viruses currently in the wild are coded to spread indiscriminately.[30] Computer network attacks that cannot distinguish between civilian and military networks and systems are inherently indiscriminate; therefore, where they cause sufficient damage to amount to an attack under the laws of armed conflict, their use is unlawful under the law of weaponry.

Subparagraph (c) prohibits the use of means and methods of combat whose effects cannot be limited as required under the Protocol and which consequently are of a nature to strike military objectives and civilians or civilian objects without distinction.[31] This is distinct from means and methods which can, depending on the circumstances of their use, have a restricted use or be completely out of the control of those using them.[32] The Stuxnet worm acted as proof of concept that this particular type of computer network attack does not fall within

[28] *Nuclear Weapons Case*, para. 78.

[29] The ICRC Commentary gives the example of the V2 rockets used by Germany at the end of World War II. Pilloud, *et al.*, *Commentary*, para. 1958.

[30] For example, the Storm worm is hard-coded with a sample set of addresses and a random dialling algorithim which enables it to spread from computer to computer.

[31] The examples given in the Commentary relate to the poisoning of drinking water or the spread of bacteriological weapons, although the Commentary notes that both means of warfare are prohibited under other provisions. Pilloud, *et al.*, *Commentary*, 165.

[32] *Ibid.*

that category. The sophisticated malware showed the increasing ability on the part of coders to take advantage of its component structure to limit the effect, rather than the targeting, of the malware. While the worm spreads indiscriminately, it does not deploy the payload component until it finds itself on a system with the correct configuration. It is also worth noting that it is the deployment of this component that will raise the impact of the computer network attack to the level of an attack under the laws of armed conflict. Thus, despite the indiscriminate spread of the worm, it does not constitute an indiscriminate attack under the Protocol. In a variant of a Stuxnet-type worm, an attacking coder could easily set the conditions of the payload deployment to ensure that the system being attacked included a piece of software or a system file that is unique to the military systems targeted.

Obviously, malicious code or computer network attacks which are capable of being directed at specific military objectives, or whose effects can be limited but are nevertheless utilised in an indiscriminate manner, are also prohibited.[33] However, they do not breach the discrimination principle as it applies to the law of weaponry.

1.1.3. Martens Clause

The Martens Clause was first included in the preambular provisions of the Hague Convention of 1899; a modern formulation is expressed in Article 1(2) of Additional Protocol I:

> In cases not covered by this Protocol or by other international agreements, civilians and combatants remain under the protection and authority of the principles of international law derived from established custom, from the principles of humanity and from the dictates of public conscience.

As the Court in the *Nuclear Weapons* case noted, the clause has proved to be an effective means of addressing the rapid evolution of military technology.[34] Thus, although there is currently no rule or agreement in international humanitarian law which expressly bans or restricts the use of computer network attacks, where a particular type of computer network attack *per se* would have results which violate the principles of humanity or dictates of public consciousness, it would contravene the Martens Clause.[35] Christopher Greenwood notes that one of the

33 See section 4.2 of Chapter 6 above discussing indiscriminate attacks.

34 *Nuclear Weapons Case*, para. 78.

35 See generally, Isabelle Daoust, Robin Coupland and Rikke Ishoey, 'New Wars, New Weapons? The Obligation of States to Assess the Legality of Means and Methods of Warfare' (2002) 84(846) *IRRC* 345, 351; ICRC, *A Guide to the Legal Review of New*

effects of the Clause is that the absence of a specific treaty provision outlawing a particular weapon does not mean that a weapon must be lawful; the Clause makes clear that the general principles embodied in customary law still apply and that the use of a weapon contrary to those principles will be unlawful.[36]

1.2. *Explicit prohibitions of weapons*

The International Court of Justice in the *Nuclear Weapons* case stated that the illegality of certain weapons is formulated in terms of prohibition rather than absence of authorisation.[37] As noted above, there is currently no rule or agreement in international humanitarian law which explicitly bans or restricts the use of computer network attacks.[38] However, because computer network attacks can cause a multiplicity of effects, some attack techniques may fall within the definitions of other weapons conventions which may restrict or ban their use. Care must be taken, however, not to misunderstand the underlying principles behind the prohibitions of specific weapons.[39] While the law of weaponry seeks to protect core humanitarian values, it is also used to affect disarmament objectives which are not directly relevant to computer network attacks, and aspects of fair dealing which are discussed in section 2 on perfidy below.[40]

Protocol II and Amended Protocol II of the Conventional Weapons Convention may restrict the use of some forms of computer network attack. The definition of a booby trap is wide enough to encompass computer network attacks which are 'designed, constructed or adapted to kill or injure'.[41] A file, device or system may be rigged to execute

Weapons, Means and Methods of Warfare: Measures to Implement Article 36 of Additional Protocol I of 1977 (2006) 17.

36 Greenwood, 'Law of Weaponry', 206.

37 *Nuclear Weapons Case*, para. 52.

38 In 1998, and every year since, the Russian Federation has tabled a draft resolution for the investigation into communications technologies and information security which would address these issues. To date, it has not been adopted. See, for example, draft resolution A/C.1/55/L. 6 introduced by the Russian Federation, 18th mtg, 19 October 1998.

39 For example, Brown has proposed that certain attacks i.e. logic bombs, are analogous to landmines and should therefore be prohibited: Brown, 'Proposal for an International Convention', 197.

40 For a discussion of the objectives of the law of weaponry see generally, Greenwood, 'Law of Weaponry', 189.

41 Art. 2 of both Protocols defines booby traps as 'any device or material which is designed, constructed or adapted to kill or injure, and which functions unexpectedly when a person disturbs or approaches an apparently harmless object or performs an apparently safe act'.

some form of malicious code on access which would cause part of the system to explode or trigger some other form of injury to the activator. For example, fooling the surge protectors into thinking there was a lightning strike and shutting off the power; on restart the additional power surge would overload the distribution node and could be rigged to cause the computer monitor to explode.[42] It is important to note, however, that the Protocols only apply to booby traps which are designed to kill or injure, thus the use of booby traps which merely destroy or prevent unauthorised access to information or render a system useless, which are fairly common in domestic computer crime, would not be covered by the Conventions. Whether the device is designed, constructed or adapted to kill or injure is a matter of fact, thus its legality or otherwise will need to be assessed on the basis of all of the available information.[43]

Booby traps are prohibited if, by their nature or employment, their use violates the legal protection accorded to a protected person or object by another customary rule of international law.[44] Thus any code which is capable of falling within the definition which disguises itself as an email from the ICRC, for instance, would automatically be banned.[45] It is also prohibited to design and manufacture certain booby traps to look like harmless portable objects, although it is not prohibited to convert an existing harmless object into a booby trap.[46] However, although the general restrictions on booby traps are broad enough to encompass a computer network attack specifically designed to cause injury, the wording of the particular provisions relating to harmless portable objects refers expressly to objects constructed to contain explosive material and thus would not be applicable to malicious code. However, in the event that the principle was applied by analogy, it would mean that rootkit malware (which inserts itself into existing code) would be a valid form of attack, whereas a separate file disguised as a harmless email attachment or other such item would not. This would be in keeping with the prohibition which applies to

[42] Whether or not a computer would in fact explode or otherwise cause injury is dependent on a number of variables, for instance the type of monitor used by the rigged system (a plasma screen would melt rather than explode), or whether backup generators were in place etc.

[43] Boothby, *Weapons*, 158–9.

[44] Henckaerts and Doswald-Beck, *Customary International Humanitarian Law*, Vol. I, 278.

[45] This would undoubtedly constitute an act of perfidy in any event, and the misuse of protected symbols is covered more fully in section 2 on perfidy below.

[46] Art. 6(1) Protocol II, Art. 7(2) Amended Protocol II.

letter bombs.[47] Such prohibitions would be better dealt with in terms of the principle against perfidy and aspects of fair dealing which are discussed in section 2 on perfidy below.[48]

1.3. *Article 36 obligations*

Article 36 of Additional Protocol I imposes an obligation on contracting parties to perform legal reviews of new weapons, means and methods of warfare:

> In the study, development, acquisition or adoption of a new weapon, means or method of warfare, a High Contracting Party is under an obligation to determine whether its employment would, in some or all circumstances, be prohibited by this Protocol or by any other rule of international law applicable to the High Contracting Party.

Although a new obligation to international humanitarian law, this obligation follows in the footsteps of the St Petersburg Declaration of 1868,[49] and is an ‘obvious and indispensable corollary’ to Article 23(e) of the Hague Regulations and Article 35(2) of Additional Protocol I.[50] The ICRC Commentary to the Additional Protocol states that the words ‘methods and means’ include weapons in the widest sense, as well as the way in which they are used.[51] Further, the meaning of the phrase ‘some or all circumstances’ is to require a determination of whether the employment for its *normal or expected use* would be prohibited under some or all circumstances, not to foresee or analyse all possible misuse of the weapon.[52]

The ICRC notes that the faithful and responsible application of its international law obligations would require a state to ensure that the new weapons, means and methods of warfare it develops or acquires will not violate these obligations.[53] This obligation to review undoubtedly also applies to new computer network attack techniques and states will be required to assess the legality of each type of attack as they are developed. As with conventional weapons' development and review, assessment of computer network attacks under this provision will

[47] See generally Dinstein, *Conduct of Hostilities* (1st edn), 65.

[48] For a discussion of the objectives of the law of weaponry see generally, Greenwood, ‘Law of Weaponry’, 189.

[49] Whereby parties were to come to an understanding about advances in armaments to maintain the principles which they had established and to ‘conciliate the necessities of war with the laws of humanity’: St Petersburg Declaration.

[50] Bothe, *et al.*, *New Rules*, 199.

[51] Pilloud, *et al.*, *Commentary*, para. 1402.

[52] Bothe, *et al.*, *New Rules*, 200.

[53] ICRC, *Legal Review of New Weapons*, 4.

remain largely a matter of trust. As noted previously in this volume, many computer network attacks will only work once before countermeasures are developed and installed; therefore any international scrutiny of the technique would render them useless before they have been used.

2. Perfidy and ruses of war

Advanced network technology provides a myriad of new opportunities for parties to deceive and mislead the adversary. While certain deceptions are entirely permissible under the laws of armed conflict, where they amount to perfidy (or treachery) they are prohibited.

2.1. Ruses of war

The Hague Regulations and Additional Protocol I both confirm that ruses of war are not prohibited.[54] Article 37(2) of Additional Protocol I states:

> Ruses of war are not prohibited. Such ruses are acts which are intended to mislead an adversary or to induce him to act recklessly but which infringe no rule of international law applicable in armed conflict and which are not perfidious because they do not invite the confidence of an adversary with respect to protection under that law. The following are examples of such ruses: the use of camouflage, decoys, mock operations and misinformation.

Advances in computer network technology have enabled twenty-first-century ruses to develop in lock-step with advancements in attack scenarios. Many of the options available to armed forces for modern day ruses of war fall into the category of electronic warfare;[55] however, militaries are increasingly adopting more sophisticated techniques which utilise computer network attack capabilities in order to mislead and deceive their adversaries. For example, basic jamming of radar signals would constitute an electronic attack; however, the recent use of an airborne network attack system by Israel in the air strike against a target in northern Syria demonstrates other possibilities. US aerospace

[54] Art. 24 Hague Regulations; Art. 37(2), Additional Protocol I.

[55] For example, jamming, electromagnetic pulse (EMP), high energy radio frequency weapons are examples of electronic warfare and while raising interesting issues for LOAC will not be dealt with in this book. For a general overview of electronic warfare techniques see Wilson, *Information Operations, Electronic Warfare, and Cyberwar.*

industry and retired military officials have indicated that the Israelis utilised a system like the US 'Suter' system to allow their fighters to approach undetected by Syrian defences.[56] The US Suter system enables users to invade communications networks, see what the enemy sees and even take over as the systems administrator so that sensors can be manipulated into positions where approaching aircraft cannot be seen.[57] An example of a permissible ruse of war would be to infiltrate the targeting data on the adversary's computer and enter false information.[58]

Routing an attack through multiple hosts in multiple countries to disguise its origin is common practice in computer network attacks outside the military context. Any number of stepping stone hosts (whether routers, servers, or individual computers) may act as conduits for an attack, effectively 'laundering' the packets of information and making tracing the path and the ultimate origin of the attack extremely difficult. Because the purpose of such tactics is to obfuscate the source of the attack, use of any stepping stone host which is part of a civilian network or a neutral state network runs the risk of the victim state retaliating against the apparent source. As Michael Schmitt points out, such retaliation may be kinetic in nature.[59] The question is whether this use of civilian networks is analogous to using civilian aircraft or vehicles to transport military cargo, or whether it amounts to feigning civilian status, which would be a prohibited act.[60] It is permissible to transmit to the enemy false messages which appear to come from their headquarters; however, such messages must not be distress signals or cause the enemy to attack civilians or civilian objects under the wrong impression that they are combatants

[56] Fulghum and Barrie, 'Israel Used Electronic Attack in Air Strike against Syrian Mystery Target'.

[57] *Ibid.* See Appendix 1 for details of the development of the Suter system.

[58] For example, all US target data for Operation Allied Force in Kosovo was stored on and accessed through a classified computer system. The information contained imagery, descriptions of the facility and its functions, analysis on impact (military advantage anticipated) if destroyed, possible collateral damage concerns and historical information on the target. Tony Montgomery, 'Legal Perspective from the EUCOM Targeting Cell' in A. E. Wall (ed.), *Legal and Ethical Lessons of NATO's Kosovo Campaign* (Naval War College, Newport, RI, 2002) 189, 192.

[59] Schmitt, 'Wired Warfare', 206.

[60] Note that this analogy is in respect of the laws of war on land. Naval vessels may fly false colours under certain circumstances without it constituting perfidy. See generally, Dieter Fleck, 'Ruses of War and Prohibition of Perfidy' (1974) 13 *Revue de Droit Penal Militare et de Droit de la Guerre* 269, 292.

or military objectives.[61] Such measures will be permitted ruses of war as long as they do not cross the line into perfidy.

2.2. *Perfidy*

The prohibition against perfidy reflects customary international law and is expressed in Article 23(b) of the Hague Regulations (where it is referred to as treachery) and expanded on by Article 37(1) of Additional Protocol I.[62]

Perfidy is defined by the Additional Protocol as:

> [a]cts inviting the confidence of an adversary to lead him to believe that he is entitled to, or is obliged to accord, protection under the rules of international law applicable in armed conflict, with intent to betray that confidence.

The application of the prohibition of perfidy to internal armed conflicts has also been confirmed by the Appeals Chamber in *Tadic*.[63] The prohibition against perfidy has two purposes: first, it seeks to protect those who genuinely wish to surrender, possess protected status or who are injured, by prohibiting the misuse of such status on the basis that abuse will erode respect for immunity in future cases. It also seeks to impose a minimum level of 'fairness' to dealings between combatants, even where the act endangers no one else.[64] Thus, manipulating information systems so that enemy forces wrongly believe that troops are surrendering rather than gathering for an attack would be perfidious, as would causing them to believe that combat vehicles were medical vehicles or those of neutrals.[65] Similarly, manipulating an enemy's targeting database so that it believed that an army division headquarters was a hospital would be prohibited.[66] Using protected symbols such as the UN Symbol or the emblem of the Red Cross would be prohibited acts of perfidy, as well as constituting a misuse of the symbols under Article 38 of Additional Protocol I.[67]

[61] Dinstein, *Conduct of Hostilities* (2nd edn), 240.

[62] Art. 23(b) Hague Regulations states that it is especially forbidden 'to kill or wound treacherously individuals belonging to the hostile nation or army'.

[63] *Tadic (Interlocutory Appeal)*, para. 125, citing the Supreme Court of Nigeria in *Pius Nwaoga* v. *The State* (1972) 52 ILR 494, 496–7 (Nig. S. Ct.).

[64] Greenwood, 'Law of Weaponry', 190.

[65] Greenberg, Goodman and Hoo, *Information Warfare and International Law*, 13.

[66] *Ibid*.

[67] As well as breaching Arts 38 and 39 on misuse of protective symbols, see below. Note that use of the UN symbol is only considered perfidy where the United Nations is *not* engaged in combat.

The feigning of civilian, non-combatant status is one of the examples given in Article 37(1) of perfidious acts. This has led some commentators to state incorrectly that the use of emails purporting to be from Microsoft support, which in actuality contain viruses or executable software patch designed to wreck the targeted computer system, would be perfidious.[68] This misunderstands the limitations of the prohibition contained in the Additional Protocol. Perfidy is only prohibited in so far as it results in the killing, injuring or capture of an adversary. Although this limitation has been criticised,[69] beyond this restrictive definition, acts such as sabotage or the destruction of property even through the use of perfidious deception are not prohibited under international law.[70] Bothe *et al.*'s commentary to the Protocol notes there must be a direct proximate causation between the act of perfidy and the killing, injury or capture of the adversary to breach Article 37.[71]

A further question arises over whether states have a legitimate expectation of authenticity of emails. Because of the long-standing view that communications may be disrupted, and because, unlike uniforms, information systems are in no way required by the laws of war but are rather combat aids, such tactics might seem less treacherous than would taking advantage of the requirement that troops wear distinct uniforms to set themselves apart from their foes and civilians.[72]

2.3. *Misuse of protected symbols and emblems*

Article 38 of Additional Protocol I prohibits any improper or deliberate misuse of internationally recognised signs, symbols or signals, including those of the Red Cross and associated symbols, the flag of truce, cultural property or of the emblem of the United Nations except as authorised by the UN.[73] An equivalent provision relating to the emblem of the Red Cross, Crescent, Lion or Sun exists in Article 12 of Additional

[68] Shulman, 'Discrimination in the Laws of Information Warfare', 959. See also Dörmann, 'Additional Protocols', 152.

[69] Oeter, 'Methods and Means of Combat', 228–9; Knut Ipsen, 'Perfidy' in R. Bernhardt (ed.), *Encyclopedia of Public International Law* (Max Planck Institute, Amsterdam; New York, 1997) 978–81.

[70] Oeter, 'Methods and Means of Combat', 229.

[71] Bothe, *et al.*, *New Rules*, 204. Although as Bothe *et al.* point out, the saboteur would not be entitled to POW status and may also be guilty of a breach of Art. 44(3) Additional Protocol I for failing to distinguish themselves from the civilian population.

[72] Greenberg, Goodman and Hoo, *Information Warfare and International Law*, 13.

[73] Note that the emblem of the UN is only protected where the UN is not a party to the conflict.

Protocol II for internal armed conflicts. This is an absolute prohibition which does not require any link to killing, injuring or capture. Thus, in the above example, if an email containing malware purported to be from a UN representative or Red Cross or Crescent society and misused the protected emblem, it would be prohibited under the broader auspices of improper use of such emblems. By extension it may well be likely that the use of a spoofed email address (i.e. joe.bloggs@icrc.org) would also be prohibited, although it does not fall within the category of protective emblems, signs or signals.

3. Destruction and seizure of property

The laws regulating the treatment of an adversary's property in times of armed conflict raise interesting issues for networked societies and knowledge economies. The basic principle is set out in Article 46 of the Hague Regulations, which provides that private property must be respected and cannot be confiscated. Further, Article 23(g) provides that it is forbidden to 'destroy or seize the enemy's property, unless such destruction or seizure be imperatively demanded by the necessities of war'.[74] Article 8(2)(b)(xiii) of the Rome Statute makes such destruction or seizure a war crime.[75] Pillage also is formally prohibited by the Hague Regulations, Geneva Convention IV, Additional Protocol II, and made a war crime under the Rome Statute.[76] However, the protections for property are subject to several exceptions, some of which are particularly relevant to communications networks and digital property.

Before turning to the exceptions, two preliminary matters call for comment. First, as indicated in Chapter 7 above in relation to the protection of cultural property, the fact that property may be intangible is not a unique issue for armed conflict involving information technology. The Nuremburg war crimes tribunals in both the *Krupp* and *I.G. Farben* trials were in no doubt that the property offences set out by the

74 Art. 23(g), Hague Regulations.

75 An almost identical provision exists in Art. 8(2)(e)(xii) of the Rome Statute in relation to armed conflicts not of an international character.

76 Arts 28 and 47, Hague Regulations; Art. 33(2), Geneva Convention IV; Art. 4(2)(g), Additional Protocol II; Art. 8(2)(b)(xvi), Rome Statute for international armed conflicts and Art. 8(2)(e)(v) for armed conflicts not of an international character. It is also an offence under Art. 4(f) of the ICTR Statute, Art. 3(e) of the ICTY Statute (referring to plunder, although the French text still refers to pillage) and Art. 3(f) of the SCSL Statute.

Hague Regulations were broad enough to encompass the acquisition of intangible property by a number of means:[77]

> Property offences recognised by modern international law are not, however, limited to offences against physical tangible possessions or to open robbery in the old sense of pillage, but include the acquisition of intangible property and the securing of ownership, use or control of all kinds of property by many ways other than open violence.

The Tribunal in the *I.G. Farben Trial* held that: 'In our view, the offences against property defined in the Hague Regulations are broad in their phraseology and do not admit of any distinction between "plunder" in the restricted sense of acquisition of physical properties, … the plunder or spoliation resulting from acquisition of intangible property such as is involved in the acquisition of stock ownership, or of acquisition of ownership or control through any other means.'[78] Further, following the invasion of Germany in World War II, Allied Forces had no difficulty with acquiring such intangibles as technical and scientific information and military expertise in varying forms including patents and documentation, both as booty and 'intellectual reparations'.[79]

The second preliminary issue to be dealt with is the concept of seizure. While destruction of property is fairly unambiguous even in respect of computer systems, networks and the information contained on them,[80] what is not clear is what actions will constitute seizure of those items. There are two separate aspects to this issue: the first is the remote appropriation of a system or network while leaving its physical components in place, and the second is the copying of information located on that network or system. The issue is relevant both for property captured by a party on the battlefield as booty of war and for property seized in occupied territory. For both questions, however, it should be noted that it is sufficient that the property in question is located in the battlespace; the location of the actor is not of primary

[77] *Krupp Trial*, 129; *I.G. Farben Trial*, 46. United Nations War Crimes Commission, *Law Reports of Trials of War Criminals* (HMSO for the United Nations War Crimes Commission, London, 1949), Vol. XV, 129, citing the *Krupp Trial* which dealt in part with transfer of shares, transfer of corporate property, contractual transfer of property rights and the like.

[78] *I.G. Farben Trial*, 46.

[79] See generally, John Farquharson, 'Governed or Exploited? The British Acquisition of German Technology, 1945–48' (1997) 32(1) *Journal of Contemporary History* 23.

[80] Although a question may be raised as to whether destruction of information can be established if it can be restored or reconstructed, albeit with effort and expenditure from backups.

importance.[81] Although the tangible aspects of a system or network may be physically seized,[82] it is possible to assume control of a network without doing so. No formal definition of seizure appears in international instruments or in decisions before international tribunals; however, the Military Tribunal in the *Krupp Trial* held that the offence of spoliation was achieved even if no definite alleged transfer of title was accomplished, stating:[83]

> However, if, for example, a factory is being taken over in a manner which prevents the rightful owner from using it and deprives him from lawfully exercising his prerogative as owner it cannot be said that his property 'is respected' under Article 46 as it must be.

William Downey has argued that effective seizure requires that the property is placed under substantial guard and is in the 'firm possession' of the capturing state.[84] Although there is conflicting case law on what is required in order to reduce property to firm possession, Downey cites an opinion by the Legal Advisor of the Office of Military Government for Germany (OMGUS) as the preferable view. The OMGUS opinion concludes:[85]

> It appears that 'firm possession' requires some manifestation of intention to seize and retain the property involved and some affirmative act or declaration of a possessory or custodial nature with respect to the property. The circumstances which will satisfy these two elements of firm possession will, of course, vary in each case. It is, however, our conclusion that the general

[81] See above discussion on location of offences in section 1.2.1 of Chapter 5, relating to sabotage.

[82] It may be questioned whether the cable element of a fibre-optic network would be considered 'movable' property for the purposes of seizure or booty. However, it would appear that the distinction is to refer to personal property or chattels (as distinguished from real property) rather than any requirement for actual movability: William Gerald Downey, Jr., 'Captured Enemy Property: Booty of War and Seized Enemy Property' (1950) 44 *AJIL* 488, 489.

[83] *The Krupp Trial*, 623–4. The tribunal was responding to a defence argument that the laws and customs of war do not prohibit seizure and exploitation of property in belligerently occupied territory, so long as no definite transfer of title is accomplished.

[84] Downey, 'Captured Enemy Property', 492.

[85] Legal Advisor of the Office of Military Government for Germany (OMGUS) IX Selected Opinions, OMGUS, 57, 60, cited *ibid.*, 493. *Cf.* an alternate opinion cited by Downey which dealt with a case of US Civil War era confederate cannon found at the bottom of a river in Arkansas during World War II. It was held in 1947 that the cannon became the property of the United States when the area in which the cannon were located was captured by Federal forces; i.e. mere seizure and occupation of the territory was enough to transfer title: 6 Bull JAG (1947) 238–89, cited in Downey, 'Captured Enemy Property', 493.

occupation of an area by a belligerent is not of itself sufficient to satisfy either of the two elements of the doctrine of firm possession.

In 1985 the Israeli Supreme Court confirmed this approach in *Al Nawar* v. *Minister of Defence et al.*, although the Court noted the practical impossibility of seizing all property at once and stated that in order to effect seizure, it would suffice to arrange for a general guarding or patrolling of the area where the property was located.[86]

Following this reasoning, in order to seize a network or other system remotely, it would be necessary to access the system or network and change the access codes in order to prevent the original owner from accessing and controlling their system or network. This already happens to a certain extent with computers which are compromised by malicious software and recruited to a botnet.[87] In the case of criminal use of infected computers, the author or controller of the botnet leaves the owner's access and control intact. However, once the controller has access to the system or network as an administrator it would be possible to change the access permissions so that the system or network is under the sole control of the accessing party and remove all access rights of the original owner. This appears to satisfy the requirements of firm possession set out above, as the act of excluding the original owner manifests the intention of the controller to retain control of the network or system. Placing the network or system under guard may be as simple as changing passwords and ensuring that the system is running up-to-date virus protection software and that all program updates and patches are installed.

Mere copying of data and information resident on systems poses a slightly more difficult analysis.[88] The nature of digital information is such that copying renders an identical copy capable of being used by the capturing party for any purpose that the original would have been,

[86] *Al Nawar* v. *Minister of Defence et al.* (1985) 39(3) Piskei Din 449, Israel Supreme Court Excerpted in English in F. Domb, 'Judgments of the Supreme Court of Israel' (1986) 16 *Israel Y.B. Hum. Rts* 321, 326.

[87] See, for example, the Storm Worm which was first detected in the wild on 17 January 2007.

[88] It should be noted that there is considerable disagreement at present in domestic jurisdictions concerning the appropriate property rights, other than intellectual property rights, to be granted over information. See, for example, the discussion on theft of information in Ian J. Lloyd, *Information Technology Law* (4th edn, Oxford University Press, 2004), 315–21, comparing England, Scotland, the United States and Canada. Database rights in particular are especially contentious.

thus the requirement of intention to seize and retain would still be met. Generally speaking, property is seized during armed conflicts for one of two purposes: either to deprive the opposing forces of its use; or to turn it to the capturing state's advantage.[89] While copied information may fulfil the latter purpose, it would not deny the use of the property to the opposing forces. This 'seizure by copying' is not a new phenomenon; following the invasion of Germany in World War II, significant documentation programmes were put in place by the Allied Forces. Vast amounts of technical information, research facilities and prototypes were confiscated as booty by the American and British authorities, with much of the documentation being acquired by way of extensive microfilming projects that left the original documents in place.[90] Historian John Farquharson notes that, by mid-1947, 5 million pages were available on microfilm in the US to businesses and academic institutions. These pages were comprised entirely of documents, patents etc. found by the occupying powers in Germany.[91] However, one of the difficulties that arises with this form of seizure is the differing treatment of seized property under the laws relating to booty and military necessity on the one hand, and that of seizure of property in an occupied area on the other. Both Article 23(g) and Article 53 of the Hague Regulations refer to property being seized, but the treatment (inferred in the case of Article 23(g) and specified in the case of Article 53) is very different. Under Article 53, ownership rights are not transferred to the occupying party; the property may be seized but must be returned when peace is made and compensation paid. However, property seized for military operations or under the law of booty becomes the property of the seizing party. Title passes on effective seizure for purposes of booty and the previous owner is completely divested of all rights in the property.[92] Logically, it would seem that for seizure under those conditions, the affirmative action or declaration must

[89] Lauterpacht, *Oppenheim's International Law*, 152.

[90] Farquharson, 'Governed or Exploited?'; John Gimbel, *Science, Technology, and Reparations: Exploitation and Plunder in Postwar Germany* (Stanford University Press, 1990), 60–74. There is controversy over the extent of the documentation programmes put in place in respect of the type of information confiscated which included large-scale acquisitions from private industry. However, much was taken of purely military usage and was nearly all taken as booty: Farquharson, 'Governed or Exploited?', 37.

[91] *Ibid.*, 23.

[92] Yoram Dinstein, 'Booty in Land Warfare' in R. Bernhardt (ed.), *Encyclopedia of Public International Law* (Max Planck Institute; North Holland, Amsterdam, 1992) 432–4, 432.

incorporate the 'right to exclude' or in some other way deprive the original owner of their rights over the information. Thus, mere copying would not constitute valid seizure as it only succeeds in depriving the original owner of the right to exclude vis-à-vis the seizing state. As this is inconsistent with state practice, it would appear that seizure may mean different things with respect to booty than its does to occupation and that any action with respect to such property would be prohibited if it entailed the party acting in a manner inconsistent with the property rights which continue to vest in the original owner.

3.1. *Booty*

In accordance with customary international law, all movable state property captured on the battlefield, with the sole exception of cultural property,[93] may be appropriated as booty of war. The seized property becomes the property of the capturing state rather than the individual soldier or unit seizing it, and title passes on seizure.[94] Title to the property is acquired automatically by the belligerent state whose armed forces have seized it, irrespective of the military character of the property (not only weapons and ammunition, but also money, food and stores).[95] This would undoubtedly apply to all government-owned systems and networks, as the computers, servers and routers would all fall in the category of movable property. Although the issue has not been formally addressed in international law, it would seem that all information resident on such computers, networks and other devices could also be lawfully appropriated in a similar way to the information, technical documents, patents and other intellectual property which were seized by the Allied powers after World War II.[96]

In *Al Nawar* v. *The Minister of Defence et al.* the Israeli Supreme Court held that Article 23(g) of the Hague Regulations does not accord

[93] Cultural property is exempted under Art. 14 of the Cultural Property Convention 1954.

[94] Downey, 'Captured Enemy Property', 500; Dinstein, *Conduct of Hostilities* (2nd edn), 247–8; Dinstein, 'Booty in Land Warfare', 432.

[95] Dinstein, *Conduct of Hostilities* (2nd edn), 247.

[96] Although there was significant discussion over the extent and type of property that could be confiscated (in terms of much of it being from private firms), none of the discussion appears to have focused on the intellectual nature of the property. Following March 1946, a narrow definition of booty was adopted by the Allied powers limiting booty to 'arms, munitions and implements of war, and all research and development facilities (including documents, material and training devices) relative thereto': Farquharson, 'Governed or Exploited?', 33; Gimbel, *Science, Technology, and Reparations*, 172–5.

protection to property used for hostile purposes; such property enjoys protection from arbitrary destruction, but it is still subject to the enemy's right of appropriation as booty.[97] While private property is generally immune from seizure on the battlefield, any private property actually used for hostile purposes may be appropriated by the belligerent state.[98] The Court also held that the distinction between state and private property should be based on the functional test applied in the 1921 Arbitral Award in the *Cession of Vessels and Tugs for Navigation on the Danube Case*, which determines the nature of the property in question based on its actual use.[99] Following this reasoning, any commercial network, system or computer that is utilised by the opposing state for military operations may also be seized, a significant concern given the large percentage of military communications that travel over civilian networks.[100]

Further, it is permissible to seize any weapons, ammunition, military equipment, military papers and the like, regardless of whether it can be used for military operations or not, even though they constitute private property.[101] In an age of computer network attack this may prove to be an extremely broad exception. Taken to its full extent, practically all networks operating in the battlespace as well as any systems that are capable of being used in a computer network attack may be liable to be seized as booty. As the average home computer or laptop with an internet connection has this capacity, and may already be leveraged as part of a botnet (with or without the owner's consent or knowledge), it would seem that any computer could be open to seizure by the armed forces as military equipment despite the provisions of Article 23(g) of the Hague Regulations. Likewise, any information resident on networks or computers would be able to be seized, insofar as it amounts to military papers (in the case of documents or databases) or

[97] Domb, 'Judgments of the Supreme Court of Israel', 324. Article 23(g) forbids the destruction or seizure of the enemy's property, unless such destruction or seizure be imperatively demanded by the necessities of war.

[98] *Al Nawar* excerpted in English in Domb, 'Judgments of the Supreme Court of Israel', 324.

[99] Domb, 'Judgments of the Supreme Court of Israel', 325, citing *Cession of Vessels and Tugs for Navigation on the Danube Case* (1921) 1 RIAA 97.

[100] In 2010, former Director of National Intelligence, Admiral Michael McConnell, estimated that 98 per cent of US government communications, including classified communications, travel over civilian-owned-and-operated networks and systems. Jensen, 'Precautions', 1533.

[101] Downey, 'Captured Enemy Property', 494.

military equipment (in respect of software). Such an exception would be unduly broad. It would be its use for computer network attacks that would mark the particular device, system or network as a weapon or military equipment, as it is not inherently military in nature.

The term 'battlefield' is commonly used in describing where property may be seized in accordance with the law of booty in warfare; however, as Dinstein points out, the term is to be understood very broadly and is perhaps better to be understood in terms of 'combat' or 'military engagement'.[102] The Supreme Court of Israel held in *Al Nawar* that the entire theatre of operations may be regarded as the battlefield for the purposes of the law of booty in land warfare.[103] This raises interesting questions for computer network attack as the battlespace is much larger than traditionally contemplated. Taken to its logical conclusion, this would mean that any network, computer, router, server or mobile satellite command station utilised in military operations may be appropriated as booty of war,[104] as well as any information resident on those devices (unless they amount to cultural property). Given the ability to seize such networks remotely, it would appear that a belligerent party could seize systems or networks in the territory of an adversary regardless of the size of, or indeed fact of, their physical presence in the adversary state. For example, it is possible that any attacks launched against Iraq from US military bases on US soil, would open US networks to seizure as booty by Iraq.

3.2. *Occupied territory*

Exceptions to the prohibition against destruction or seizure of an adversary's property also exist for occupied territories, which are of particular relevance for computer networks. Article 53 of the Hague

[102] Dinstein, 'Booty in Land Warfare', 433; Dinstein, *Conduct of Hostilities* (2nd edn), 247.

[103] *Al Nawar* excerpted in English in Domb, 'Judgments of the Supreme Court of Israel', 324.

[104] An interesting issue arises over whether the satellite's space architecture would be considered 'movable' property for the purposes of seizure. Where it is possible to access and obtain control of the thrusters and inclination of the satellite, it follows logically that it may be moved. As the traditional battlefield is now thought of in terms of battlespace, any satellite which is actually used by the parties to the conflict must necessarily form part of that battlespace. Note, however, the exclusion of this right for occupied territory. See generally, Thomas C. Wingfield, 'Legal Aspects of Offensive Information Operations in Space' (1998) 9 *J. Legal Stud.* 121; Michel Bourbonnière, 'Law of Armed Conflict (LOAC) and the Neutralisation of Satellites or Ius in Bello Satellitis' (2004) 9 *JC&SL* 43; Michael N. Schmitt, 'International Law and Military Operations in Space' (2006) 10 *Max Planck UNYB* 89.

Regulations allows parties to take possession of communications equipment in occupied territory, stating:

An army of occupation can only take possession of cash, funds, and realizable securities which are strictly the property of the State, depots of arms, means of transport, stores and supplies, and, generally, all movable property belonging to the State which may be used for military operations.

All appliances, whether on land, at sea, or in the air, adapted for the transmission of news, or for the transport of persons or things, exclusive of cases governed by naval law, depots of arms, and, generally, all kinds of munitions of war, may be seized, even if they belong to private individuals, but must be restored and compensation fixed when peace is made.

Computer networks and other IT systems undoubtedly fall within this provision. However, where these means of communication take the form of submarine cables or satellites, specific laws apply. Submarine cables (including fibre-optic cables) connecting the occupied territory to a neutral territory may only be seized or destroyed in cases of absolute necessity and must be restored and compensation fixed when peace is restored.[105] Further, the space architecture of satellite communications systems would also fall outside the Article 53 exception. Outer space is not subject to appropriation by any state though occupation,[106] therefore the laws regulating seizure of property in occupied territories does not apply. This would not prevent the seizure of the ground-based control centres, however, so the point may be moot; subject to prohibited interference with the physical aspects of the satellite (for example the altitude and orbit control subsystems which control the thrusters and inclination of the satellite itself), exclusive control of information passing through the satellite would still be possible.

Article 53 of Geneva Convention IV relating to protection of civilian persons in occupied territory prohibits the destruction of property,

[105] Art. 54, Hague Regulations. See, for example, *Cuba Submarine Telegraph Co.* (1923) 6 RIAA 118; *Eastern Extension, Australasia and China Telegraph Co. Claim* (1923) 6 RIAA 112 cited in Leslie C. Green, *The Contemporary Law of Armed Conflict* (2nd edn, Manchester University Press, 2000), 152.

[106] Art. 2, Treaty on Principles Governing the Activities of States in the Exploration and Use of Outer Space, including the Moon and Other Celestial Bodies 1967, Vol. 610 No. 8843. Note that the Hague Regulations do not apply *per se* to space-based conflict as they are only applicable to the law of war on land. Their application is based on their status as customary international law; the International Court of Justice has indicated on several occasions that the principles contained in the Hague Regulations constitute 'intransgressible principles of international customary law': see, for example, *Nuclear Weapons Case*, para. 79.

both state-owned and private, except where it is absolutely necessary for military operations.[107] It must be noted, however, that this provision applies only to destruction of property, not seizure. The occupying authorities have a recognised right, under certain circumstances, to dispose of property within the occupied territory – namely the right to requisition private property, the right to confiscate any movable property belonging to the state which may be used for military operations and the right to administer and enjoy the use of real property belonging to the occupied state.[108] Extensive destruction is considered a grave breach of the Convention and may be prosecuted as a war crime under Article 8(2)(a)(iv) of the Rome Statute.[109]

3.3. *Pillage and plunder*

The prohibition against pillage is firmly rooted in both customary international law and treaty law.[110] Traditionally, pillage meant the looting or plundering of enemy property (public or private) by individuals for private ends,[111] and incorporated an element of violence in the appropriation of such property.[112] However, a number of these elements have been brought into question by recent judgments. The ICTY has determined that the Statute proscribes plunder committed on the entire territory of the parties to a conflict.[113] There are

[107] Art. 53, Geneva Convention IV provides: 'Any destruction by the Occupying Power of real or personal property belonging individually or collectively to private persons, or to the state, or to other public authorities, or to social or cooperative organizations, is prohibited, except where such destruction is rendered absolutely necessary by military operations.'

[108] ICRC 'Article 53' Commentary to Geneva Convention IV, 301; Pictet, *Commentary*.

[109] Art. 147, Geneva Convention IV; 8(2)(a)(iv) Rome Statute.

[110] *'Celebici' Judgment*, para. 315. Arts 28, 47 (concerning occupied territory), Hague Regulations 1907; Art. 33(2), Geneva Convention IV; Art. 4(2) of Additional Protocol II; Art. 8(2)(b)(xvi) and Art. 8(2)(e)(v), Rome Statute, make 'pillaging a town or place, even when taken by assault' a war crime; the ICTR and SCSL Statutes both include the crime of pillage, while the ICTY Statute prohibits plunder. Interestingly the official French version of both the ICTR and ICTY Statutes use the term 'le pillage'.

[111] Dinstein, *Conduct of Hostilities* (1st edn), 214; Armin A. Steinkamm, 'Pillage' in R. Bernhardt (ed.), *Encyclopedia of International Law* (Max Planck Institute; North Holland, Amsterdam, 1982) 1029–30, 1029.

[112] *Trial of Alois and Anna Bommer and Their Daughters* (1947) IX *Law Reports of Trials of War Criminals* 62, Permanent Military Court at Metz.

[113] *Prosecutor* v. *Mladen Naletilic and Vinko Martinovic* (2003) Case No. IT-98–34-T, International Criminal Court for the Former Yugoslavia, para. 615, noting that Geneva Convention IV indicates that the prohibition of pillage is not limited to acts committed in occupied territories.

two particular questions which arise in respect of pillage of digital property. First, must the owner of the property be dispossessed of it entirely, or is it sufficient that certain of the owner's property rights, namely the right to exclude, and the right to control the use made of the property and the profit from it, are infringed? And secondly, with regard to pillage, must the property be acquired through threats or use of violence?

In the *Flick Trial*, Flick was found guilty of war crimes for, *inter alia*, the plunder of public and private property and spoliation in the countries and territories occupied by Germany. However, the Tribunal stated that 'no defendant is shown by the evidence to have been responsible for any act of pillage as that word is commonly understood'.[114] It is not clear from the judgment whether this was a result of the lack of violence involved, or the fact that the property was returned to the owners in a better condition than when it was appropriated. Other courts have not separated the offences in the same manner.

The trial chamber of the Special Court for Sierra Leone in the *Fofana and Kondewa* case were of the view that the inclusion of the requirement that the appropriation be for personal or private use is an unwarranted restriction on the application of the offence of pillage.[115] However, this statement seems to stem from the conflation of the terms 'pillage', 'plunder' and 'spoliation', and the fact that the Statute for the International Criminal Tribunal for the Former Yugoslavia contains the offence of plunder rather than pillage.

Computer network attacks designed to appropriate property are generally non-violent in nature. Noting that the concept of pillage in the traditional sense implied an element of violence,[116] the Court in the *Celebici* judgment declined to decide whether the terms pillage and plunder were synonymous, stating that the term plunder embraces 'all forms of appropriation of property in armed conflict for which individual criminal liability attaches under international law, including those acts traditionally described as 'pillage'.[117] However, courts have been happy to consider property seized by enemy forces as pillage even

[114] *Flick Trial*, 21.

[115] *Prosecutor* v. *Moinina Fofana and Allieu Kondewa (Decision on Preliminary Motion Based on Lack of Jurisdiction (Child Recruitment))* (2004) Case No. SCSL-04–14-T, Special Court for Sierra Leone, 49.

[116] The requirement of violence in pillage is brought into question by the case of *Bommer*.

[117] *'Celebici' Judgment*, § 591, citing *Law Reports*, Vol. IX, 64.

when violence is not used, such as the seizure of bonds and shares abandoned by the owner in occupied territory.[118]

3.4. *Enemy-owned property on the territory of a belligerent*

Enemy-owned public property on the territory of a belligerent is subject to seizure, although diplomatic buildings are placed under protection.[119] Thus, any state-owned websites or other digital assets hosted in the adversary state would be liable to be seized in the event of an armed conflict breaking out between the states. For example, the government websites of East Timor (now Timor-Leste) were hosted on sites in the Republic of Ireland prior to independence and would have been subject to seizure or freezing in the unlikely event that the two states engaged in an armed conflict with one another. Likewise, any work that has been outsourced offshore to a belligerent state would be subject to seizure. In an age of increased civilian outsourcing of government work (including defence acquisitions), and offshore outsourcing in the commercial sector, knowledge and control over where the actual work will be performed will be an important consideration. While it may also be possible to argue that information merely passing over a server in a belligerent state is sufficient to qualify as being on the territory of a belligerent and therefore subject to seizure, most information accessed in this manner could presumably be seized as intelligence gathering which is expressly permitted under Article 24 of the Hague Regulations.

The situation with regard to private property is more complex. The protections for civilian property provided by Article 23(g) of the Hague Regulations apply equally to property on the territory of both belligerent parties. In addition, Article 38 of Geneva Convention IV provides that 'the situation of protected persons shall continue to be regulated, in principle, by the provisions governing aliens in time of peace' with the exception of measures of internment, assigned residence or other exceptional measures for control and security necessitated by the war.[120] These measures have been considered by the Eritrea Ethiopia Claims Commission in its partial awards in respect of some of *Eritrea's*

[118] See, for example, *Re Otto Wallemar* [1948] ADIL 619 (removal of goods during occupation), *Mazzoni* v. *Ministry of Finance* [1927–1928] AD Case No. 384 (on seizure of bonds and shares abandoned by the owner in occupied territory).

[119] Green, *The Contemporary Law of Armed Conflict*, 155.

[120] Art. 27, Geneva Convention IV.

Civilian Claims and its claim for *Loss of Property in Ethiopia*.[121] The decision of the Commission in its partial award for *Eritrea's Civilian Claims* noted that belligerents have 'substantial latitude to place freezes or other discriminatory controls on the property of the nationals of the enemy state or otherwise to act in ways contrary to international law in time of peace'.[122] While observing that the control measures were necessary to deny the enemy access to economic resources which might otherwise be potentially available to support its conduct of the war, the Commission commented that states have not consistently done so, and that where states have vested the assets of enemy nationals it has been done under controlled conditions and for reasons directly tied to higher state interests.[123] The Commission went on to find:[124]

> a belligerent is bound to ensure insofar as possible that the property of protected persons and of other enemy nationals are not despoiled and wasted. If private property of enemy nationals is to be frozen or otherwise impaired in wartime, it must be done by the State, and under conditions providing for the property's protection and its eventual disposition by return to the owners or through post-war agreement.

The Commission noted that such limitations on the vesting of property have been emphasised by commentators.[125] Digital property or assets in the territory of the belligerent may be frozen by that party in the same way that tangible property or funds may be frozen, and would be subject to the same conditions to ensure protection of the assets and for reasons of state interest.

4. Conclusion

Of all of the principles that govern modern armed conflict, it is those relating to the means and methods of warfare which have the most uneasy fit in respect of computer network attacks. One thing is certain,

[121] *Partial Award, Loss of Property in Ethiopia Owned by Non-Residents, Eritrea's Claim 24 between the State of Eritrea and the Federal Democratic Republic of Ethiopia* (2005), Eritrea Ethiopia Claims Commission; *Partial Award, Civilians Claims, Eritrea's Claims 15, 16, 23 & 27–32 between the State of Eritrea and the Federal Democratic Republic of Ethiopia* (2004), Eritrea Ethiopia Claims Commission.

[122] *Partial Award in Eritrea's Civilians Claims*, para. 124.

[123] *Ibid.*, para. 127–8.

[124] *Ibid.*, para. 151.

[125] *Ibid.*, para. 128, citing Lauterpacht, *Oppenheim's International Law*, 326–31; Ian Brownlie, *Principles of Public International Law* (6th edn, Oxford University Press, 2003), 514.

however: the general principles of the law of weaponry will continue to apply, but their specific application will only become apparent with the details of particular computer network attacks. The application of Article 36 obligations to assess individual attack techniques should mitigate against these problems.

Perfidy is one of the most difficult concepts to translate into an online environment where anonymity is the norm. However, it must be borne in mind that it only applies to attacks that result in the killing, injuring or capture of the adversary. Because standard practice in computer network attacks is to disguise the origin of the attack, states will need to take particular care to ensure that such ruses do not cross the line into perfidy.

The changing conceptions of property and the increasing reliance of economies on intellectual property and intangibles also raise serious issues in terms of the rules regarding the protection of property in armed conflict. Some issues do not readily fit within the current framework; others reflect the difficulties found in international criminal law in respect of computer crime generally. Concepts such as pillage or plunder raise difficulties with the possible requirement that the owner must be permanently deprived of the property, thus the copying of files (and even their deletion, in circumstances where backup copies exist) may mean that liability is avoided. However, the frequent conflation of property offences by international courts may mean that this is less of a difficulty in practice.

Concluding remarks

The advent of computer network attacks poses new challenges for the international law regulating force and for international humanitarian law. Not only do computer network attacks represent a fundamentally different method of warfare, they come at a time when the laws of armed conflict are struggling to meet the challenges of greater than ever civilian participation in conflict, increased asymmetry and technological advance. However, despite these challenges, the author believes that the underlying framework and general principles of the laws of armed conflict remain applicable to conflicts involving computer network attacks. Some adaptations in detail will be required, as always happens with law over time;[1] however, the underlying principles of the laws of armed conflict are aimed at the fundamental nature of war, which remains unchanging. The exact content and contours of the laws will be determined by the prevailing principles of the societies that shape them.

At present, examination of the legality of computer network attacks under international law results in a complex picture. Under the *jus ad bellum*, the need for a physical effect, namely death, injury or destruction of physical property, appears to remain constant, although it may be achieved indirectly. The perceived intent of the attacker will play a large role in determining the victim state's response. Certainly for a computer network attack to qualify as an armed attack triggering the right of self-defence under international law, the attack must result in consequences of a sufficient scale and effect. Adopting a restrictive approach to computer network attacks in respect of the *jus ad bellum*

[1] Oeter, 'Comment: Is the Principle of Distinction Outdated?', with regard to the changes wrought by increasing asymmetry.

also serves to act as a restraint on the right to resort to traditional force in self-defence.

It is perhaps the *jus in bello* which shows more clearly the complex relationship between fundamental principles and specific applications of the laws of armed conflict with regard to computer network attacks. Principles such as the distinction between combatants and civilians, proportionality in attack, and the prohibition against causing unnecessary suffering, remain at the core of the commitment to the law regardless of the technology employed. It is in the specific applications of the laws where the effects of technology and the changing values and conceptions brought about by the information revolution are seen. One of the most significant is the increased value that information societies place on intangible property and information. It will have an impact on the application of the laws governing the conduct of hostilities in relation to targeting analysis, protection of cultural property and property offences generally.

Despite the attempt to adopt a comprehensive approach, as always, there remain a few areas which require further research. As stated in the introduction, this volume does not examine the law of neutrality, or look in any meaningful way at the effects of computer network attacks on the law of naval warfare (other than a brief mention to draw a comparison in the text). There are other questions which have been raised in the course of this work and these will require future investigation. Some are raised by the nature of digital property and the protection to be provided to it; for example, in a format capable of perfect digital copies, what criteria will be used to determine which copies of cultural property are worthy of protection as reproductions?

Appendix 1 Selected computer network attack examples

1982 **Trans-Siberian Pipeline – the 'Farewell Dossier'.**[1] Following the theft of technology from Western powers by the Soviet KGB, the Central Intelligence Agency (CIA) of the United States and a Canadian firm planted a Trojan horse in the control software for the pipeline system which a KGB operative had been sent to steal. Once installed 'the pipeline software that was to run the pumps, turbines, and valves was programmed to go haywire, after a decent interval, to reset pump speeds and valve settings to produce pressures far beyond those acceptable to pipeline joints and welds. The result was the most monumental non-nuclear explosion and fire ever seen from space.'[2] The explosion happened in a remote area of Siberia and there were no physical casualties from the attack; however, it is not clear from the documents publicly available whether this was by fortunate happenstance or design. A former KGB officer has called into question whether this incident happened as reported, blaming the explosion on poor construction.[3]

1997 **Eligible Receiver.** Eligible Receiver is the code name of a 1997 internal exercise initiated by the US Department of Defense. A 'red team' of hackers from the National Security Agency (NSA) was organised to infiltrate the Pentagon systems. The red team was only allowed to use publicly available computer equipment and hacking software. Although many details about Eligible Receiver are still classified, it is known that the red team was

[1] See Weiss, 'The Farewell Dossier', 269; Reed, *At the Abyss*, 266–70; Matthew French, 'Tech Sabotage During the Cold War', *Federal Computer Week*, 26 April 2004, http://fcw.com/articles/2004/04/26/tech-sabotage-during-the-cold-war.aspx (last accessed 16 January 2012).

[2] Reed, *At the Abyss*, 269.

[3] Anatoly Medetsky, 'KGB Veteran Denies CIA Caused '82 Blast', *Moscow Times*, 18 March 2004, 4.

able to infiltrate and take control of the Pacific Command Center computers, as well as power grids and 911 systems in nine major US cities.[4] The red team intruded computer networks; denied services; changed, removed and read emails; and disrupted phone services. They also gained super-user access in over thirty-six computer systems, which meant they could have created new accounts, deleted accounts, turned the system off or reformatted the server hard drives.

In October 2002, a subsequent 'no-notice' mock attack against military computers, titled 'Eligible Receiver 2003' indicated a need for greater coordination between US military and non-military organisations to deploy a rapid military computer counter-attack.[5]

1998 **Solar Sunrise.** In February 1998, a number of US Department of Defense networks were attacked using a well-known vulnerability in the operating system (the UNIX-based Solaris). The attacks were widespread and appeared to come from multiple servers in the US as well as the United Arab Emirates, Israel, France, Taiwan and Germany. Although the targeted systems were all reported as unclassified, many key support systems reside on unclassified networks (for example the global transportation system, defence finance system, medical personnel, logistics and email).[6] The attacks came at a time when the US was preparing for potential military action against Iraq due to UN weapons inspection disputes and raised fears that the attacks were aimed at disrupting deployments and operations. Investigators eventually traced the attacks to two Californian teenage high-school students directed by an Israeli teenaged mentor, who were subsequently arrested and charged.[7]

1998 **Roosevelt Dam, Arizona.** In 1998 a twelve-year-old US hacker, exploring for fun, broke into the computer system that

[4] PBS Frontline, *Cyberwar! The Warnings*, www.pbs.org/wgbh/pages/frontline/shows/cyberwar/warnings/ (last accessed 7 July 2011); William M. Arkin, *Code Names: Deciphering US Military Plans, Programs, and Operations in the 9/11 World* (Steerforth Press, Hanover, NH, 2005), 358.

[5] Clay Wilson, 'Information Operations and Computer Network Attack Capabilities of Today' (Paper presented at the International Expert Conference on Computer Network Attacks and the Applicability of International Humanitarian Law, Stockholm, 17–19 November 2004) 28–79, 64. Citing a US Department of Defense Briefing for the Congressional Research Service, 9 January 2003.

[6] 'Solar Sunrise' GlobalSecurity.org, www.globalsecurity.org/military/ops/solar-sunrise.htm (last accessed 8 July 2011).

[7] *Ibid.*; John A. Serabian, Jr, 'Cyber Threats and the US Economy: Statement for the Record before the Joint Economic Committee on Cyber Threats and the US Economy' (23 February 2000) (transcript available from https://www.cia.gov/news-information/speeches-testimony/2000/cyberthreats_022300.html (last accessed 8 July 2011).

runs Arizona's Roosevelt Dam.[8] Although he was unaware of the fact, federal authorities claim he had complete control of the SCADA system which controls the dam's massive floodgates and the 489 billion gallons of water which it contains. Unleashed, the water would have coursed down the Salt River and over a downstream flood plain (home to an estimated population of one million people) before reaching the state capital, Phoenix.

The facts of this incident have since been disputed.[9] An alternative report claims that although a hacker did break into the computers of an Arizona water facility, the Salt River Project in the Phoenix area, the hacker was twenty-seven not twelve, and the incident occurred in 1994, not 1998. Moreover, the article claims that while clearly trespassing in critical areas, the hacker never had control of any dams.

1998 **Electronic Disturbance Theatre.** Electronic Disturbance Theatre launched a series of denial of service attacks on the US Department of Defense in support of Mexico's Zapatista rebels. The attack was in protest at alleged support by the US government for the Mexican government who were accused of serious human rights abuses in the Chiapas region of Mexico. When users logged on to an EDT website, the Zapatista Floodnet software was downloaded to their computer. As with most DDoS attacks, the software is designed to initiate automatic and repeated requests to reload an IP address, in this case the Pentagon's website DefenceLink. As Floodnet performs automatic reloads of the site, it slows or halts access to the targeted server and clogs bandwidth.[10] The attack is interesting primarily for the Pentagon's response to an attack that they knew was coming. On 9 September 1998, the Pentagon responded in kind by sending a java applet back to the originating computer and disabling the browser of the computer initiating the attack.[11] The action

[8] Gellman, 'Cyber-Attacks by Al Qaeda Feared'.

[9] ZDNet, 'Cyberterrorism: The Real Risks', 27 August 2002, http://news.zdnet.co.uk/internet/0,1000000097,2121358,00.htm (last accessed 8 July 2011).

[10] The Electronic Disturbance Theatre views this act as 'performance art', hence the term theatre in their title, and classifies the FloodNet action as virtual or electronic civil disobedience. Coco Fusco, 'Performance Art in a Digital Age: A Conversation with Ricardo Dominguez', 25 November 1999; see also Karl J. Shawhan 'Vital Interests, Virtual Threats: Reconciling International Law with Information Warfare and United States Security' (Unpublished Thesis) Maxwell Air Force Base, May 1999, 48 www.dtic.mil/cgi-bin/GetTRDoc?AD=ADA391632 (last accessed 8 July 2011).

[11] Friel, 'DoD Launches Internet Counterattack'. According to the EDT only two computers of the 80,000-plus who participated were crashed by the DoD counter-attack.

caused a storm of controversy on the internet as it involved an offensive attack on civilian servers. The incident sparked a joint task force to investigate the legalities involved.[12]

1998 **Indonesia and East Timor.**[13] East Timor (now Timor-Leste), occupied by Indonesia since 1975, declared its virtual independence in 1998 and established its own Country Code Top Level Domain and website hosted by an Irish Internet Service Provider (ISP), Connect-Ireland. Following the launch of the domain, the Indonesian embassy relayed its concerns regarding the launch to the *Irish Times*, complaining that the site represented misuse of computer freedom to campaign against Indonesia. In January 1999, Connect-Ireland, because the focus of a coordinated attack on its servers. Martin Maguire, founder and managing director of Connect-Ireland, believed the attacks were perpetrated by the Indonesian government and complained to the embassy. Maguire asserted 'the attacks were systematic and took place over a long period of time, from 18 different locations, and were targeted at the .tp domain name'. Attacks took the form of buffer overflow attacks, defacement of web pages and denial of service attacks. A spokesperson for the Indonesian embassy denied the claims.

1999 **Moonlight Maze.** Moonlight Maze was the code name for the investigation into a highly classified incident of early 1999: an electronic assault involving hackers based in Russia. In this attack, US officials accidentally discovered a pattern of probing of computer systems at the Pentagon, NASA, Energy Department, private universities and research labs that had begun in March 1998 and had been going on for nearly two years. Intruders accessed unclassified but highly sensitive DoD science and technology information, systematically marauding through tens of thousands of files – including maps of military installations, troop configurations and military hardware designs. The Defense Department traced the trail back to a mainframe computer in the Russian Academy of Science, a government organisation that interacts closely with the Russian military,[14] but the sponsor of the attacks is unknown and Russia

[12] Seffers, 'Legalities Cloud Pentagon's Cyber Defence', cited in Shawhan 'Vital Interests, Virtual Threats', 48.

[13] 'Indonesia, Ireland in Info War?', *Wired News* 27 January 1999; Arthur, 'The Day East Timor Was Deleted'; Nuttall, 'Virtual Country "Nuked" on Net'.

[14] Inside Defense, Defense Information and Electronics Report 1 (22 Oct. 1999). In James P. Terry, 'The Lawfulness of Attacking Computer Networks in Armed Conflict and in Self-Defense in Periods Short of Armed Conflict: What Are the Targeting Constraints?' (2001) 169 *Mil. L. Rev.* 70.

denies any involvement. Moonlight Maze is still being actively investigated by US intelligence.[15]

1999 **UK Ministry of Defence Satellite Hack**. In March 1999, a news report surfaced claiming a group of hackers had seized control of a British military communications satellite.[16] The hackers alledgedly intercepted the link between the Skynet satellite's control centre and the ground station and accessed the control system of the satellite, using it to change the characteristics of the channels used to convey military communications, satellite television and telephone calls. As there was only one news report of this incident, care should be taken in relying on the report.

1999 **Kosovo – Operation Allied Force.** The United States used computer network attacks to confuse and disable the Yugoslav air defence systems. Used in combination with jamming and other electronic warfare techniques, computer network attacks were used to insert misleading messages and false targets into the Yugoslav computer systems. While it is not clear what method was used by the US Air Force to do so, they have aircraft capable of intercepting and amending data before retransmission to the enemy system.[17]

2000 **Queensland Water Supplies – Vitek Boden.** In April 2000, Vitek Boden was arrested after being caught using a stolen computer and radio transmitter to gain access to a sewerage treatment system in Queensland, Australia. Over the previous two-month period Boden had accessed the system forty-six times, gaining complete control of treatment of the region's sewerage and drinking water facilities and dumping 250 million tonnes of putrid sludge into the area's rivers and parks, killing wildlife and plants.[18]

2001 **Code Red.** Code Red was a worm with multiple variants that first appeared in July 2001 and ultimately affected nearly 300,000 computers in the US. Exploiting a hole in Microsoft's IIS web servers, it was time-sensitive based on the date: from days 1–19 of the month the worm would propagate; from days 20–27 it

[15] PBS Frontline, *Cyberwar! The Warnings.*

[16] 'British Hackers Attack MOD Satellite', *Telegraph* (London), 4 March 1999.

[17] For more details see David A. Fulghum, 'Yugoslavia Successfully Attacked by Computers', *Aviation Week and Space Technology* 8, 23 August 1999, 31; David A. Fulghum, 'Data Link, EW Problems Pinpointed by Pentagon', *Aviation Week and Space Technology* 10, 6 September 1999, 87; William M. Arkin, 'The Cyber Bomb in Yugoslavia', *Washington Post* (Washington DC), 25 October 1999, www.washington-post.com/wp-srv/national/dotmil/arkin.htm (last accessed 15 July 2011); Arkin and Windrem, 'The Other Kosovo War'.

[18] *R* v. *Boden*; Gellman, 'Cyber-Attacks by Al Qaeda Feared'.

would launch a denial of service attack against a particular site; and from day 27 to the end of the month the worm would 'sleep', dormant in the computer. In Code Red's first variation, the affected computers were programmed to launch a denial of service attack against the White House website at a certain date and time. If the assault had worked, the hundreds of thousands of pings would have overwhelmed the internet in nanoseconds. Richard Clarke, the president's adviser for cyberspace security, worked with the nation's internet providers to thwart the attack by blocking traffic to the White House site. Other websites were shut down, however, and replaced by a message that read 'Hacked by Chinese'.[19]

2001 **Mountain View.** In the summer of 2001, the coordinator of the website for the city of Mountain View, California, noticed a suspicious pattern of intrusions. The FBI investigated and found similar 'multiple casings of sites' in other cities throughout the US. The probes were seemingly emanating from the Middle East and South Asia, and the visitors were looking up information about the cities' utilities, government offices and emergency systems. This information took on a new significance when US intelligence officials examined computers seized from Al Qaeda operatives after the 11 September 2001 attacks and discovered what appeared to be a broad pattern of surveillance of US infrastructure.[20]

According to former White House adviser Richard Clarke, 'I think the bottom line on the Mountain View case is the ease with which people can do virtual reconnaissance from overseas on our physical infrastructure and on our cyber infrastructure, and the difficulty that we have in knowing what is being done. We were lucky in the case of Mountain View that there were good people watching. It's probably occurring in lots of other places around the country, and we don't have people who are catching it.'[21]

2001 **Houston Port Authority.** Aaron Caffrey, 19, was accused of launching a denial of service attack that hampered operations at the Port of Houston, Texas, on 20 September 2001, by crippling its web-based systems which contained crucial information on navigation, tides, water depths and weather. Caffrey, who allegedly launched the attack against a female internet chatroom user who had insulted his American girlfriend, used a list of unpatched servers downloaded from the internet to hijack the machines and launch a denial of service attack. But it almost ended in disaster

[19] PBS Frontline, *Cyberwar! The Warnings.* [20] *Ibid.*

[21] PBS Frontline, *Interview with Richard Clarke – Cyberwar!* www.pbs.org/wgbh/pages/frontline/shows/cyberwar/interviews/clarke.html (last accessed 8 July 2011).

when it crashed the Port of Houston's systems under the weight of 100,000 requests to ping data at the girl's computer, leaving vital navigation and weather data inaccessible.[22]

2002 **Al-Qaeda Laptops.** US troops clearing the cave system in the Tora Bora region of Afghanistan uncovered an Al Qaeda laptop which indicated a strong interest in computer network attacks. Computer forensics indicated that the laptop had made multiple visits to sites offering sabotage handbooks, software and programming instructions on SCADA systems, and other 'cracking' tools. In January 2002, another computer was seized at an Al Qaeda office in Kabul, Afghanistan. The computer contained models of a dam, made with structural architecture and engineering software and included geological soil identification software which would enable the planners of an attack to simulate the dam's catastrophic failure and plot the consequences of a breach.[23]

2002 **Attack on DNS Root Servers.** A massive denial of service attack of unknown origin briefly interrupted traffic on nine of the thirteen DNS 'root' servers that control the internet, but the overall threat was dismissed as 'minimal'.[24] The attack took place over a one-hour window and appeared to be the work of experts. The attack is primarily of interest as it was an attack on the internet itself, rather than particular websites.

2000–2006 **Suter Systems.**[25] In 2000, 2002 and 2004, the US military tested the capability for US forces to secretly enter an enemy computer network and monitor what their radar systems could detect. Further experiments tested the added capability for US Forces to take over the enemy computers and start manipulating their radar to show false images. Suter 1, 2 and 3 progressively enabled information warfare experts to penetrate anti-aircraft defense networks, using radar and radio antennas and microwave relays as portals. Once inside, Suter operators could see what the enemy radars saw, then jam and spoof the flow of information or even take over as system administrator to control movement

[22] Andy McCue, '"Revenge" Hack Downed U.S. Port Systems', *ZDNet News*, 7 October 2003, http://news.zdnet.co.uk/security/0,1000000189,39116978,00.htm (last accessed 15 July 2011); Rebecca Allison, 'Youth Cleared of Crashing American Port's Computer', *Guardian* (London), 18 October 2007, Guardian Home Pages 7; Rebecca Allison, 'Hacker Attack Left Port in Chaos: Busiest US Port Hit after Dorset Teenager Allegedly Launched Electronic Sabotage against Chatroom User', *Guardian* (London), 7 October 2003, Home 7.

[23] Gellman, 'Cyber-Attacks by Al Qaeda Feared'.

[24] Naraine, *Massive DDoS Attack Hit DNS Root Servers*.

[25] David Bond, 'The Dog That Didn't Bark', *Aviation Week & Space Technology*, 8 May 2006, 19; David Fulghum, 'Sneak Attack', *Aviation Week & Space Technology*, 28 June 2004, 34.

of radar antennas. By the 2006 joint forces exercise, the Suter series of communications network invasion and exploitation capabilities were absent. Senior Air Force officials stated that this change in emphasis was because the technology was no longer experimental and had been moved into operational use in Iraq and Afghanistan.

2003– **Titan Rain.** Titan Rain is the code name given to a series of computer intrusions originating in Guangdong province, China. The attacks were significant for the high speed and technical skill of the intrusions, although the motivation behind them remains unknown, as does the identity of the perpetrators. It has been speculated that these were Chinese military attacks.[26] The code name has since been changed from Titan Rain but the new designation remains classified.

2004 **Mydoom and variants.** The original Mydoom virus began circulating around email systems and peer-to-peer networks at the end of January 2004. The original virus contained a mass-mailing worm,[27] which set up a backdoor into the infected computer by opening TCP ports. These backdoors potentially allow an attacker to connect to the computer either to gain access to its network resources, or to make the computer follow remote commands from the attacker to launch attacks on other computers. In the case of the Mydoom virus, it was the latter. The original virus was programmed to launch denial of service attacks at US company SCO over a period of twelve days, apparently as part of an ongoing battle over control of Unix source code.[28] The Mydoom variant, Doomjuice, launches a similar attack at the Microsoft website. However, unlike the original virus, Doomjuice does not travel by email. Instead, both Doomjuice and its counterpart Deadhat[29] randomly scan net addresses and upload themselves to any infected machines they find, through the backdoor opened by the original virus.[30]

2007 **Attack on DNS Root Servers.**[31] On 6 February 2007 a distributed denial of service attack was launched against three

[26] Graham, 'Hackers Attack Via Chinese Web Sites'.

[27] A mass-mailing worm arrives as an email attachment which sends itself out to all other addresses in the compromised computer's address book.

[28] 'Mydoom Cripples US Firm's Website', *BBC News*, 1 February 2004.

[29] The Deadhat virus is designed to find machines infected with the Mydoom virus, it removes any copies of Mydoom.A and Mydoom.B that are resident, installs itself and then attempts to stop the computer running anti-virus software or getting updates to protect itself against future infections.

[30] Security firms suspect that Doomjuice was written by the author of the original Mydoom virus. 'Mydoom Mutants Launch New Attacks', *BBC News*, 10 February 2004.

[31] 'Hackers Attack Heart of the Net', *BBC News*, 7 February 2007.

DNS root servers including one operated by the US Defense Department (the others were operated by the internet's oversight body ICANN (Internet Corporation for Assigned Names and Numbers) and UltraDNS (which manages traffic for websites ending in 'org' and some other suffixes) respectively. There was no evidence of damage to the servers.

2007 **'Aurora'.** In March 2007, researchers from the Idaho National Laboratory launched an experimental cyber attack, hacking into a replica of a power plant's control system and changing the operating cycle of a generator.[32] The attack sent the generator out of control and ultimately caused it to self-destruct, alarming the federal government and electrical industry about what might happen if such an attack were carried out on a larger scale.[33]

2007 **DDoS Attacks against Estonia.**[34] In May 2007, Estonia became the victim of a prolonged series of denial of service attacks which brought its banking system, many government services and much of its media to a halt. Although no critical infrastructure was compromised, for a highly technology-dependent state like Estonia, that depends on the internet for everything from parking to banking to voting, the attacks caused serious disruption and damage estimated at tens of millions of euros.[35] Despite earlier explicit accusations that Russia was behind the offensive, the Estonian government backed away from directly accusing the Kremlin of launching the attacks.[36] For an excellent summary of the events and findings of various reports into the incident, see the 2010 report produced by the NATO Cooperative Cyber Defence Centre of Excellence.[37]

2007 **Israeli Attack on Suspected Syrian Nuclear Site.** On 6 September 2007, Israel penetrated Syrian air defence networks in order to bomb a suspected nuclear site at Dayr az-Zawr,

[32] Meserve, 'Staged Cyber Attack Reveals Vulnerability in Power Grid'.

[33] *Ibid.*, footage of the generator is also available at www.youtube.com/watch?v=fJyWngDco3g (last accessed 8 July 2011).

[34] See generally, 'The Cyber Raiders Hitting Estonia'; AFP, 'Cyber Attacks on Estonia Are Security Issue: NATO Chief'; Halpin, 'Putin Accused'; Traynor, 'Russia Accused of Unleashing Cyberwar to Disable Estonia'; Traynor, 'Web Attackers Used a Million Computers, Says Estonia'.

[35] Traynor, 'Web Attackers Used a Million Computers, Says Estonia'.

[36] *Ibid.* Russia categorically denies any involvement and no concrete evidence has been found to substantiate those claims. While technical data shows that some of the attacks came from IP addresses allocated to the Russian government, there is no evidence that these computers were involved in initiating the attacks, or that they had not been compromised or spoofed.

[37] Tikk, Kaska and Vihul, *Cyber Incidents*, 14–34.

without being engaged or even detected.[38] The attack combined electronic attack techniques in the form of brute-force jamming, precision missiles to eliminate the facility itself, and computer network attack techniques. The ability of non-stealthy Israeli aircraft to carry out the attack without interference rests in part on technology, carried on board modified aircraft, that allowed specialists to hack into Syria's networked air defence system.[39] 'Network raiders can conduct their invasion from an aircraft into a network and then jump from network to network until they are into the target's communications loop.'[40] To date, Israel has never publicly discussed this attack.

2008 **DDoS Attacks against Georgia during Russia/Georgia Conflict.** The August 2008 cyber attacks against Georgia are significant in that they fall within the timeframe and context of a broader armed conflict between the Russian Federation and Georgia over South Ossetia. Prior to the physical invasion, Georgian government websites were defaced and came under sustained DDoS attacks which continued several weeks after the ceasefire and cessation of traditional military operations. The attacks were directed against numerous public and private targets, including the website of the Georgian President, central government sites, the Ministry of Foreign Affairs, Ministry of Defence, media sites, pro-Georgian discussion forums, and national and commercial banks.[41] The methods used were similar to those used in attacks against Estonia in 2007. Instructions and DDoS attack tools were made available on several Russian language blogs, forums and websites such as stopgeorgia.ru, as well as identifying targets for hackers. As was the case with the attacks against Estonia, there is no conclusive evidence of who was behind the Georgian attacks. Despite widespread suspicion pointed at Russia, which has denied any involvement with the attacks, it appears that the attacks were carried out by 'patriotic hackers' with no direct involvement by the Russian authorities.[42] Some reports have pointed to the high level of coordination between the cyber attacks and military operations on the ground; however, evidence of possible involvement of some officials within the administration is only circumstantial.[43]

[38] Fulghum, Wall and Butler, 'Israel Shows Electronic Prowess'.
[39] *Ibid.*
[40] *Ibid.*
[41] Tikk, Kaska and Vihul, *Cyber Incidents*, 70.
[42] US Cyber Consequences Unit, *Overview*, 2.
[43] See in particular Project Grey Goose, *Phase I Report*, 75; Tikk, Kaska and Vihul, *Cyber Incidents*.

2008 **US DoD USB worm and Operation Buckshot Yankee.** In 2008, the US Department of Defense suffered a significant compromise of its classified military computer networks, when an infected USB flashdrive was inserted into a military laptop.[44] The worm (agent.btz) spread undetected on both classified and unclassified networks run by US Central Command and allowed for data to be transferred to servers under foreign control. The process of removing the malware, dubbed Operation Buckshot Yankee, was long and painful, taking almost fourteen months of stop-go effort by military network administrators. US Deputy Secretary of Defense, William Lynn, has stated that the worm was the work of a foreign intelligence agency; however, other reports quoting officers involved in the removal operation have suggested that this may be premature as computer forensics are still being carried out.[45] Whatever the attribution, the intrusion and protracted removal process resulted in the formation of the US Cyber Command to integrate cyber defence operations across the US military.

2010 **Stuxnet – Iran Nuclear Enrichment Programme.** The Stuxnet worm was discovered in June 2010. The worm targets industrial control systems, and is widely believed to be aimed at the Iranian nuclear programme; Iran was the most heavily hit target of the worm with approximately 60 per cent of the infected hosts,[46] including its Bashir and Natanz nuclear facilities. The complexity and sophistication of the worm and its multi-pronged attack vectors have caused suspicion that the most likely authors are state-backed professionals,[47] although at the time of writing no party has officially claimed responsibility and computer forensics have not conclusively attributed the attack to any party. Iranian President Mahmoud Ahmadinejad has publicly accused Israel and the West of being behind the attacks.[48] Other reports also suggest a joint effort by Israel and the United States;

44 Lynn, 'Defending a New Domain', 97.

45 Noah Shachtman, 'Insiders Doubt 2008 Pentagon Hack Was Foreign Spy Attack (Updated)', *Wired*, 25 August 2010, www.wired.com/dangerroom/2010/08/insiders-doubt-2008-pentagon-hack-was-foreign-spy-attack/#more-29819 (last accessed 12 April 2011).

46 Falliere, OMurchu and Chien, *Stuxnet Dossier*.

47 Stuxnet exploits four different vulnerabilities, a previously unheard of level of complexity in malware. *Ibid*., version 1.4.

48 The Iranian President announced that unspecified malicious software sent by Western enemies had affected Iran's centrifuges at its Natanz enrichment plant and 'succeeded in creating problems for a limited number of our centrifuges'. 'Iran says cyber foes caused centrifuge problems,' Reuters, 29 November 2010; 'Iran Says Nuclear Programme Was Hit by Sabotage'. Iranian officials have also admitted that Stuxnet infected some computers at the Bashir nuclear power station but denied that any of the facility's control systems were affected.

however, all are unsubstantiated.[49] At the time of writing there are no official reports of physical damage in the public domain; however, it is known that Iran decommissioned and replaced approximately 1,000 IR-1 centrifuges in a short space of time around the same time Stuxnet is believed to have attacked the plant.[50] Footage from IAEA installed surveillance cameras tends to suggest that the work was carried out 'feverishly' in response to damage, implying larger than expected levels of breakage.[51] The worm is believed to have been introduced into computers at the Iranian nuclear facilities via an infected USB device into a gateway target. The worm is designed to seek out its final target and cause damage by making quick changes in the rotational speed of motors and sabotaging the normal operation of control systems.[52] By adjusting these speeds the virus could ultimately cause centrifuges (such as those used in nuclear enrichment facilities) to fly apart.

49 See, for example, Zetter, 'Did a U.S. Government Lab Help Israel Develop Stuxnet?'.
50 Albright, Brannan and Walrond, *ISIS Stuxnet Report*; Zetter, 'Report Strengthens Suspicions that Stuxnet Sabotaged Iran's Nuclear Plant' (ISIS Stuxnet Report).
51 Zetter, 'Surveillance Footage and Code Clues Indicate Stuxnet Hit Iran'.
52 Broad and Sanger, 'Worm Was Perfect for Sabotaging Centrifuges'. The worm targets frequency converters, a type of power supply that changes its output frequency to control the speed of a motor.

Appendix 2 Glossary of selected computing terms[1]

Applet	A program designed to be executed from within another application (they cannot be executed directly from the operating system). Web browsers, which are often equipped with Java virtual machines, can interpret applets from Web servers. Because applets are small in file size, cross-platform compatible, and highly secure (cannot be used to access users' hard drives), they are ideal for small internet applications accessible from a browser.
Backdoor	Also called a trapdoor. An undocumented way of gaining access to a program, online service or an entire computer system.
Bitstream	A bitstream is a contiguous sequence of bits (0s and 1s), representing a stream of data, transmitted continuously over a communications path, serially (one at a time).[2]
Botnet	Also called a Zombie net. A group of compromised computers controlled by a master computer and used primarily for distributed denial of service attacks or spam. According to figures in the *Independent*, a reasonable-sized botnet of 8,000–10,000 computers may be rented for £200 an hour.[3]
Computer network attack	Operations designed to disrupt, deny, degrade, or destroy information resident in computers or computer networks, or the computer networks themselves.

1 Definitions are primarily taken and adapted from the *Webopedia: Online Dictionary of Computing and Internet*, www.webopedia.com (last accessed 8 July 2011) and the *Department of Defense Dictionary of Military and Associated Terms* (Government Reprints Press, Washington DC, 2001)

2 SearchNetworking.com Definitions http://searchnetworking.techtarget.com/sDefinition/0,,sid7_gci213496,00.html (last accessed 13 August 2008).

3 Sarah Arnott, 'How Cyber Crime Went Professional', *Independent* (London), 13 August 2008.

Denial of service attack	A type of attack on a network that is designed to bring the system or network to its knees by flooding it with useless traffic.
Distributed denial of service attack	An attack where multiple compromised systems (which are usually infected with a Trojan) are used to target a single system causing a denial of service attack. Victims of a DDoS attack consist of both the end-targeted system and all systems maliciously used and controlled by the hacker in the distributed attack.
DNS server	Domain Name Service server. DNS is an internet service that translates domain names into IP addresses. For example, the domain name www.example.com might translate to *198.105.232.4*. If one DNS server in the network doesn't know how to translate a particular domain name, it asks another one, and so on, until the correct IP address is returned.
Electronic warfare	Military action involving the use of electromagnetic and directed energy to control the electromagnetic spectrum or to attack the enemy.
Hard coded	A feature which is built into the hardware or software in such a way that it cannot be modified.
In the wild	In order for a virus to be considered 'in the wild', 'it must be spreading as a result of normal day-to-day operations on and between the computers of unsuspecting users'. Although there are an estimated 47,000 computer viruses, fewer than 600 are said to be circulating outside of laboratories and research facilities – hence, 'in the wild'. Experts say these wild viruses pose the most significant threat to computers.[4]
Information	Facts, data or instructions in any medium or form.
Information operations	The integrated employment of the core capabilities of electronic warfare, computer network operations, psychological operations, military deception and operations security, in concert with specified supporting and related capabilities, to influence, disrupt, corrupt or usurp adversarial human and automated decision making while protecting one's own systems and decision making.
IP address	An identifier for a computer or device on a TCP/IP network. Networks using the TCP/IP protocol route messages based on the IP address of the destination. The format of an IP address is a 32-bit numeric address written as four numbers separated by points. Each number can be zero to 255. For example, 1.160.10.240 could be an IP address.

4 SearchSecurity.com definitions http://searchsecurity.techtarget.com/sDefinition/0,,sid14_gci511204,00.html (last accessed 2 December 2007).

ISP Internet Service Provider. A company that provides access to the internet.

Logic bomb Malicious programming code which is inserted into application software or an operating system. The code lies dormant until a predetermined period of time has elapsed, or a triggering event (or series of events) occurs, at which time the code activates.

Malware Malicious software, i.e. software designed specifically to damage or disrupt, e.g. Trojan horse, virus etc.

Peer-to-peer Peer-to-peer (P2P) architecture is a type of network in which each workstation has equivalent capabilities and responsibilities. Often used to describe one user linking with another user to transfer information and files through the use of a common P2P client to download MP3s, videos, images, games and other software. This, however, is only one type of P2P networking. Generally, P2P networks are used for sharing files, but a P2P network can also mean grid computing or instant messaging.

Rootkit Type of malicious software that is activated each time a system boots up. Rootkits are difficult to detect because they are activated before the system's Operating System (OS) has completely booted up. A rootkit often allows the installation of hidden files, processes, hidden user accounts, and more in the system's OS. Rootkits are able to intercept data from terminals, network connections and the keyboard.

Shareware Software distributed on the basis of an honour system. Usually free of charge, but with the request that if the user likes the program and uses it regularly the author is paid a small fee. This may entitle the user to service assistance and updates. Shareware may be copied and shared with the same fee expectation.

Spoof To fool or deceive. Although many things can be spoofed, it generally refers to IP spoofing which allows the sender of data to forge the source address in the header of the IP packet. The receiving computer will then send all replies to the forged address rather than the actual computer. This technique is often used in denial of service attacks.

Trapdoor See 'Backdoor'.

Trojan horse Malicious programming code disguised to look like a harmless application. Trojans are further broken down in classification based on how they breach systems and the damage they cause.

URL Universal Resource Locator. A global address of documents and other resources on the world wide web. A web address.

Virus Malicious programming code written to replicate itself and attach to another file or program in order to spread from one computer to another. They may also cause damage or destruction as they travel. Unlike a worm, a virus cannot travel without human intervention.

Worm A sub-class of virus which does not require human intervention to spread from host to host. A worm takes advantage of file or information transport features on the computer system to allow it to travel unaided.

Bibliography

Books

Alberts, David S., Garstka, John and Stein, Frederick P., *Network Centric Warfare: Developing and Leveraging Information Superiority* (2nd edn, National Defense University Press, Washington DC, 1999).

Alexandrov, Stanimir A., *Self-Defense against the Use of Force in International Law* (Kluwer Law International, The Hague; London, 1996).

Arkin, William M., *Code Names: Deciphering US Military Plans, Programs, and Operations in the 9/11 World* (Steerforth Press, Hanover, NH, 2005).

Arquilla, John, Ronfeldt, David F. and United States Department of Defense, Office of the Secretary of Defense, *Networks and Netwars: The Future of Terror, Crime, and Militancy* (RAND, Santa Monica, 2001).

Best, Geoffrey, *Humanity in Warfare: The Modern History of the International Law of Armed Conflicts* (Weidenfeld and Nicolson, London, 1980).

Black, Jeremy, *War in the New Century* (Continuum, London, 2001).

Boothby, William H., *Weapons and the Law of Armed Conflict* (Oxford University Press, Oxford; New York, 2009).

Bothe, Michael, *et al.*, *New Rules for Victims of Armed Conflicts* (Martinus Nijhoff Publishers, Leiden, 1982).

Bowett, D. W., *Self-Defence in International Law* (University of Manchester Press, 1958).

Boylan, Patrick J., *Review of the Convention for the Protection of Cultural Property in the Event of Armed Conflict* (UNESCO, Paris, 1993).

Brownlie, Ian, *International Law and the Use of Force by States* (Oxford University Press, 1963).

International Law and the Use of Force by States: Revisited (Europaeum, Oxford, 2001).

Principles of Public International Law (6th edn, Oxford University Press, 2003).

Cassese, Antonio, *International Criminal Law* (Oxford University Press, 2003).

Castrén, Erik, *The Present Law of War and Neutrality* (Suomalaisen Tiedeakatemian Toimituksia, Helsinki, 1954).

Chamberlain, Kevin, *War and Cultural Heritage: An Analysis of the 1954 Convention for the Protection of Cultural Property in the Event of Armed Conflict and Its Two Protocols* (Institute of Art & Law, Leicester, 2004).

Clausewitz, Carl von, Graham, J. J. and Maude, F. N., *On War* (new and rev. edn, Kegan Paul, Trench, Trubner & Co., London, 1940).

De Wet, Erika, *The Chapter VII Powers of the United Nations Security Council* (Hart, Oxford, 2004).

Deibert, Ronald J., *Parchment, Printing, and Hypermedia: Communication in World Order Transformation* (Columbia University Press, New York, 1995).

Dinstein, Yoram, *The Conduct of Hostilities under the Law of International Armed Conflict* (Cambridge University Press, 2004).

The Conduct of Hostilities under the Law of International Armed Conflict (2nd edn, Cambridge University Press, 2010).

War, Aggression, and Self-Defense (3rd edn, Cambridge University Press, New York, 2001).

Dunnigan, James F., *The Next War Zone: Confronting the Global Threat of Cyber Terrorism* (Citadel Press Books, New York, 2003).

Franck, Thomas M., *Recourse to Force: State Action against Threats and Armed Attacks* (Cambridge University Press, 2002)

Gimbel, John, *Science, Technology, and Reparations: Exploitation and Plunder in Postwar Germany* (Stanford University Press, 1990).

Gray, Christine D., *International Law and the Use of Force* (2nd edn, Oxford University Press, 2004).

International Law and the Use of Force (3rd edn, Oxford University Press, 2008).

Gray, Colin S., *Another Bloody Century: Future Warfare* (Weidenfeld & Nicolson, London, 2005).

Green, Leslie C., *The Contemporary Law of Armed Conflict* (2nd edn, Manchester University Press, 2000).

Harris, D. J., *Cases and Materials on International Law* (6th edn, Sweet & Maxwell, London, 2004).

Heintschel von Heinegg, Wolff and Epping, Volker, *International Humanitarian Law Facing New Challenges: Symposium in Honour of Knut Ipsen* (Springer, Berlin; New York, 2007).

Henckaerts, Jean-Marie and Doswald-Beck, Louise, *Customary International Humanitarian Law,* Vols. I and II (Cambridge University Press, 2005).

Higgins, Rosalyn, *Problems and Process: International Law and How We Use It* (Clarendon Press, Oxford, 1993).

Hulme, Karen, *War Torn Environment: Interpreting the Legal Threshold* (Martinus Nijhoff Publishers, Leiden, 2004).

Jennings, Robert Y. and Watts, C. Arthur H. (eds.), *Oppenheim's International Law* (9th edn, Longman, Harlow, 1992).

Kalshoven, Frits and Zegveld, Liesbeth, *Constraints on the Waging of War* (ICRC, Geneva, 2001).

Kidder, Tracy, *The Soul of a New Machine* (Little, Brown, Boston, 1981).

Klare, Michael T., *Resource Wars: The New Landscape of Global Conflict* (Metropolitan Books, New York, 2001).
Lauterpacht, Hersch (ed.), *Oppenheim's International Law* (7th edn, Longmans, Green & Co., London, 1952).
Lessig, Lawrence, *Code and Other Laws of Cyberspace* (Basic Books, New York, 1999).
Libicki, Martin C., *What Is Information Warfare?* (Center for Advanced Concepts and Technology, Institute for National Strategic Studies, Washington DC, 1995).
Lloyd, Ian J., *Information Technology Law* (4th edn, Oxford University Press, 2004).
McDougal, Myres Smith and Feliciano, Florentino P., *Law and Minimum World Public Order: The Legal Regulation of International Coercion* (Yale University Press, New Haven, 1961).
McLuhan, Marshall, *Understanding Media: The Extensions of Man* (McGraw-Hill, New York, 1964).
Melzer, Nils, *Targeted Killing in International Law* (Oxford University Press, 2008).
Moir, Lindsay, *The Law of Internal Armed Conflict* (Cambridge University Press, New York, 2002).
Reappraising the Resort to Force: International Law, Jus Ad Bellum and the War on Terror (Hart, Oxford, 2010).
Murray, Andrew, *Information Technology Law: The Law and Society* (Oxford University Press, Oxford; New York, 2010).
Nye, Joseph S., Jr, *The Paradox of American Power: Why the World's Only Superpower Can't Go It Alone* (Oxford University Press, 2002).
O'Keefe, Roger, *The Protection of Cultural Property in Armed Conflict* (Cambridge University Press, Cambridge; New York, 2006).
Pictet, Jean S., *The Geneva Conventions of 12 August 1949: Commentary* (International Committee of the Red Cross, Geneva, 1952).
Pilloud, Claude, *et al.*, *Commentary on the Additional Protocols of 8 June 1977* (Martinus Nijhoff, Geneva, 1987).
Reed, Thomas C., *At the Abyss: An Insider's History of the Cold War* (Presidio, New York, 2004).
Roberts, Adam and Guelff, Richard, *Documents on the Laws of War* (3rd edn, Oxford University Press, 1999).
Rodes, Jean-Michel, Piejut, Geneviève and Plas, Emmanuèle, *Memory of the Information Society* (UNESCO, Paris, 2003).
Rogers, A. P. V., *Law on the Battlefield* (2nd edn, Manchester University Press, 2004).
Saxby, Stephen, *The Age of Information: The Past Development and Future Significance of Computing and Communications* (Macmillan, London, 1990).
Sharp, Walter G., *Cyberspace and the Use of Force* (Aegis Research Corp., Falls Church, VA, 1999)
Singer, P. W., *Corporate Warriors: The Rise of the Privatized Military Industry* (Cornell University Press, Ithaca, 2003).

Smith, Edward A., *Effects Based Operations: Applying Network Centric Warfare to Peace, Crisis, and War* (DOD-CCRP, Washington DC, 2002).
Smith, Rupert, *The Utility of Force: The Art of War in the Modern World* (Penguin, London, 2005).
Solis, Gary D., *The Law of Armed Conflict: International Humanitarian Law in War* (Cambridge University Press, 2010).
Toffler, Alvin and Toffler, Heidi, *War and Anti-War: Survival at the Dawn of the 21st Century* (Warner Books, London, 1994).
Toman, Jirí, *The Protection of Cultural Property in the Event of Armed Conflict* (Dartmouth; UNESCO, Aldershot; Brookfield, VT, 1996).
UK Ministry of Defence, *The Manual of the Law of Armed Conflict* (Oxford University Press, Oxford; New York, 2004).
United Nations, International Law Commission and Crawford, James, *The International Law Commission's Articles on State Responsibility : Introduction, Text, and Commentaries* (Cambridge University Press, Cambridge; New York, 2002).
US Department of Defense, *Dictionary of Military and Associated Terms (as Amended through 20 March 2006)* (Washington DC, 2001).
Dictionary of Military and Associated Terms, 8 November 2010 (as Amended through 31 January 2011) (Washington DC, 2010).
Vandewiele, Tiny, *Commentary on the United Nations Convention on the Rights of the Child, 46 Optional Protocol: The Involvement of Children in Armed Conflicts* (Martinus Nijhoff Publishers, Leiden; Boston, 2005).
Wall, Andru E., *Legal and Ethical Lessons of NATO's Kosovo Campaign* (Naval War College, Newport, RI, 2002)
Weisburd, A. Mark, *Use of Force: The Practice of States Since World War II* (Pennsylvania State University Press, 1997)
Wingfield, Thomas C., *The Law of Information Conflict: National Security Law in Cyberspace* (Aegis Research Corp., Falls Church, VA, 2000).

Chapters

'Discussion' in Wall, A. E. (ed.), *Legal and Ethical Lessons of Nato's Kosovo Campaign* (Naval War College, Newport, RI, 2002) 211–22.
Abi-Saab, Georges, 'The Specificities of Humanitarian Law' in Swinarski, C. (ed.), *Studies and Essays on International Humanitarian Law and Red Cross Principles in Honour of Jean Pictet* (Martinus Nijhoff Publishers, Geneva, The Hague, 1984) 265–80.
Arquilla, John and Ronfeldt, David, 'The Advent of Netwar (Revisited)' in Arquilla, J. and Ronfeldt, D. (eds.), *Networks and Netwars* (RAND, Santa Monica, 2001) 1–24.
'Information, Power and Grand Strategy: In Athena's Camp – Section 1' in Arquilla, J. and Ronfeldt, D. (eds.), *In Athena's Camp: Preparing for Conflict in the Information Age* (RAND, Santa Monica, 1997) 141–71.
Baxter, Richard R., 'The Duties of Combatants and the Conduct of Hostilities (the Law of the Hague)' in UNESCO (ed.), *International Dimensions of Humanitarian Law* (Martinus Nijhoff, Dordrecht, 1988) 93–134.

Blix, Hans, 'Means and Methods of Combat' in UNESCO (ed.), *International Dimensions of Humanitarian Law* (Martinus Nijhoff, Dordrecht, 1998) 135–51.
Combacau, Jean, 'The Exception of Self Defence in U.N Practice' in Cassese, A. (ed.), *The Current Legal Regulation of the Use of Force* (Martinus Nijhoff, Dordrecht, 1986) 9–38.
Denning, Dorothy E., 'Cyber-Security as an Emergent Infrastructure' in Latham, R. (ed.), *Bombs and Bandwidth: The Emerging Relationship between Information Technology and Security* (Manas Publications, New Delhi, 2004) 25.
Dinstein, Yoram, 'Booty in Land Warfare' in Bernhardt, R. (ed.), *Encyclopedia of Public International Law* (Max Planck Institute; North Holland, Amsterdam, 1992) 432–4.
'Computer Network Attacks and Self-Defense' in Schmitt, M. N. and O'Donnell, B. T. (eds.), *Computer Network Attack and International Law* (Naval War College, Newport, RI, 1999) 99–119.
'Legitimate Military Objectives under the Current Jus in Bello' in Wall, A. E. (ed.), *Legal and Ethical Lessons of NATO's Kosovo Campaign* (Naval War College, Newport, RI, 2002) 139–73.
'Unlawful Combatancy' in Borch, F. and Wilson, P. (eds.), *International Law and the War on Terror* (Naval War College, Newport, RI, 2003) 151–74.
Doswald-Beck, Louise, 'Some Thoughts on Computer Network Attack and the International Law of Armed Conflict' in Schmitt, M. N. and O'Donnell, B. T. (eds.), *Computer Network Attack and International Law* (Naval War College, Newport, RI, 2002) 163–86.
Farer, Tom, 'Political and Economic Aggression in Contemporary International Law' in Cassese, A. (ed.), *The Current Legal Regulation of the Use of Force* (Martinus Nijhoff, Dordrecht, 1986) 121–32.
Fleck, Dieter, 'The Law of Non-International Armed Conflicts' in Fleck, D. (ed.), *The Handbook of International Humanitarian Law* (2nd edn, Oxford University Press, 2008) 605–33.
Frowein, Jochen, 'Article 39' in Simma, B. (ed.), *The Charter of the United Nations: A Commentary* (2nd edn, Oxford University Press, 2002) 717.
Green, Leslie C., 'The Status of Mercenaries in International Law' in Green, L. C. (ed.) *Essays on the Modern Law of War* (Transnational Publishers, Dobbs Ferry, NY, 2000).
Greenwood, Christopher, 'Customary International Law and the First Geneva Protocol of 1977 in the Gulf Conflict' in Rowe, P. J. (ed.), *The Gulf War 1990–1991 in International and English Law* (Routledge, London, 1993) 63–88.
'The Law of War (International Humanitarian Law)' in Evans, M. D. (ed.), *International Law* (Oxford University Press, 2003) 789–821.
'The Law of Weaponry at the Start of the New Millennium' in Schmitt, M. N. and Green, L. C. (eds.), *The Law of Armed Conflict: Into the Next Millennium* (Naval War College, Newport, RI, 1998) 185–231.
'Scope of Application of Humanitarian Law' in Fleck, D. (ed.), *The Handbook of Humanitarian Law in Armed Conflict* (2nd edn, Oxford University Press, 2008) 45–78.

Hampson, Françoise J., 'Means and Methods of Warfare in the Conflict in the Gulf' in Rowe, P. J. (ed.), *The Gulf War of 1990–1991 in International and English Law* (Routledge, London, 1993) 89–110.

Held, David and McGrew, Anthony, 'Introduction' in Held, D. and McGrew, A. (eds.), *Governing Globalisation: Power, Authority and Global Governance* (Polity Press, Cambridge, 2002) 1–21.

Hulme, Karen, 'Natural Environment' in Wilmshurst, E. and Breau, S. (eds.), *Perspectives on the ICRC Study on Customary International Humanitarian Law* (Cambridge University Press, 2007) 204–37.

Ipsen, Knut, 'Combatants and Non-Combatants' in Fleck, D. (ed.), *The Handbook of Humanitarian Law in Armed Conflicts* (Oxford University Press, 1999) 65–104.

'Perfidy' in Bernhardt, R. (ed.), *Encylopedia of Public International Law* (Max Planck Institute, Amsterdam; New York, 1997) 978–81.

Kalshoven, Frits, 'The Soldier and His Golf Clubs' in Swinarski, C. (ed.), *Studies and Essays on International Humanitarian Law and Red Cross Principles in Honour of Jean Pictet* (Martinus Nijhoff, The Hague; Boston, 1984) 369–85.

Mészáros, Ernö, 'Techniques for Manipulating the Atmosphere' in Westing, A. H. (ed.), *Environmental Warfare: A Technical, Legal and Policy Appraisal* (Taylor & Francis, London, 1984).

Montgomery, Tony, 'Legal Perspective from the EUCOM Targeting Cell' in Wall, A. E. (ed.), *Legal and Ethical Lessons of NATO's Kosovo Campaign* (Naval War College, Newport, RI, 2002) 189.

Nichiporuk, Brian and Builder, Carl H., 'Societal Implications' in Arquilla, J. and Ronfeldt, D. (eds.), *In Athena's Camp* (RAND, Santa Monica, 1997) 295.

Noltimier, Hallan C., 'Techniques for Manipulating the Geosphere' in Westing, A. H. (ed.), *Environmental Warfare: A Technical, Legal and Policy Appraisal* (Taylor & Francis, London, 1984) 25–31.

Oeter, Stefan, 'Comment: Is the Principle of Distinction Outdated?' in Heinegg, W. H. von and Epping, V. (eds.), *International Humanitarian Law Facing New Challenges* (Springer, Berlin; New York, 2007) 53–65.

'Methods and Means of Combat' in Fleck, D. (ed.), *The Handbook of Humanitarian Law in Armed Conflicts* (2nd edn, Oxford University Press, 2008) 120–235.

Parks, W. Hays, 'Asymmetries and the Identification of Legitimate Military Objectives' in Heinegg, W. H. von and Epping, V. (eds.), *Interntional Humanitarian Law Facing New Challenges* (Springer, Berlin, 2007) 65–116.

Randelzhofer, Albrecht, 'Article 2(4)' in Simma, B. (ed.), *The Charter of the United Nations: A Commentary* (2nd edn, Oxford University Press, 2002).

'Article 51' in Simma, B. (ed.), *The Charter of the United Nations: A Commentary* (2nd edn, Oxford University Press, 2002) 788.

Robertson, Horace B., 'Self-Defense against Computer Network Attack under International Law' in Schmitt, M. N. and O'Donnell, B. T. (eds.), *Computer Network Attack and International Law* (Naval War College, Newport, RI, 2002) 122–45.

Sassòli, Marco, 'Targeting: The Scope and Utility of the Concept Of "Military Objectives" for the Protection of Civilians in Contemporary Armed Conflicts' in Wippman, D. and Evangelista, M. (eds.), *New Wars, New Laws? Applying the Laws of War in 21st Century Conflicts* (Transnational Publishers, Ardsley, NY, 2005) 181–210.

Schmitt, Michael N., 'Asymmetrical Warfare and International Humanitarian Law' in Heintschel von Heinegg, W. and Epping, V. (eds.), *International Humanitarian Law Facing New Challenges: Symposium in Honour of Knut Ipsen* (Springer, Berlin; New York, 2007) 11–48.

'Direct Participation in Hostilities and 21st Century Armed Conflict' in Fischer, H *et al.* (eds.), *Crisis Management and Humanitarian Protection* (Berliner Wissenschafts Verlag, Berlin, 2004) 505–29.

'Fault Lines in the Law of Attack' in Breau, S. and Jachec-Neale, A. (eds.), *Testing the Boundaries of International Humanitarian Law* (British Institute of International and Comparative Law, London, 2006) 277–307.

'Wired Warfare: Computer Network Attack and the *Jus in Bello*' in Schmitt, M. N. and O'Donnell, B. T. (eds.), *Computer Network Attack and International Law* (US Naval War College, Newport, RI, 2002) 187–218.

Steinkamm, Armin A., 'Pillage' in Bernhardt, R. (ed.), *Encyclopedia of International Law* (Max Planck Institute; North Holland, Amsterdam, 1982) 1029–30.

Toffler, Alvin and Toffler, Heidi, 'Foreword: The New Intangibles' in Arquilla, J. *et al.* (eds.), *In Athena's Camp: Preparing for Conflict in the Information Age* (RAND, Santa Monica, 1997) xiii–xxiv.

Wedgwood, Ruth G., 'Proportionality, Cyberwar, and the Law of War' in Schmitt, M. N. and O'Donnell, B. T. (eds.), *Computer Network Attack and International Law* (Naval War College, Newport, RI, 2002) 219–32.

Zanardi, Pierluigi L., 'Indirect Military Aggression' in Cassese, A. (ed.), *The Current Legal Regulation on the Use of Force* (Martinus Nijhoff, Dordrecht, 1986) 111–19.

Journal Articles

Ago, Roberto, 'Addendum to the Eighth Report on State Responsibilty' (1980) II *UN Y.B. Int'l L. Comm'n* 13.

'State Responsibility' (1980) Vol. 1 *UN Y.B. Int'l L. Comm'n* 18826 June 1980.

Aldrich, Richard W., 'How Do You Know You Are at War in the Information Age?' (2000) 22 *Hous. J. Int'l L.* 223.

'The International Legal Implications of Information Warfare' (1996) *Airpower* 99.

Alvey, Ruth, 'Russian Hackers for Hire: The Rise of the E-Mercenary' (2001) 13(7) *Jane's Intelligence Review* 52.

American Society of International Law, 'The Jurisprudence of the Court in the Nicaragua Decision' (1987) 81 *ASIL Proc.* 258.

Arkin, William M., 'Cyber Warfare and the Environment' (2001) 25 *Vermont Law Review* 779.

Barkham, Jason, 'Information Warfare and International Law on the Use of Force' (2001) 34 *New York University Journal of International Law* 57.
Basu, Sandra, 'Military Electronic Medical Records Support Quality Treatment Abroad', *US Medicine* (Washington DC), February 2006.
Baxter, Richard R., 'So-Called "Unprivileged Belligerency": Spies, Guerrillas, and Saboteurs' (1951) 28 *BYBIL* 323.
Belknap, Margaret H., 'The CNN Effect: Strategic Enabler or Operational Risk?' (2002) Fall *Parameters* 100.
Belt, Stuart Walters, 'Missiles over Kosovo: Emergence, Lex Lata, of a Customary Norm Requiring the Use of Precision Munitions in Urban Areas' (2000) 47 *Naval Law Review* 115.
Bothe, Michael, 'Legal Restraints on Targeting: Protection of Civilian Population and the Changing Faces of Modern Conflict' (2001) 31 *Israel Y.B. Hum. Rts* 35.
'Terrorism and the Legality of Pre-Emptive Force' (2003) 14(2) *EJIL* 227.
Bourbonnière, Michel, 'Law of Armed Conflict (LOAC) and the Neutralisation of Satellites or Ius in Bello Satellitis' (2004) 9 *JC&SL* 43.
Bouvier, Antoine, 'Protection of the Natural Environment in Time of Armed Conflict' (1991) 285 *IRRC* 567 www.icrc.org/web/eng/siteeng0.nsf/html/57JMAU (last accessed 19 September 2006).
Brown, Davis, 'A Proposal for an International Convention to Regulate the Use of Information Systems in Armed Conflict' (2006) 47 *Harv. Int'l L.J.* 179.
Brownlie, Ian, 'International Law and the Activities of Armed Bands' (1958) 7 *International and Comparative Law Quarterly* 712.
Cassese, Antonio, 'The Status of Rebels under the 1977 Geneva Protocol on Non-International Armed Conflicts' (1981) 30 *International and Comparative Law Quarterly* 416.
Corten, Olivier, 'The Controversies over the Customary Prohibition on the Use of Force: A Methodological Debate' (2005) 16 *EJIL* 803.
Coupland, Robin M, 'The Effects of Weapons and the Solferino Cycle: Where Disciplines Meet to Prevent or Limit the Damage Caused by Weapons' (1999) 319(7214) *BMJ* 864.
Damrosch, Lori Fisler, 'Politics across Borders: Nonintervention and Nonforcible Influence over Domestic Affairs' (1989) 83 *AJIL* 1.
Daoust, Isabelle, Coupland, Robin and Ishoey, Rikke, 'New Wars, New Weapons? The Obligation of States to Assess the Legality of Means and Methods of Warfare' (2002) 84(846) *IRRC* 345.
Domb, F., 'Judgments of the Supreme Court of Israel' (1986) 16 *Israel Y.B. Hum. Rts* 321.
Doswald-Beck, Louise, 'Vessels, Aircraft and Persons Entitled to Protection During Armed Conflicts at Sea' (1994) 65 *BYBIL* 211.
Downey, William Gerald, Jr, 'Captured Enemy Property: Booty of War and Seized Enemy Property' (1950) 44 *AJIL* 488.
Dunlap, Charles J, 'The End of Innocence: Rethinking Non-Combatancy in the Post-Kosovo Era' (2000) *Summer Strategic Review* 9.

Evron, Gadi, 'Battling Botnets and Online Mobs' (2008) 9(1) *GJIA* 121.
Farquharson, John, 'Governed or Exploited? The British Acquisition of German Technology, 1945–48' (1997) 32(1) *Journal of Contemporary History* 23.
'Ruses of War and Prohibition of Perfidy' (1974) 13 *Revue de Droit Penal Militare et de Droit de la Guerre* 269.
Frigo, Manlio, 'Cultural Property v. Cultural Heritage: A "Battle of Concepts" in International Law?' (2004) 86(854) *IRRC* 367.
Gardam, Judith G., 'Proportionality and Force in International Law' (1993) 87(3) *AJIL* 391.
Gordon, Edward, 'Article 2(4) in Historical Context' (1985) 10 *Yale J. Int'l L.* 271.
Gray, Christine D., 'The Bush Doctrine Revisited: The 2006 National Security Strategy of the USA' (2006) 5(3) *Chi. J. Int' L.* 555.
Greenwood, Christopher, 'International Law and the Pre-Emptive Use of Force: Afganistan, Al-Qaida, and Iraq' (2003) 4 *San Diego Int'l L.J.* 7.
'International Law and the War against Terrorism' (2002) 78 *International Affairs* 301.
Guillory, Michael E., 'Civilianising the Force: Is the United States Crossing the Rubicon?' (2001) 51 *AFL Rev.* 111.
Hanseman, Robert G., 'The Realities and Legalities of Information Warfare' (1997) 42 *AFL Rev.* 173.
Happold, Matthew, 'Child Soldiers in International Law: The Legal Regulation of Children's Participation in Hostilities' (2000) 47 *NIL Rev.* 27.
Harbom, Lotta and Wallensteen, Peter, 'Armed Conflict, 1989–2006' (2007) 44(5) *Journal of Peace Research* 623.
Hargrove, John, 'The Nicaragua Judgment and the Future of the Law of Force and Self-Defense' (1987) 81 *AJIL* 135.
Haslam, Emily, 'Information Warfare: Technological Changes and International Law' (2000) 5(2) *JC&SL* 157.
Henderson, Christian, 'The 2010 United States National Security Strategy and the Obama Doctrine of "Necessary Force"' (2010) 15 *JC&SL* 403.
Henderson, Christian M., 'The 2006 National Security Strategy of the United States: The Pre-Emptive Use of Force and the Persistent Advocate' (2007) 15 *Tulsa J. Comp. & Int'l L.* 1.
Hollis, David, 'Cyberwar Case Study: Georgia 2008' (2011) 7(1) *Small Wars Journal* http://smallwarsjournal.com/blog/journal/docs-temp/639-hollis.pdf (last accessed 6 January 2011).
Hollis, Duncan B., 'Why States Need an International Law for Information Operations' (2007) 11(4) *Lewis & Clark L. Rev.* 1023.
ICRC, 'External Activities: September-October 1987' (1987) 27(261) *IRRC* 650.
Intoccia, Gregory F. and Moore, Joe Wesley, 'Communications Technology, Warfare, and the Law: Is the Network a Weapons System?' (2006) 28 *Hous. J. Int'l L.* 467.
Jacobsen, Walter L., 'A Juridical Examination of the Israeli Attack on the USS *Liberty*' (1986) 36 (Winter) *Naval Law Review* 69.

Jacobson, Mark, 'War in the Information Age: International Law, Self Defence and the Problem of "Non-Armed" Attacks' (1998) 21(3) *Journal of Strategic Studies* 1.

Jensen, Eric Talbot, 'Computer Attacks on Critical National Infrastructure: A Use of Force Invoking the Right of Self-Defense' (2002) 38 *Stan. J. Int'l L.* 207.

'Cyber Warfare and Precautions against the Effects of Attacks' (2010) 88 *Tex. L. Rev.* 1533.

'The International Law of Environmental Warfare: Active and Passive Damage During Armed Conflict' (2005) 38 *Vand. J. Transnat'l L.* 145.

Kanuck, Sean P., 'Information Warfare: New Challenges for Public International Law' (1996) 37 *Harv. Int'l L.J.* 272.

Kritsiotis, Dino, 'Mercenaries and the Privatisation of Warfare' (1998) 22 *Fletcher Forum of World Affairs* 11.

Ku, Raymond, 'Foreword: A Brave New Cyberworld' (2000) 22 *T. Jefferson L. Rev.* 125.

Kuehl, Daniel T., 'Airpower Vs Electricity' (1995) 18 *Journal of Strategic Studies* 28.

LaRae-Perez, Cassandra, 'Economic Sanctions as a Use of Force: Re-Evaluating the Legality of Sanctions from an Effects-Based Perspective' (2002) 20 *BU Int'l L.J.* 161.

Lev, Izhar, 'E-Intifada: Political Disputes Cast Shadows in Cyberspace' (2000) 12(12) *Jane's Intelligence Review* 16.

Lynn, William F., III, 'Defending a New Domain – the Pentagon's Cyberstrategy' (2010) 89(5) *Foreign Affairs* 97.

Mégret, Frédéric, '"War"? Legal Semantics and the Move to Violence' (2002) 13(2) *EJIL* 361.

Melzer, Nils, 'Keeping the Balance between Military Necessity and Humanity: A Response to Four Critiques of the ICRC's Interpretive Guidance on the Notion of Direct Participation in Hostilities' (2010) 42 *NYU J. Int'l L. & Pol.* 831.

Meyer, Jeanne M., 'Tearing Down the Facade: A Critical Look at the Current Law on Targeting the Will of the Enemy and Air Force Doctrine' (2001) 51 *AFL Rev.* 143.

Milliard, Todd S., 'Overcoming Post-Colonial Myopia: A Call to Recognise and Regulate Private Military Companies' (2003) 176 *Mil. L. Rev.* 1.

Morth, Todd A., 'Considering Our Position: Viewing Information Warfare as a Use of Force Prohibited by Article 2(4) of the U.N. Charter' (1998) 30 *Case W. Res. J. Int'l L.* 567.

Murphy, Sean D., 'Self-Defense and the Israeli Wall Advisory Opinion: An Ipse Dixit from the Court?' (2005) 99 *AJIL* 62.

O'Keefe, Roger, 'The Meaning of "Cultural Property" under the 1954 Hague Convention' (1999) 46 *NIL Rev.* 26.

Parks, W. Hays, 'Air War and the Law of War' (1990) 32 *AFL Rev.* 1.

Reisman, W. Michael, 'Criteria for the Use of Force in International Law' (1985) 10 *Yale J. Int'l L.* 279.

Roach, J. Ashley, 'The Law of Naval Warfare at the Turn of Two Centuries' (2000) 94 *AJIL* 64.
Robbat, Michael J., 'Resolving the Legal Issues Concerning the Use of Information Warfare in the International Forum: The Reach of the Existing Legal Framework, and the Creation of a New Paradigm' (2000) 6 *BU J. Sci. & Tech. L.* 10.
Robertson, Horace B., 'The Principle of the Military Objective' (1997–1998) 8 *USAF Acad. J. Legal Stud.* 35.
Romanych, Marc J. and Krumm, Kenneth, 'Tactical Information Operations in Kosovo' (2004) September–October *Military Review* 56.
Schmitt, Michael N., 'Bellum Americanum Revisited: US Security Strategy and the Jus Ad Bellum, Transcript of Sixteeth Waldemar A. Solf Lecture in International Law to the Judge Advocate General School, US Army, Charlottesville, Virginia, 28 February 2003' (2003) 176 *Mil. L. Rev.* 364.
'Computer Network Attack and the Use of Force in International Law: Thoughts on a Normative Framework' (1999) 37 *Col. J. Trans. L.* 885.
'Deconstructing Direct Participation in Hostilities: The Constitutive Elements' (2010) 42 *NYU J. Int'l L. & Pol.* 697.
'Green War: An Assessment of the Environmental Law of International Armed Conflict' (1997) 22 *Yale J. Int'l L.* 1.
'Humanitarian Law and Direct Participation by Private Contractors or Civilian Employees' (2004) 5(2) *Chi. J. Int'l L.* 511.
'International Law and Military Operations in Space' (2006) 10 *Max Planck UNYB* 89.
'The Interpretive Guidance on the Notion of Direct Participation in Hostilities: A Critical Analysis' (2010) 1 *Harvard National Security Journal* 55.
'The Principle of Discrimination in 21st Century Warfare' (1999) 2 *Yale Hum. Rts & Dev. L.J.* 143.
'The Resort to Force in International Law: Reflections on Positivist and Contextualist Approaches' (1994) 37 *AFL Rev.* 105.
'Targeting and Humanitarian Law: Current Issues' (2004) 34 *Israel Y.B. Hum. Rts* 59.
Scott, Roger D., 'Legal Aspects of Information Warfare: Military Disruption of Telecommunications' (1998) 45 *Naval Law Review* 57.
Shulman, Mark R., 'Discrimination in the Laws of Information Warfare' (1999) 37 *Col. J. Trans. L.* 939.
Simma, Bruno and Paulus, Andreas L., 'The Responsibility of Individuals for Human Rights Abuses in Internal Conflicts: A Positivist View' (1999) 93 *AJIL* 302.
Singer, Peter W., 'Outsourcing War' (2005) 84(2) *Foreign Affairs* 119.
'War, Profits, and the Vacuum of Law: Privatized Military Firms and International Law' (2004) 42 *Col. J. Trans. L.* 521.
Solf, Waldemar A., 'Comment: Non-International Armed Conflicts' (1981–1982) 31 *Am. U. L. Rev.* 927.

Swiney, Gabriel, 'Saving Lives: The Principle of Distinction and the Realities of Modern War' (2005) 39 *International Lawyer* 733.

Terry, James P., 'The Lawfulness of Attacking Computer Networks in Armed Conflict and in Self-Defense in Periods Short of Armed Conflict: What Are the Targeting Constraints?' (2001) 169 *Mil. L. Rev.* 70.

Thomas, Timothy L., 'Russian Views on Information Based Warfare' (1996) special edn *Airpower* 25.

Watts, Sean, 'Combatant Status and Computer Network Attack' (2010) 50(2) *Virginia Journal of International Law* 391.

Weiss, Gus W., 'The Farewell Dossier: Duping the Soviets' (1996) 35(5) *Studies in Intelligence* 121 https://www.cia.gov/library/center-for-the-study-of-intelligence/csi-publications/csi-studies/studies/96unclass/farewell.htm (last accessed 12 April 2011).

Wingfield, Thomas C., 'Legal Aspects of Offensive Information Operations in Space' (1998) 9 *J. Legal Stud.* 121.

Zamparelli, Steven J., 'What Have We Signed up For? Competitive Sourcing and Privatization – Contractors on the Battlefield' (1999) 23(3) *AFJ Log.* 9.

Declarations and Treaties

Charter of the Organisation of American States, 30 April 1948, 119 UNTS 3.

Convention for the Protection of Cultural Property in the Event of Armed Conflict, 14 May 1954, 249 UNTS 240.

Convention on the Prohibition of Military or Any Other Hostile Use of Environmental Modification Techniques, 18 May 1977.

Convention Respecting the Laws and Customs of War on Land and Its Annex: Regulations Concerning the Laws and Customs of War on Land, 18 October 1907.

Declaration on Principles of International Law Concerning Friendly Relations and Cooperation Among States in Accordance with the Charter of the United Nations, 24 October 1970.

Declaration Renouncing the Use, in Time of War, of Explosive Projectiles under 400 Grammes Weight (St Petersburg Declaration), 29 November/11 December 1868.

North Atlantic Treaty, 4 April 1949, 34 UNTS 243.

OAU Convention for the Elimination of Mercenaries in Africa, OAU Doc. CM/433/Rev.L Annex 1.

Reports and Working Papers

Albright, David, Brannan, Paul and Walrond, Christina, *Did Stuxnet Take out 1,000 Centrifuges at the Natanz Enrichment Plant?*, Institute for Science and International Security (2010).

Stuxnet Malware and Natanz: Update of ISIS December 22, 2010 Report, Institute for Science and International Security (2011).

Bank for International Settlements, *Trienniel Central Bank Survey: Foreign Exchange and Derivitaves Market Activity in 2007*, Bank for International Settlements (2007) www.bis.org/publ/rpfxf07t.pdf (last accessed 12 April 2011).

Boivin, Alexandra, *The Legal Regime Applicable to Targeting Military Objectives in the Context of Contemporary Warfare*, University Centre for International Humanitarian Law, 2 (2006) www.cudih.org/recherche/objectif_militaire_recherche.pdf (last accessed 7 July 2011).

Bond, James, *Peacetime Foreign Data Manipulation as One Aspect of Offensive Information Warfare: Questions of Legality under the United Nations Charter Article 2(4)* (unpublished thesis) Naval War College (1996) http://handle.dtic.mil/100.2/ADA310926 (last accessed 5 July 2011).

Brookings Institution, *Iraq Index*, Brookings Institute (2006) www.brookings.edu/iraqindex (last accessed 5 July 2011).

Center for Strategic and International Studies, *Significant Cyber Incidents since 2006* (2011) http://csis.org/files/publication/110309_Significant_Cyber_Incidents_Since_2006.pdf (last accessed 12 April 2011).

Dorman, Andrew M., *Transforming to Effects-Based Operations: Lessons from the United Kingdom Experience*, Strategic Studies Institute (2008).

Dörmann, Knut, 'Applicability of the Additional Protocols to Computer Network Attacks' (Paper presented at the International Expert Conference on Computer Network Attacks and the Applicability of International Humanitarian Law, Stockholm, 17–19 November 2004) 139–54.

Falliere, Nicolas, OMurchu, Liam and Chien, Eric, *W32.Stuxnet Dossier*, Symantec (2011) www.symantec.com/content/en/us/enterprise/media/security_response/whitepapers/w32_stuxnet_dossier.pdf (last accessed 12 February 2011).

Gerhard, William D. and Millington, Henry W., *Attack on a Sigint Collector, the U.S.S. Liberty* (National Security Agency, 1981).

Greenberg, Lawrence T., Goodman, Seymour E. and Hoo, Kevin J. Soo, *Information Warfare and International Law* (CCRP, Washington DC, 1998).

Heise Security, *Estonian DDoS – a Final Analysis* http://h-online.com/-732971 (last accessed 12 April 2011).

Holz, Thorsten, *et al.*, 'Measurements and Mitigation of Peer-to-Peer-Based Botnets: A Case Study on Storm Worm' (Paper presented at the First USENIX Workshop on Large-Scale Exploits and Emergent Threats (LEET '08), San Francisco, http://honeyblog.org/junkyard/paper/storm-leet08.pdf) (last accessed 7 July 2011).

Hosmer, Stephen T., *The Conflict over Kosovo: Why Milosevic Decided to Settle When He Did* (RAND, Santa Monica, 2001).

Hundley, Richard O., *et al.*, *The Global Course of the Information Revolution: Recurring Themes and Regional Variations* (National Defense Research Institute, RAND, Santa Monica, 2003).

ICRC, *A Guide to the Legal Review of New Weapons, Means and Methods of Warfare: Measures to Implement Article 36 of Additional Protocol I of 1977*, ICRC (2006).

'Guidelines for Military Manuals and Instructions on the Protection of the Environment in Times of Armed Conflict' (1996) 311 *IRRC* 230 www.icrc.org/eng/resources/documents/misc/57jn38.htm (last accessed 15 July 2011).

ICRC Clarification Process on the Notion of Direct Participation in Hostilities under International Humanitarian Law (Proceedings) www.icrc.org/eng/resources/documents/article/other/direct-participation-article-020709.htm (last accessed 26 February 2011).

Interpretive Guidance on the Notion of Direct Participation in Hostilities (ICRC, Geneva, 2009).

Third Expert Meeting on the Notion of Direct Participation in Hostilities – Summary Report (ICRC, Geneva, 2005).

ICTY, *Final Report to the Prosecutor of the Committee Established to Review the NATO Bombing Campaign against the Federal Republic of Yugoslavia*, ICTY (2000) www.icty.org/x/file/About/OTP/otp_report_nato_bombing_en.pdf (last accessed 15 July 2011).

Information Warfare Monitor, *Tracking Ghostnet: Investigating a Cyber Espionage Network*, Information Warfare Monitor (2009) www.infowar-monitor.net/ (last accessed 12 April 2010).

Intergovernmental Conference on the Protection of Cultural Property in the Event of Armed Conflict, *Records of the Conference Convened by the United Nations Educational, Scientific and Cultural Organisation and Held at the Hague from 21 April to 14 May 1954* (Government of the Netherlands, The Hague Staatsdrukkerij en Uitgeverijbedrijf, 1961).

Krekal, Bryan, *et al.*, *Capability of the People's Republic of China to Conduct Cyber Warfare and Computer Network Exploitation Prepared for the US-China Eonomic and Security Review Commission*, Northrop Grumman (2009).

Lambeth, Benjamin S., US Air Force and Project Air Force, *NATO's Air War for Kosovo: A Strategic and Operational Assessment* (RAND, Santa Monica, 2001).

Livingston, Steven, *Clarifying the CNN Effect: An Examination of Media Effects According to Type of Military Intervention*, The Joan Shorenstein Center on the Press, Politics and Public Policy, John F. Kennedy School of Government, Harvard University, R-18 (1997) www.hks.harvard.edu/presspol/publications/papers/research_papers/r18_livingston.pdf (last accessed 15 July 2011).

Lyman, Peter and Varian, Hal R., *How Much Information?*, University of California at Berkley (2003) www.sims.berkeley.edu/research/projects/how-much-info-2003/ (last accessed 5 July 2011).

Malhotra, Yogesh, 'Measuring the Knowledge Assets of a Nation: Knowledge Systems for Development' (Paper presented at the United Nations Advisory Meeting of the Department of Economic and Social Affairs, Division for Public Administration and Development Management, Ad hoc Group of Experts Meeting – Knowledge Systems for Development, United Nations Headquarters, New York, 4–5 September 2003).

McAfee, *Virtual Criminology Report*, McAfee Inc. (2007).

Office of General Counsel, *An Assessment on International Legal Issues in Information Operations*, United States Department of Defense (1999) www.au.af.mil/au/awc/awcgate/dod-io-legal/dod-io-legal.pdf (last accessed 5 July 2011).

OpenNet Initiative, *Pulling the Plug: A Technical Review of the Internet Shutdown in Burma*, OpenNet Initiative, (2007) http://opennet.net/sites/opennet.net/files/ONI_Bulletin_Burma_2007.pdf (last accessed 12 April 2011).

Pace, Scott, *et al.*, *The Global Positioning System: Assessing National Policies* (RAND, Santa Monica, 1995).

Project Grey Goose, *Phase I Report: Russia/Georgia Cyber War – Findings and Analysis* (2008).

Quéguiner, Jean-François, *Direct Participation in Hostilities under International Humanitarian Law*, Program on Humanitarian Policy and Conflict Research at Harvard University (2003) http://ihl.ihlresearch.org/_data/n_0002/resources/live/briefing3297.pdf (last accessed 5 July 2011).

Royal Academy of Engineering, *Global Navigation Space Systems: Reliance and Vulnerabilities*, Royal Academy of Engineering (2011) www.raeng.org.uk/gnss (last accessed 8 March 2011).

Schmitt, Michael N., 'CNA and the Jus in Bello: An Introduction' (Paper presented at the International Expert Conference on Computer Network Attacks and the Applicability of International Humanitarian Law, Stockholm, 17–19 November 2004) 101–25.

The Impact of High and Low-Tech Warfare on the Principle of Distinction, Program on Humanitarian Policy and Conflict Research at Harvard University (2003) http://ihl.ihlresearch.org/_data/n_0002/resources/live/briefing3296.pdf (last accessed 5 July 2011).

Schwartz, Moshe and Swain, Joyprada, *Department of Defense Contractors in Afghanistan and Iraq: Background and Analysis*, Congressional Research Service (2011) www.fas.org/sgp/crs/natsec/R40764.pdf (last accessed 12 April 2011).

Shawhan, Karl J., 'Vital Interests, Virtual Threats: Reconciling International Law with Information Warfare and United States Security' (2001) (unpublished thesis) School of Advanced Airpower Studies, Air University, Maxwell Air Force Base, Alabama, www.dtic.mil/cgi-bin/GetTRDoc?AD=ADA391632 (last accessed 8 July 2011).

Shulman, Mark R., *Legal Constraints on Information Warfare*, Center for Strategy and Technology, Air War Center, Occasional Paper No. 7 (1999).

Stewart, Joe, *Storm Worm DDoS Attack*, Secure Works (1997) www.secureworks.com/research/threats/storm-worm (last accessed 7 July 2011).

Streltsov, Anatolij, 'Threat Analysis in the Post Cold-War Order' (Paper presented at the International Expert Conference on Computer Network Attacks and the Applicability of International Humanitarian Law, Stockholm, 17–19 November 2004) 21–7.

Symantec, *Symantec Global Internet Security Threat Report*, Symantec, Vol. 13 (2008).

Symmetricon, *Why Convert to a SAASM Based Global Positioning System (GPS)?*, (2006) www.symmetricom.com/resources/downloads/white-papers/GPS-Solutions/Why-Convert-to-a-SAASM-based-Global-Positioning-System-GPS/ (last accessed 7 July 2011).
Tikk, Eneken, Kaska, Kadri and Vihul, Liis, *International Cyber Incidents: Legal Considerations*, Cooperative Cyber Defence Centre of Excellence (2010).
Tsymbal, V. I., 'Kontseptsiya "Informatsionnoy Voyny"' (Concept of Information Warfare) (Paper presented at the Evolving post Cold War National Security Issues, Moscow, 12–14 September 1995) 7.
UK Cabinet Office, *The National Security Strategy of the United Kingdom: Security in an Interdependent World*, UK Cabinet Office, Cm 7291 (2008).
A Strong Britian in an Age of Uncertainty: The National Security Strategy (Cm 7935 edn, The Stationery Office, London, 2010).
UN Economic and Social Council, *Report of the Working Group on a Draft Optional Protocol to the Convention on the Rights of the Child on Involvement of Children in Armed Conflicts in Its Sixth Session*, UN Doc. E/CN.4/2000/74 (2000).
UN Secretary General, *In Larger Freedom: Towards Development, Security and Human Rights for All*, United Nations, UN Doc. A/59/2005 (2005).
US Cyber Consequences Unit, *Overview by the US-CCU of the Cyber Campaign against Georgia in August of 2008*, US Cyber Consequences Unit (2009) www.registan.net/wp-content/uploads/2009/08/US-CCU-Georgia-Cyber-Campaign-Overview.pdf (last accessed 25 February 2011).
US Department of Defense, *Conduct of the Persian Gulf War: Final Report to Congress*, US Department of Defense (1992).
Information Operations, Joint Chiefs of Staff, Joint Publication 3–13 (2006).
Strategy for Operating in Cyberspace, July 2011, US Department of Defense (2011) www.defense.gov/news/d20110714cyber.pdf (last accessed 18 January 2012).
US Department of Homeland Security, *National Infrastructure Protection Plan*, US Department of Homeland Security (2006) www.dhs.gov/xlibrary/assets/NIPP_Plan.pdf (last accessed 4 July 2011).
US Joint Chiefs of Staff, *Information Warfare: A Strategy for Peace … The Decisive Edge in War* (1996) http://handle.dtic.mil/100.2/ADA318379 (last accessed 5 July 2011).
United Nations, *A More Secure World: Our Shared Responsibility Report of the Secretary-General's High-Level Panel on Threats Challenges and Change*, United Nations, UN Doc. A/59/565 (2004).
Watkin, Kenneth, *Combatants, Unprivileged Belligerents and Conflicts in the 21st Century* (HPCR, Cambridge, MA, 2003).
White House, *National Security Strategy of the United States of America* (White House, 2010), www.whitehouse.gov/sites/default/files/rss_viewer/national_security_strategy.pdf (last accessed 15 July 2011).
National Security Strategy of the United States of America (White House, 2002).
National Security Strategy of the United States of America (White House, 2006).

Wilson, Clay, 'Information Operations and Computer Network Attack Capabilities of Today' (Paper presented at the International Expert Conference on Computer Network Attacks and the Applicability of International Humanitarian Law, Stockholm, 17–19 November 2004) 28–79.

Information Operations, Electronic Warfare, and Cyberwar: Capabilities and Related Policy Issues, Congressional Research Service, RL31787 (2007).

World Bank, *World Development Report: Knowledge for Development*, World Bank (1998).

News Reports, Magazines and Other Materials

'British Hackers Attack MOD Satellite', *Telegraph* (London), 4 March 1999.

'China Denies Hacking Dalai Lama Computer', CNN, 25 September 2002, http://europe.cnn.com/2002/TECH/internet/09/25/dalailama.hacking.ap/ (last accessed 28 September 2002).

'The Cyber Raiders Hitting Estonia', *BBC News*, 17 May 2007, http://news.bbc.co.uk/1/hi/world/europe/6665195.stm (last accessed 15 July 2011).

'Hackers Attack Heart of the Net', *BBC News*, 7 February 2007, http://news.bbc.co.uk/1/hi/technology/6338261.stm (last accessed 15 July 2011).

'Hezbollah Takes over West Beirut', *BBC News*, 9 May 2008, http://news.bbc.co.uk/1/hi/world/middle_east/7391600.stm (last accessed 15 July 2011).

'Indonesia, Ireland in Info War?', *Wired News*, 27 January 1999, www.wired.com/politics/law/news/1999/01/17562 (last accessed 15 July 2011).

'Information', *A Dictionary of Computing* (Oxford University Press, 2004).

'Internet Access Is "a Fundamental Right"', *BBC News*, 8 March 2010, http://news.bbc.co.uk/1/hi/technology/8548190.stm (last accessed 12 April 2011).

'Internet Blackouts: Reaching for the Kill Switch', *The Economist*, 10 February 2011, www.economist.com/node/18112043 (last accessed 16 March 2011).

'Iran Says Cyber Foes Caused Centrifuge Problems', Reuters, 29 November 2010.

'Iran Says Nuclear Programme Was Hit by Sabotage', *BBC News*, 29 November 2010, www.bbc.co.uk/news/world-middle-east-11868596 (last accessed 10 April 2011).

'Iran to Take Pre-Emptive Action against Cyber Terrorism: General', *Tehran Times*, 27 February 2011, www.tehrantimes.com/Index_view.asp?code=236455 (last accessed 12 April 2011).

'Military Eyes Electronic Warfare', Associated Press, *South China Morning Post*, 28 September 2002, http://china.scmp.com/chimain/ZZZH3UK2F6D.html (last accessed 30 September 2002).

'Mydoom Cripples US Firm's Website', *BBC News*, 1 February 2004, http://news.bbc.co.uk/1/hi/technology/3449931.stm (last accessed 15 July 2011).

'Mydoom Mutants Launch New Attacks', *BBC News*, 10 February 2004, http://news.bbc.co.uk/1/hi/technology/3475235.stm (last accessed 15 July 2011).
'NATO Denies Targeting Water Supplies', *BBC News*, 24 May 1999, http://news.bbc.co.uk/1/hi/world/europe/351780.stm (last accessed 15 July 2011).
'President's News Conference on Foreign and Domestic Issues', *New York Times*, 21 December 1983, A22.
Private Warriors, 2005, Frontline, PBS, Airdate: 21 June 2005.
'Rice Condemns Ongoing Cyber-Attacks as Estonian Embassy Siege Ends', earthtimes.org, 4 May 2007.
'Solar Sunrise', GlobalSecurity.org, www.globalsecurity.org/military/ops/solar-sunrise.htm (last accessed 8 July 2011).
'Taiwan Plays Cyber War Games', *BBC News*, 7 August 2000, http://news.bbc.co.uk/1/hi/world/asia-pacific/870386.stm (last accessed 15 July 2011).
'U.S. Gov IP Addresses You Should Not Scan', Hellbound Hackers 21 June 2007, www.hellboundhackers.org/articles/721-US-GOV-IP-ADDRESSES-YOU-SHOULD-NOT-SCAN.html (last accessed 5 July 2011).
'UK Hacker "Should Be Extradited"', *BBC News*, 10 May 2006, http://news.bbc.co.uk/1/hi/technology/4757375.stm (last accessed 15 July 2011).
Webopedia: Online Dictionary of Computing and Internet, www.webopedia.com (last accessed 15 July 2011).
AFP, 'Cyber Attacks on Estonia Are Security Issue: NATO Chief', *Age* (Melbourne), 26 May 2007.
Allison, Rebecca, 'Hacker Attack Left Port in Chaos: Busiest US Port Hit after Dorset Teenager Allegedly Launched Electronic Sabotage against Chatroom User', *Guardian* (London), 7 October 2003, 7.
'Youth Cleared of Crashing American Port's Computer', *Guardian* (London), 18 October 2007, 7.
Arkin, William M., 'The Cyber Bomb in Yugoslavia', *Washington Post* (Washington DC), 25 October 1999, www.washingtonpost.com/wp-srv/national/dotmil/arkin.htm (last accessed 15 July 2011).
Arkin, William M. and Windrem, Robert, 'The Other Kosovo War', *MSNBC News*, 29 August 2001.
Arnott, Sarah, 'How Cyber Crime Went Professional', *Independent* (London), 13 August 2008, www.independent.co.uk/news/business/analysis-and-features/how-cyber-crime-went-professional-892882.html (last accessed 15 July 2011).
Arquilla, John, 'The Great Cyberwar of 2002', *Wired* 6.02, February 1998, www.wired.com/wired/archive/6.02/cyberwar.html (last accessed 12 April 2011).
Arthur, Charles, 'The Day East Timor Was Deleted', *Independent* (London), 28 February 1999, 8.
Beaumont, Peter, 'The Truth About Twitter, Facebook and the Uprisings in the Arab World', *Guardian* (London), 25 February 2011, www.guardian.co.uk/world/2011/feb/25/twitter-facebook-uprisings-arab-libya (last accessed 12 April 2011).

Bond, David, 'The Dog That Didn't Bark', *Aviation Week & Space Technology* 19, 8 May 2006, 19.

Boyes, Roger, 'China Accused of Hacking into Heart of Merkel Administration', *The Times* (London), 27 August 2007, www.timesonline.co.uk/tol/news/world/europe/article2332130.ece (last accessed 15 July 2011).

Broad, William J. and Sanger, David E., 'Worm Was Perfect for Sabotaging Centrifuges', *New York Times*, 18 November 2010, A1.

Cieslak, Mark, 'Bridging an African Digital Divide' *BBC Click Online*, 7 September 2007 http://news.bbc.co.uk/1/hi/programmes/click_online/6983397.stm (last accessed 16 January 2012).

Clover, Charles, 'Kremlin-Backed Group Behind Estonia Cyber Blitz', *Financial Times* (London), 11 March 2009, 8, www.ft.com/techblog (last accessed 27 October 2010).

Cody, Edward, 'Chinese Official Accuses Nations of Hacking', *Washington Post* (Washington DC), 13 September 2007, A16.

Cowie, James, 'What Libya Learned from Egypt', Renesys Blog, 5 March 2011, www.renesys.com/blog/2011/03/what-libya-learned-from-egypt.shtml (last accessed 12 April 2011).

Davis, Joshua, 'If We Run out of Batteries, This War Is Screwed', *Wired* 6, June 2003, www.wired.com/wired/archive/11.06/battlefield.html (last accessed 12 April 2011).

French, Matthew, 'Tech Sabotage During the Cold War', *Federal Computer Week*, 26 April 2004, http://fcw.com/articles/2004/04/26/tech-sabotage-during-the-cold-war.aspx.

Friedman, Thomas L., 'Global Is Good', *Guardian* (London), 21 April 2005,G2.

'Syria Says Airman Seized in US Raid Will Not Be Freed', *New York Times*, 6 December 1983, A1.

'Widened Cabinet Sought in Beirut', *New York Times*, 8 December 1983, A18.

Friel, Brian, 'DoD Launches Internet Counterattack', *Government Executive*, 18 September 1998, http://govexec.com/dailyfed/0998/091898b1.htm (last accessed 12 April 2011).

Fulghum, David A., 'Data Link, Ew Problems Pinpointed by Pentagon', *Aviation Week & Space Technology* 10, 6 September 1999, 87.

'Frustrations and Backlogs', *Aviation Week & Space Technology* 10, 10 March 2003, 33.

'Sneak Attack', *Aviation Week & Space Technology* 28 June 2004, 34.

'Yugoslavia Successfully Attacked by Computers', *Aviation Week & Space Technology* 8, 23 August 1999, 31.

Fulghum, David A. and Barrie, Douglas, 'Israel Used Electronic Attack in Air Strike against Syrian Mystery Target', *Aviation Week & Space Technology* 8 October 2007, www.aviationweek.com/aw/generic/story_channel.jsp?channel=defense&id=news/aw100807p2.xml (last accessed 12 April 2011).

Fulghum, David A., Wall, Robert and Butler, Amy, 'Cyber-Combat's First Shot: Attack on Syria Shows Israel Is Master of the High-Tech Battle' (2007) 167(21) *Aviation Week & Space Technology* 28, 26 November 2007.

'Israel Shows Electronic Prowess', *Aviation Week & Space Technology* 25 November 2007, www.aviationweek.com/aw/generic/story_channel.jsp?channel=defense&id=news/aw112607p2.xml&headline=Israel%20Shows%20Electronic%20Prowess (last accessed 12 April 2011).

Fusco, Coco, 'Performance Art in a Digital Age: A Conversation with Ricardo Dominguez', Unpublished interview, The Institute of Visual Arts, London, 25 November 1999.

Gallagher, David F., 'Hackers; Government Tells Vigilantes Their "Help" Isn't Necessary', *New York Times*, 20 February 2003, 5.

Gellman, Barton, 'Cyber-Attacks by Al Qaeda Feared', *Washington Post* (Washington DC), 27 June 2002, A01.

Goodin, Dan, 'Electrical Supe Charged with Damaging California Canal System', *The Register*, 30 November 2007, www.theregister.co.uk/2007/11/30/canal_system_hack/ (last accessed 15 July 2011).

Gorman, Siobhan, 'Georgia States Computers Hit by Cyberattack', *Wall Street Journal* (New York), 12 August 2008, A9.

'Hackers Stole IDs for Attacks', *Wall Street Journal*, 17 August 2009, A525 February 2011.

Graham, Bradley, 'Bush Orders Guidelines for Cyber-Warfare', *Washington Post* (Washington DC), 7 February 2003, A1.

'Hackers Attack Via Chinese Web Sites', *Washington Post* (Washington DC), 25 August 2005, A1.

Grimes, Roger A., 'Security Adviser: DNS Attack Puts Web Security in Perspective', *InfoWorld* 8, 19 February 2007, www.infoworld.com/article/07/02/16/08OPsecadvise_1.html (last accessed 15 July 2011).

Halpin, Tony, 'Putin Accused of Launching Cyber War', *The Times* (London), 18 May 2007, 46.

Hart, Kim, 'Longtime Battle Lines Are Recast in Russia and Georgia's Cyberwar', *Washington Post* (Washington DC), 14 August 2008, D01.

Heickerö, Roland, 'Electronic Warriors Use Mail Order Equipment', *Framsyn Magazine*, April 2005, www.foi.se/FOI/templates/Page____4554.aspx# (last accessed 12 April 2011).

Hoagland, Jim, 'U.S., Iraq to Confer on Air War', *Washington Post* (Washington DC), 25 May 1987, 1.

Holland, Jesse J., 'Bush Advisor Warns Cyberterrorists', *Washington Post* (Washington DC), 13 February 2002, www.washingtonpost.com/wp-dyn/articles/A6846–2002Feb13.htm (last accessed 30 September 2002).

Hulme, George V., 'Taiwan Accuses China of Launching Cyberattack', *Information Week*, 16 June 2004, www.informationweek.com/news/22100221 (last accessed 3 January 2012).

IRIN News Report, 'Africa: Getting Connected at Last' 24 January 2006, www.irinnews.org/report.aspx?reportid=57903 (last accessed 16 January 2012).

Keizer, Gregg, 'Iran Admits Stuxnet Worm Infected PCs at Nuclear Reactor', *Computerworld*, 27 September 2010, www.computerworld.com/s/article/9188147/Iran_admits_Stuxnet_worm_infected_PCs_at_nuclear_reactor (last accessed 10 April 2011).

Kinver, Mark, '"Damage Is Done" to Lebanon Coast', BBC News, 8 August 2006, http://news.bbc.co.uk/1/hi/sci/tech/5255966.stm (last accessed 15 July 2011).

Krulak, Charles C., 'The Strategic Corporal: Leadership in the Three Block War', *Marines Magazine* 1, January 1999, 28–34.

Lague, David, 'Chinese See Military Dependence on Computers as Weakness', *International Herald Tribune* (Paris), 29 August 2007, www.iht.com/articles/2007/08/29/news/cyber.php (last accessed 15 July 2011).

Landler, Mark and Markoff, John, 'Digital Fears Emerge after Data Seige in Estonia', *New York Times*, 29 May 2007, www.nytimes.com/2007/05/29/technology/29estonia.html (last accessed 15 July 2011).

Leyden, John, 'France Blames China for Hack Attacks', *The Register* (London), 12 September 2007, www.theregister.co.uk/2007/09/12/french_cyberattacks/ (last accessed 15 July 2011).

Linzer, Dafna and Witte, Griff, 'U.S. Airstrike Targets Al Qaeda's Zawahiri', *Washington Post* (Washington DC), 14 January 2006, A09 www.washingtonpost.com/wp-dyn/content/article/2006/01/13/AR2006011302260_pf.html (last accessed 15 July 2011).

Lopez, C. Todd, 'Military Students Get Lesson in Cyberwarfare', *Air Force Print News*, 3 May 2006, http://searchsecurity.techtarget.com/originalContent/0,289142,sid14_gci1186049,00.html (last accessed 15 July 2011).

Maguire, Kevin, 'Guard Tried Sabotage at Nuclear Reactor: Security Checks Tightened after High-Level Alert', *Guardian* (London), 9 January 2001, 2.

Markoff, John, 'Before the Gunfire, Cyberattacks', *New York Times*, 13 August 2008, www.nytimes.com/2008/08/13/technology/13cyber.html (last accessed 15 July 2011).

McChrystal, Stanley A., 'It Takes a Network' (2011) 185 *Foreign Policy* 66–70.

McCue, Andy, '"Revenge" Hack Downed U.S. Port Systems', *ZDNet News*, 7 October 2003, http://news.zdnet.co.uk/security/0,1000000189,39116978,00.htm (last accessed 15 July 2011).

Medetsky, Anatoly, 'KGB Veteran Denies CIA Caused '82 Blast', *Moscow Times*, 18 March 2004, 4.

Mertl, Steve, 'Cyberspace Experts Await Full-Scale Attack', *Globe & Mail* (Canada), 27 December 2002, A11.

Meserve, Jeanne, 'Staged Cyber Attack Reveals Vulnerability in Power Grid', CNN.com, 26 September 2007, http://edition.cnn.com/2007/US/09/26/power.at.risk/index.html (last accessed 15 July 2011).

Moore, David, *et al.*, *The Spread of the Sapphire/Slammer Worm* (2003) www.caida.org/publications/papers/2003/sapphire/sapphire.html (last accessed 7 July 2011).

Morozov, Evgeny, 'An Army of Ones and Zeroes: How I Became a Soldier in the Georgia-Russia Cyberwar', *Slate,* 14 August 2008, www.slate.com/id/2197514/ (last accessed 12 April 2011).

Myrli, Sverre, *NATO and Cyber Defence*, 173 DSCFC 09 E bis (2009), www.nato-pa.int/default.asp?SHORTCUT=1782 (last accessed 12 April 2011).

Nagashima, Ellen, 'U.S. Eyes Pre-emptive Cyber-Defense Strategy', *Washington Post* (Washington DC), 29 August 2010, A05.

Naraine, Ryan, *Massive DDoS Attack Hit DNS Root Servers,* www.internetnews.com/bus-news/article.php/1486981 (last accessed 6 September 2007).

National Archives, *Dambusters: The Legacy,* www.nationalarchives.gov.uk/dambusters/legacy.htm (last accessed 15 July 2011).

Nuttall, Chris, 'Virtual Country "Nuked" on Net', *BBC News*, 26 January 1999, http://news.bbc.co.uk/1/hi/sci/tech/263169.stm (last accessed 14 December 2011.

PBS Frontline, *Cyberwar! The Warnings,* www.pbs.org/wgbh/pages/frontline/shows/cyberwar/warnings/ (last accessed 5 July 2011).

Interview with Hacker – Cyberwar! 24 April 2003, www.pbs.org/wgbh/pages/frontline/shows/cyberwar/interviews/hacker.html (last accessed 16 January 2012).

Interview with James Lewis – Cyberwar! 24 April 2003, www.pbs.org/wgbh/pages/frontline/shows/cyberwar/interviews/lewis.html (last accessed 16 January 2012).

Interview with John Hamre – Cyberwar! 24 April 2003, www.pbs.org/wgbh/pages/frontline/shows/cyberwar/interviews/hamre.html (last accessed 15 July 2011

Interview with Peter Singer – Private Warriors, www.pbs.org/wgbh/pages/frontline/shows/warriors/interviews/singer.html (last accessed 5 July 2011).

Interview with Richard Clarke – Cyberwar!, www.pbs.org/wgbh/pages/frontline/shows/cyberwar/interviews/clarke.html (last accessed 16 January 2012).

Interview with Stephen Schooner – Private Warriors, www.pbs.org/wgbh/pages/frontline/shows/warriors/interviews/schooner.html (last accessed 5 July 2011).

Pessach, Guy, *Digital Art Museums – Legal Perspectives* (2006), http://islandia.law.yale.edu/isp/writing%20paper/digital_art.htm (last accessed 9 June 2006).

Poulsen, Kevin, 'Slammer Worm Crashed Ohio Nuke Plant Network', *Security Focus,* 19 August 2003, www.securityfocus.com/news/6767 (last accessed 7 July 2011).

Priest, Dana, 'Private Guards Repel Attack on US Headquarters', *Washington Post*, 6 April 2004, A01.

Reuters, 'Cyber-War Rages over Iraq', *ZDNet News*, 31 March 2003, www.zdnet.com/newstech/security/stoty/0,2000024985,20273268,00.htm (last accessed 31 March 2003).

Robertson, Jordan, 'The Day Part of the Internet Died: Egypt Goes Dark', *Washington Times*, 28 January 2011, www.washingtontimes.com/news/2011/jan/28/day-part-internet-died-egypt-goes-dark/ (last accessed 12 April 2011).

Rojas, Peter, 'The Paranoia That Paid Off', *Guardian* (London), 24 April 2003, 27.

Schouten, Hank, 'China Denies Role in NZ Cyber Attack', *Dominion Post* (New Zealand), 12 September 2007, B4.

Seffers, George I., 'Legalities Cloud Pentagon's Cyber Defence', *DefenceNews*, 25 January 1999, 3.

Serabian, John A., Jr, 'Cyber Threats and the US Economy: Statement for the Record before the Joint Economic Committee on Cyber Threats and the US Economy' (23 February 2000), https://www.cia.gov/news-information/speeches-testimony/2000/cyberthreats_022300.html (last accessed 8 July 2011).

Shachtman, Noah, 'Insiders Doubt 2008 Pentagon Hack Was Foreign Spy Attack (Updated)', *Wired*, 25 August 2010, www.wired.com/danger-room/2010/08/insiders-doubt-2008-pentagon-hack-was-foreign-spy-attack/#more-29819 (last accessed 12 April 2011).

'Under Worm Assault, Military Bans Disks, USB Drives', *Wired*, 19 November 2008, www.wired.com/dangerroom/2008/11/army-bans-usb-d/ (last accessed 12 April 2011).

Smith, Charles R, 'U.S. Information Warriors Wrestle with New Weapons', NewsMax.com, 13 March 2003, www.newsmax.com/archives/art-icles/2003/3/12/134712.shtml (last accessed 7 July 2011).

The Environment Agency, *The Thames Barrier: Flood Defence for London*, www.environment-agency.gov.uk/regions/thames/323150/335688/341764/ (last accessed 29 November 2006).

Thornburgh, Nathan, *et al.*, 'The Invasion of the Chinese Cyberspies (and the Man Who Tried to Stop Them)' (2005) 166(10) *Time* 34.

Todd, Bennett, *Distributed Denial of Service Attacks*, (2000) www.linuxsecurity.com/resource_files/intrusion_detection/ddos-faq.html (last accessed 12 April 2011).

Traynor, Ian, 'Russia Accused of Unleashing Cyberwar to Disable Estonia', *Guardian* (London), 17 May 2007, 1.

'Web Attackers Used a Million Computers, Says Estonia', *Guardian* (London), 18 May 2007, 30.

United Nations War Crimes Commission, *Law Reports of Trials of War Criminals* (HMSO for the United Nations War Crimes Commission, London, 1949).

University of Southern California, 'How Much Information Is There in the World?' *ScienceDaily*, 11 February 2011, www.sciencedaily.com/releases/2011/02/110210141219.htm (last accessed 12 April 2011).

Verton, Dan, 'Navy Opens Some IT Ops to Vendors' (2000) *Federal Computer Week*, www.fcw.com/fcw/articles/2000/0821/pol-navy-08–21–00.asp (last accessed 21 August 2000).

'The Prospect of Iraq Conflict Raises New Cyber Attack Fears', *Computerworld*, 30 September 2002, www.computerworld.com/s/art-icle/74688/Prospect_of_Iraq_conflict_raises_new_cyberattack_fears (last accessed 5 July 2011).

Williams, Christopher, 'Israeli Security Chief Celebrates Stuxnet Cyber Attack', *Daily Telegraph* (London), 16 February 2011, www.telegraph.co.uk/technology/news/8326274/Israeli-security-chief-celebrates-Stuxnet-cyber-attack.html# (last accessed 10 April 2011).

ZDNet, 'Cyberterrorism: The Real Risks', *ZDNet News*, 27 August 2002, http://news.zdnet.co.uk/internet/0,1000000097,2121358,00.htm (last accessed 8 July 2011).

Zetter, Kim, 'Did a U.S. Government Lab Help Israel Develop Stuxnet?', *Wired*, 17 January 2011, www.wired.com/threatlevel/2011/01/inl-and-stuxnet/ (last accessed 10 April 2011).

'Report Strengthens Suspicions that Stuxnet Sabotaged Iran's Nuclear Plant', *Wired*, 27 December 2010, www.wired.com/threatlevel/2010/12/isis-report-on-stuxnet/# (last accessed 13 February 2011).

'Report: Stuxnet Hit 5 Gateway Targets on Its Way to Iranian Plant', *Wired*, 11 February 2011, www.wired.com/threatlevel/2011/02/stuxnet-five-main-target/# (last accessed 21 March 2011).

'Surveillance Footage and Code Clues Indicate Stuxnet Hit Iran', *Wired*, 16 February 2011, Threat Level, www.wired.com/threatlevel/2011/02/isis-report-stuxnet/# (last accessed 10 April 2011).

Index

CAMBRIDGE STUDIES IN INTERNATIONAL AND COMPARATIVE LAW

Books in the series

Cyber Warfare and the Laws of War
Heather Harrison Dinniss

The Right to Reparation in International Law for Victims of Armed Conflict: The Role of the UN in Advocating for State Responsibility
Christine Evans

Global Public Interest in International Investment Law
Andreas Kulick

State Immunity in International Law
Xiaodong Yang

Reparations and Victim Support in the International Criminal Court
Conor McCarthy

Reducing Genocide to Law: Definition, Meaning, and the Ultimate Crime
Payam Akhavan

Decolonizing International Law: Development, Economic Growth and the Politics of Universality
Sundhya Pahuja

Complicity and the Law of State Responsibility
Helmut Philipp Aust

State Control over Private Military and Security Companies in Armed Conflict
Hannah Tonkin

'Fair and Equitable Treatment' in International Investment Law
Roland Kläger

The UN and Human Rights: Who Guards the Guardians?
Guglielmo Verdirame

Sovereign Defaults before International Courts and Tribunals
Michael Waibel

Making the Law of the Sea: A Study in the Development of International Law
James Harrison

Science and the Precautionary Principle in International Courts and Tribunals: Expert Evidence, Burden of Proof and Finality
Caroline E. Foster

Legal Aspects of Transition from Illegal Territorial Regimes in International Law
Yaël Ronen

Access to Asylum: International Refugee Law and the Globalisation of Migration Control
Thomas Gammeltoft-Hansen

Trading Fish, Saving Fish: The Interaction between Regimes in International Law
Margaret Young

The Individual in the International Legal System: State-Centrism, History and Change in International Law
Kate Parlett

The Participation of States in International Organisations: The Role of Human Rights and Democracy
Alison Duxbury

'Armed Attack' and Article 51 of the UN Charter: Evolutions in Customary Law and Practice
Tom Ruys

Science and Risk Regulation in International Law: The Role of Science, Uncertainty and Values
Jacqueline Peel

Theatre of the Rule of Law: The Theory, History and Practice of Transnational Legal Intervention
Stephen Humphreys

The Public International Law Theory of Hans Kelsen: Believing in Universal Law
Jochen von Bernstorff

Vicarious Liability in Tort: A Comparative Perspective
Paula Giliker

Legal Personality in International Law
Roland Portmann

Legitimacy and Legality in International Law: An Interactional Account
Jutta Brunnée and Stephen J. Toope

The Concept of Non-International Armed Conflict in International Humanitarian Law
Anthony Cullen

The Challenge of Child Labour in International Law
Franziska Humbert

Shipping Interdiction and the Law of the Sea
Douglas Guilfoyle

International Courts and Environmental Protection
Tim Stephens

Legal Principles in WTO Disputes
Andrew D. Mitchell

War Crimes in Internal Armed Conflicts
Eve La Haye

Humanitarian Occupation
Gregory H. Fox

The International Law of Environmental Impact Assessment: Process, Substance and Integration
Neil Craik

The Law and Practice of International Territorial Administration: Versailles, Iraq and Beyond
Carsten Stahn

Cultural Products and the World Trade Organization
Tania Voon

United Nations Sanctions and the Rule of Law
Jeremy Farrall

National Law in WTO Law: Effectiveness and Good Governance in the World Trading System
Sharif Bhuiyan

The Threat of Force in International Law
Nikolas Stürchler

Indigenous Rights and United Nations Standards
Alexandra Xanthaki

International Refugee Law and Socio-Economic Rights
Michelle Foster

The Protection of Cultural Property in Armed Conflict
Roger O'Keefe

Interpretation and Revision of International Boundary Decisions
Kaiyan Homi Kaikobad

Multinationals and Corporate Social Responsibility: Limitations and Opportunities in International Law
Jennifer A. Zerk

Judiciaries within Europe: A Comparative Review
John Bell

Law in Times of Crisis: Emergency Powers in Theory and Practice
Oren Gross and Fionnuala Ní Aoláin

Vessel-Source Marine Pollution: The Law and Politics of International Regulation
Alan Tan

Enforcing Obligations Erga Omnes *in International Law*
Christian J. Tams

Non-Governmental Organisations in International Law
Anna-Karin Lindblom

Democracy, Minorities and International Law
Steven Wheatley

Prosecuting International Crimes: Selectivity and the International Law Regime
Robert Cryer

Compensation for Personal Injury in English, German and Italian Law: A Comparative Outline
Basil Markesinis, Michael Coester, Guido Alpa, Augustus Ullstein

Dispute Settlement in the UN Convention on the Law of the Sea
Natalie Klein

The International Protection of Internally Displaced Persons
Catherine Phuong

Imperialism, Sovereignty and the Making of International Law
Antony Anghie

Necessity, Proportionality and the Use of Force by States
Judith Gardam

International Legal Argument in the Permanent Court of International Justice: The Rise of the International Judiciary
Ole Spiermann

Great Powers and Outlaw States: Unequal Sovereigns in the International Legal Order
Gerry Simpson

Local Remedies in International Law
C. F. Amerasinghe

Reading Humanitarian Intervention: Human Rights and the Use of Force in International Law
Anne Orford

Conflict of Norms in Public International Law: How WTO Law Relates to Other Rules of Law
Joost Pauwelyn

Transboundary Damage in International Law
Hanqin Xue

European Criminal Procedures
Edited by Mireille Delmas-Marty and John Spencer

The Accountability of Armed Opposition Groups in International Law
Liesbeth Zegveld

Sharing Transboundary Resources: International Law and Optimal Resource Use
Eyal Benvenisti

International Human Rights and Humanitarian Law
René Provost

Remedies Against International Organisations
Karel Wellens

Diversity and Self-Determination in International Law
Karen Knop

The Law of Internal Armed Conflict
Lindsay Moir

International Commercial Arbitration and African States: Practice, Participation and Institutional Development
Amazu A. Asouzu

The Enforceability of Promises in European Contract Law
James Gordley

International Law in Antiquity
David J. Bederman

Money Laundering: A New International Law Enforcement Model
Guy Stessens

Good Faith in European Contract Law
Reinhard Zimmermann and Simon Whittaker

On Civil Procedure
J. A. Jolowicz

Trusts: A Comparative Study
Maurizio Lupoi

The Right to Property in Commonwealth Constitutions
Tom Allen

International Organizations Before National Courts
August Reinisch

The Changing International Law of High Seas Fisheries
Francisco Orrego Vicuña

Trade and the Environment: A Comparative Study of EC and US Law
Damien Geradin

Unjust Enrichment: A Study of Private Law and Public Values
Hanoch Dagan

Religious Liberty and International Law in Europe
Malcolm D. Evans

Ethics and Authority in International Law
Alfred P. Rubin

Sovereignty Over Natural Resources: Balancing Rights and Duties
Nico Schrijver

The Polar Regions and the Development of International Law
Donald R. Rothwell

Fragmentation and the International Relations of Micro-States: Self-determination and Statehood
Jorri Duursma

Principles of the Institutional Law of International Organizations
C. F. Amerasinghe